Skin Microbiology
Relevance to Clinical Infection

Skin Microbiology

Relevance to Clinical Infection

Edited by
Howard I. Maibach and Raza Aly

Springer-Verlag
New York   Heidelberg   Berlin

HOWARD I. MAIBACH, M.D.
Professor, Department of Dermatology, University of California, School of
Medicine, San Francisco, California, USA

RAZA ALY, Ph.D.
Associate Professor, Department of Dermatology, Department of Micro-
biology, University of California, School of Medicine, San Francisco,
California, USA

Library of Congress Cataloging in Publication Data
Maibach, Howard and Aly, Raza
  Skin microbiology, relevance to clinical infection
  Includes index
  1. Skin—infections.   2. Skin—Microbiology.
RL201.S58   1981   616.5'01   80-28395

9 8 7 6 5 4 3 2 1

ISBN-13: 978-1-4612-5870-4   e-ISBN-13: 978-1-4612-5868-1
DOI: 10.1007/978-1-4612-5868-1

Preface

Not since the 1965 publication of *Skin Bacteria and Their Role in Infection* has our knowledge of clinical skin microbiology been reviewed and summarized. In the more than a decade and a half since that publication, we have seen a careful reevaluation of the ideas and information current in 1965 and the development of important new discoveries and information.

This volume, *Skin Microbiology: Relevance to Clinical Infection*, reviews developments in the field since 1965 and summarizes the current state of the art in thirty-six carefully prepared chapters. Emphasis is on the clinical perspective rather than straight microbiology, although we include enough of the latter to put the clinical aspects in a proper scientific context.

The authors contributing to this volume represent a cross section of authorities in the many specialty areas that contribute to our knowledge of skin microbiology. They include investigators in microbiology, infectious disease, epidemiology, surgery, pediatrics, and dermatology. Significant efforts have been made to minimize repetition and overlap in the various chapters. In some cases, however, information is deliberately repeated in order to provide for the reader a necessary frame of reference.

We hope that this volume will be of value to dermatologists, microbiologists, pediatricians, surgeons, public health workers, nurses, and others involved in the diagnosis and treatment of dermatologic problems caused by bacteria.

The editors acknowledge with appreciation the assistance of Drs. A. Allen, F. Marzulli, F. Engley, G. Hildick-Smith, A. Kligman, M. Bruch, H. Eiermann, and D. Taplin.

HOWARD MAIBACH
RAZA ALY

Contents

Contributors

ALFRED M. ALLEN, Preventive Medicine Division, U.S. Army Health Services Command, Fort Sam Houston, San Antonio, TX, USA

RAZA ALY, Department of Dermatology, University of California, School of Medicine, San Francisco, CA, USA

INGRID ANDERSSON, Hospital Infection Control Laboratory, Goteborg, Sweden

BO R. BERGMAN, Department of Orthopaedic Surgery, Ostra sjukhuset, Goteborg, Sweden

AKE BRANDBERG, Hospital Infection Control Laboratory, Goteborg, Sweden

MARY BRUCH, Bureau of Drugs, U.S. Food and Drug Administration, Washington, DC, USA

JOHN F. BURKE, Department of Surgery, Harvard Medical School; Surgical Services, Shriner's Burns Institute and Massachusetts General Hospital, Boston MA, USA

M. W. CASEWELL, St. Thomas Hospital Medical School, London, UK

STEPHEN N. COHEN, Departments of Laboratory Medicine, Medicine, and Microbiology, University of California, School of Medicine, San Francisco, CA, USA

SYLVIA F. DIAS, Evans Department of Medicine, University Hospital, Boston University Medical Center, Boston, MA, USA

JOHN M. DRISCOLL, Department of Pediatrics, Columbia University, College of Physicians and Surgeons, New York, NY, USA

DAVID J. DRUTZ, Department of Medicine, University of Texas Health Science Center, San Antonio, TX, USA

W. CHRISTOPHER DUNCAN, Department of Dermatology, Baylor College of Medicine, Houston, TX, USA

HEINZ J. EIERMANN, Bureau of Foods, U.S. Food and Drug Administration, Washington, DC, USA

PETER M. ELIAS, Dermatology Service, The Veterans Administration Medical Center, San Francisco, CA, USA

PETER FRITSCH, Department of Dermatology, University of Innsbruck, Innsbruck, Austria

JOHN R. GRAYBILL, Department of Medicine, University of Texas Health Sciences Center, and Memorial Veterans Hospital, San Antonio, TX, USA

JOSEPH GREENBERG, Department of Dermatology, University of California at San Francisco, San Francisco, CA, USA

JAN HAMMARSTEN, Department of Surgery, Sahlgrenska Hospital, Goteborg, Sweden

JAN HOLM, Department of Surgery, Sahlgrenska Hospital, Goteborg, Sweden

PETER J. H. JACKMAN, Department of Bacteriology, Institute of Dermatology, St. John's Hospital for Diseases of the Skin, London, UK

WESLEY E. KLOOS, Department of Genetics, North Carolina State University, Raleigh, NC, USA

RICHARD B. KOHLER, Department of Medicine, Indiana University School of Medicine, and Veterans Administration Hospital, Indianapolis, IN, USA

ALBERT M. KLIGMAN, Department of Dermatology, Duhring Laboratories, University of Pennsylvania, School of Medicine, Philadelphia, PA, USA

JOHN M. KNOX, Department of Dermatology, Baylor College of Medicine, Houston, TX, USA

JAMES J. LEYDEN, Department of Dermatology, Duhring Laboratories, University of Pennsylvania, School of Medicine, Philadelphia, PA, USA

ANDERS LINDBERG, Institute of Medical Microbiology, Hospital Infection Control Laboratory, Goteborg, Sweden

JOSE A. LOPEZ, Evans Department of Medicine, Boston University Medical Center and University Hospital, Boston, MA, USA

E. J. L. LOWBURY, Department of Microbiology, University of Aston, Birmingham, UK

HOWARD I. MAIBACH, Department of Dermatology, University of California School of Medicine, San Francisco, CA, USA

RICHARD R. MARPLES, Central Public Health Laboratory, London, UK

FRANCIS N. MARZULLI, Bureau of Foods, U.S. Food and Drug Administration, Washington, DC

MOLLIE E. MCBRIDE, Department of Dermatology, Baylor College of Medicine, Houston, TX, USA

WILLIAM C. NOBLE, Institute of Dermatology, London, UK

JANE O'NEILL, Department of Pediatrics, Columbia University, College of Physicians and Surgeons, New York, NY, USA

DAVID G. PITCHER, Institute of Dermatology, St. John's Hospital for Diseases of the Skin, London, UK

S. M. PUHVEL, Department of Medicine, Center for Health Sciences, University of California, Los Angeles, CA, USA

HERBERT S. ROSENKRANZ, Case Western Reserve University School of Medicine, Cleveland OH, USA

STAFFAN SEEBERG, Institute of Medical Microbiology, Hospital Infection Control Laboratory, Goteborg, Sweden

SYDNEY SELWYN, Department of Medical Microbiology, Westminster Medical School, University of London, London, UK

TORE SCHERSTEN, Department of Surgery, Sahlgrenska Hospital, Goteborg, Sweden

H. R. SHINEFIELD, Department of Pediatrics, University of California School of Medicine, San Francisco, CA, USA

WILLIAM T. SPECK, Department of Pediatrics, School of Medicine, Case Western Reserve University, School of Medicine Cleveland, OH, USA

ALEXANDER S. D. SPIERS, Division of Oncology, Albany Medical College, Albany, NY, USA

DAVID TAPLIN, Division of International Community Medicine, University of Miami School of Medicine, Miami, FL, USA

JOHN A. ULRICH, Department of Microbiology, University of New Mexico School of Medicine, Albuquerque, NM, USA

GUY F. WEBSTER, Department of Pathology, School of Dental Medicine, University of Pennsylvania, Philadelphia, PA, USA

L. JOSEPH WHEAT, Department of Medicine, Indiana University School of Medicine; Veterans Administration Hospital, Indianapolis, IN, USA

ARTHUR WHITE, Department of Medicine, Indiana University School of Medicine; Veterans Administration Hospital, Indianapolis, IN, USA

THE CUTANEOUS FLORA
AND ITS CONTROL

Chapter 1

The Identification of *Staphylococcus* and *Micrococcus* Species Isolated from Human Skin

Wesley E. Kloos

Staphylococci and micrococci represent major groups of bacteria inhabiting human skin. The identity of each genus and currently recognized species offers a challenge to the skin microbiologist that should be rewarding in uncovering the population structure of cutaneous microbial communities and in providing a starting point from which to explore adaptive and evolutionary mechanisms. I believe that most human *Staphylococcus* and *Micrococcus* species have been discovered and given an appropriate taxonomic status. Perhaps others will be found, especially from unique niches or specially isolated human populations. Schleifer and coworkers[20] (K. H. Schleifer, personal communication) have shown that strains identified as *Peptococcus saccharolyticus* are actually anaerobic staphylococci which might constitute a new species. Preliminary ecological studies have indicated that the human forehead may carry rather large populations of this organism.[14]

In this paper, an attempt will be made to familarize the reader with the current status of *Staphylococcus* and *Micrococcus* taxonomy including the choice of simple characters that may be tested in the routine laboratory for genus and species identification.

Separation of Staphylococci and Micrococci

Staphylococci and micrococci can be separated by several key tests easily performed in the routine laboratory. One such test proposed by Schleifer and Kloos[46] is based on the ability of staphylococci to produce acid, aerobically, from glycerol in the presence of 0.4 μg erythromycin per ml and on their susceptibility to lysostaphin at a concentration of 200 μg/ml. Curry and Borovian[9] have shown that a nitrofuran-containing medium (FTO agar) permits the growth of micrococci but not staphylococci. Most *Staphylococcus* species produce detectable growth in a semisolid thioglycolate medium described by Evans and Kloos,[16] whereas most *Micrococcus* species will not grow in this medium except near the surface.[25,47] The species usually not conforming to this test include *S. hominis*, some strains of *S. haemolyticus*, *S. cohnii*, and *S. xylosus*, and *M. kristinae*; however, the test provides more reliable results for the separation of these genera than the standard anaerobic glucose utilization test proposed by the International Subcommittee on Staphylococci and Micrococci (1965). Seidl and Schleifer[52] introduced a serological test for separating staphylococci from micrococci based on different peptidoglycan types. Preliminary identification of these organisms can, in most instances, be accomplished on the basis of colony morphology and pigment and growth rate. Colonies of micrococci develop much slower than those of staphylococci (except for *Staphylococcus sciuri* subsp. *lentus*, which has been isolated from sheep and goats). Human *Staphylococcus* species produce detectable colonies (> 1 mm in diameter) on most nonselective plating agars, within 18 h at 34–37°C; whereas, micrococci require 36–48 h to produce detectable colonies. Most micrococci produce colonies typically more convex than staphylococcus colonies. Occasionally colonies of micrococci may be confused with those of corynebacteria, though these two groups can be resolved on the basis of cell morphology.

Micrococci and staphylococci are included currently in the family Micrococcaceae[5] representing gram-positive, catalase-positive cocci; however, major differences in their DNA base composition[1,6,25,30,33,44,47] and 16S ribosomal RNA sequences[53] indicate clearly that these organisms bear no special relationship to one another. It is expected that staphylococci and micrococci will be eventually separated and moved into other existing families.

Characterization of *Staphylococcus* Species Inhabiting Human Skin

Staphylococci are currently classified according to specific combinations of phenotypic characters[10,19,25,31,47] and DNA relatedness.[10,22,28,39,43,50] Extensive systematic and ecological studies have identified at least ten different *Staphylococcus* species living on human skin.[8,11,12,15,18,23,24,25,26,31,47,50,57,58] These species include *S. aureus*, *S. epidermidis*, *S. hominis*, *S. haemolyticus*, *S. capitis*, *S. warneri*, *S. saprophyticus*, *S. cohnii*, *S. xylosus*, and *S. simulans*. In the routine laboratory, they can be identified with reasonably good accuracy (> 80 to > 95 percent, depending upon the species) using specific combinations of selected characters that are relatively simple to test and have a high predictive value in resolving DNA homology groups.[24,27,32,55] The task of resolving at least ten different species is a relatively large one in any study. Here the investigator must determine whether or not it is important to resolve all species or only certain ones. The widely accepted Baird-Parker taxonomic schemes cannot identify most of the species inhabiting human skin, though they can identify with reasonable accuracy (> 70 percent) the more medically important species *S. aureus*

(*Staphylococcus* subgroup I), *S. epidermidis (S.* subgroup II or *S. epidermidis* biotype 1), and *S. saprophyticus (Micrococcus* subgroup 3 or *S. saprophyticus* biotype 3).[2,3,4] Some microbiologists adhere to Baird-Parker's schemes, but in doing so, must realize that they are limiting their ability to recognize the variety of widely divergent populations present on skin, some of which demonstrate niche and host preferences.

S. aureus is considered the most potentially pathogenic species and for this reason has received the most attention. Several years ago, all coagulase-positive staphylococci were identified as belonging to this species. Today, we recognize the three coagulase-positive species *S. aureus, S. intermedius,* and *S. hyicus;* the latter two species are found primarily on lower mammals and birds.[10,19,49] A moderate percentage of *S. hyicus* strains are coagulase-negative and were classified earlier as *Staphylococcus* subgroup III or *S. epidermidis* biotype 2 according to Baird-Parker's schemes. *S. intermedius* and *S. hyicus* are probably of minor concern to one studying the human cutaneous microflora, though they may be isolated on occasion from human skin if contact has been made with natural animal hosts (e.g. with pets or farm animals). Although these species may produce infections in a variety of animals and birds, it is not known if they are capable of producing infections in man. All three coagulase-positive species produce higher deoxyribonuclease activity than usually found with coagulase-negative species. Coagulase-positive staphylococci that produce weak, delayed acid or no acid, areobically, from maltose are probably not *S. aureus. S. intermedius* strains isolated from carnivora (e.g. dogs, mink, raccoons, etc.) typically produce very weak acid from maltose in 48 to 72h and *S. hyicus* strains usually fail to produce acid from this carbohydrate. *S. intermedius* and *S. hyicus* do not produce acid from D-mannitol anaerobically and do not produce acetylmethylcarbinol (acetoin), except, perhaps, in small quantity. *S. aureus* on the other hand, is usually positive for these characters. Each of the coagulase-positive species are in different, widely divergent DNA homology groups exhibiting less than 20 percent relatedness under restrictive binding conditions.[10,39,42]

The nine coagulase-negative species found living on human skin may be organized into three species groups.[26,32] The major one of human skin is the *S. epidermidis* species group composed of the species *S. epidermidis, S. hominis, S. haemolyticus, S. warneri,* and *S. capitis.* These species are related slightly more to one another than to species in other groups. They exhibit a DNA relatedness of about 45 ± 6 percent at non-restrictive conditions and 15 ± 3 percent at restrictive conditions.[22,28,50] *S. hominis* is more related to *S. haemolyticus,* and *S. epidermidis* is more related to *S. capitis* than to other species of this group.

S. epidermidis and *S. capitis* may be identified easily and with considerable accuracy (>95 percent). On a peptone-yeast extract-salt (P agar) medium,[30] colonies of *S. epidermidis* are small and are about 3.5 to 4 mm in diameter after incubation at 34°C for 3 days, followed by storage at room temperature for an additional 2 days. Colonies are slightly raised, rather sticky, and usually gray or gray-white. With crowding or prolonged incubation the center of colonies becomes dark and nearly transparent. Rare strains may develop a violet or yellowish pigment. *S. capitis* colonies are small, averaging 2 to 3 mm in diameter and more convex in appearance. They are opaque, chalk white, and rather striking. Colonies of both species grow significantly larger on P agar supplemented with 5 percent bovine blood, which is the opposite of the response observed with *S. hominis* colonies. *S. epidermidis* and *S. capitis* do not produce acid, aerobically, from D-(+)-trehalose. Unlike *S. epidermidis*, *S. capitis* does not produce acid from maltose or α-lactose, but often produces acid from D-mannitol. *S. epidermidis* has usually strong phosphatase activity and produces a dense uniform growth under anaerobic conditions in a thioglycolate medium. *S. capitis*, on the other hand, has usually weak or no phosphatase activity and produces large discrete colonies or moderate growth in thioglycolate.

S. hominis, *S. haemolyticus*, and *S. warneri* are sometimes difficult to distinguish from one another by simple biochemical tests. Some advantage will be gained in the recognition of these species if the investigator becomes familiar with colonial characteristics. *S. hominis* and *S. warneri* produce colonies usually smaller than those of *S. haemolyticus*. Their diameter is about 3.5 to 5 mm; whereas *S. haemolyticus* colonies are usually about 5 to 8 mm. On P agar the more common pigmented *S. hominis* colonies demonstrate a yellow or orange-yellow center with two or more alternating light and dark (e.g., whitish and grayish) concentric rings proceeding out from near the center to the edge. Unpigmented *S. hominis* colonies exhibit the light and dark concentric ring pattern. *S. hominis* colonies are usually smaller on P agar supplemented with 5 percent bovine blood than on unsupplemented P agar. Pigmented *S. warneri* colonies often develop a deeper golden-yellow pigment along the edge than in the center, especially after prolonged storage at room or refrigeration temperatures. They are usually glistening, sticky, and translucent; whereas, *S. hominis* and *S. haemolyticus* colonies are usually smooth in texture and opaque. *S. haemolyticus* can be distinguished from other species of the group on the basis of its relatively large colony size and moderate to strong hemolysis of bovine blood (within 48 to 72 h). A small percentage of strains in the other species produce weak hemolysis. Unlike *S. hominis* many strains of *S. haemolyticus* and *S. warneri* produce acid, aerobically, from D-mannitol. Most strains of *S. hominis* grow poorly under anaerobic conditions in thioglycolate; whereas, *S. warneri* usually produces uniform dense growth and *S. haemolyticus* produces many large discrete colonies or moderate growth in the anaerobic portion of this medium. Most strains of *S. warneri* do not reduce nitrates or produce acid from α-lactose. Some strains of this species produce only weak acid from maltose.

The *S. saprophyticus* species group composed of the species *S. saprophyticus*, *S. cohnii*, and *S. xylosus* usually produces relatively small populations on human skin that often appear to be temporary residents. Species of this group exhibit a DNA relatedness of about 50 ± 5 percent at non-restrictive conditions and 19 ± 2 percent at restrictive conditions (W. E. Kloos and J. F. Wolfshohl, unpublished data). Each can be identified easily and with considerable accuracy (> 90 percent). Members of the *S. saprophyticus* species group are resistant to novobiocin (MIC ≥ 1.6 µg/ml) and, in practice, this property serves as a useful taxonomic character for the presumptive identification of this and the *S. sciuri* species group. Only rarely are strains in other species groups novobiocin resistant. Members of the *S. sciuri* species group are rarely isolated from human skin and can be distinguished from other species on the basis of their aerobic acid production from D-(+)-cellobiose or fucose.

    S. xylosus can be distinguished from other species of the *S. saprophyticus* species group by its ability to produce acid from D-(+)-xylose and/or L-(+)-arabinose. It often produces unique colonies that are large with an undulate to crenate edge. Growth of these spreads rapidly over the surface of soft agar. However, *S. xylosus* demonstrates a variety of other colony types, some of which may be confused occasionally with *S. saprophyticus* or *S. cohnii*. In contrast to the other species of the group, *S. xylosus* usually demonstrates phosphatase activity and nitrate reduction, but usually produces weak or no detectable acetoin. *S. cohnii* can be distinguished from *S. saprophyticus* by it failure to produce acid, aerobically, from sucrose. *S. saprophyticus* does not produce acid from D-(+)-mannose; whereas the other species often produce acid from this carbohydrate. The colonies of *S. cohnii* may vary considerably in size but are almost always entire, rather convex in profile, and glistening in lustre. Human strains of this species are rarely pigmented. *S. saprophyticus* colonies are usually entire and relatively large (5 to 8 mm in diameter), often pigmented, and have a low convex profile. Many strains of *S. saprophyticus*, and much fewer strains of the other species, produce acid, aerobically, from xylitol.

    The remaining species group is composed currently of the single species *S. simulans*. This species, though demonstrating a variety of positive characters rather similar to *S. aureus*, has diverged widely from this and other recognized species. The DNA relatedness of this species to others is less than 25 percent at non-restrictive conditions and 10 percent at restrictive conditions (Kloos and Wolfshohl 1980; unpublished data). This species can be recognized primarily by its relatively large colony size (usually 6 to 8 mm in diameter), no or only weak acid produced, aerobically, from maltose, and dense uniform growth in a thioglycolate medium. It usually reduces nitrates, demonstrates weak to moderate phosphatase activity, and produces acid, aerobically, from D-mannitol, but not from D-(+)-melezitose. The occasional hemolytic strains of this species may be initially confused with *S. haemolyticus* but very few strains of the latter species fail to produce acid from maltose or produce dense uniform growth in thioglycolate like *S. simulans*. *S. simulans* is also typically much more susceptible to lysostaphin than members of the *S. epidermidis* species group. The taxonomic status of currently recognized *Staphylococcus* species is on sound footing but additional research is still needed to find more expedient ways of identifying these organisms.

Characterization of *Micrococcus* Species Found on Human Skin

Micrococci are usually found in much smaller populations on human skin than staphylococci, though they may account for a large proportion of the aerobic bacteria in cutaneous habitats supporting relatively small numbers of aerobic bacteria (e.g. arms and legs), on certain individuals.[17,23,40] Micrococci are currently classified on the basis of a combination of key phenotypic characters,[30,37,35,38,34,48,56] genetic transformation,[21,30] and, to some extent, DNA relatedness.[41,51] Eight different *Micrococcus* species have been isolated from human skin, including *M. luteus, M. varians, M. lylae, M. nishinomiyaensis, M. kristinae, M. roseus, M. sedentarius,* and *M. agilis,* in order of prevalence. Temporal studies have suggested that certain strains of *M. luteus, M. varians, M. kristinae,* and *M. sedentarius* probably maintain a resident status; whereas others appear to be temporary residents or transients.[23]

M. luteus is the major species of human skin and can be identified with reasonably good accuracy (>90 percent) using a simple character analysis. Its colonies are usually convex and uniformly pigmented from the center to the edge. The pigment is usually different shades of cream white to yellow. *M. luteus* does not produce acid, aerobically, from D-(+)-glucose, β-D-(−)-fructose, D-(+)-galactose, D-(+)-mannose, L-(+)-rhamnose, D-(+)-xylose, or D-sorbitol, produce acetoin, or grow on Simmons citrate agar. Most strains are oxidase-positive, susceptible to lysozyme and penicillin G, and fail to reduce nitrates. This species can be confused occasionally with *M. lylae.* Transformation crosses and the immunological relationships of catalases have suggested that *M. luteus* and *M. lylae* are more closely related to one another than to other species.[30,45] *M. lylae* can usually be distinguished from *M. luteus* on the combined features of unpigmented or cream-white colonies and resistance to lysozyme. *M. luteus* does not demonstrate a close DNA relatedness to *M. varians, M. roseus,* or *M. kristinae.*[41,51] DNA relatedness to *M. lylae* is significant (30−34% under restrictive conditions).[51]

M. varians produces colonies that usually have a lower convex profile than that of colonies of *M. luteus* and young pigmented colonies demonstrate a lighter pigmentation near their edge. Most strains produce colonies that are pigmented different shades of yellow and have a sticky consistency. Unlike *M. luteus* or *M. lylae* strains, *M. varians* produces acid, aerobically, from D-(+)-glucose and β-D-(−)-fructose and occasionally from D-(+)-xylose, and usually grows on Simmons citrate agar, reduces nitrates, and is oxidase-weak or -negative. *M. varians* is resistant to lysozyme and susceptible to penicillin G, like many *M. lylae* strains. This species does not demonstrate a close DNA relatedness to *M. lylae, M. luteus* or *M. roseus.*[41,51] DNA relatedness to other species has not been reported.

M. roseus can be easily distinguished from other species by the pinkish to orange-red color of its colonies that usually have a low convex profile and glistening appearance. Pigmentation is somewhat lighter near the edge of young colonies. On prolonged storage at room or refrigeration temperatures, growth will become very wet in appearance and run down the surface of agar under normal gravitational force. Like *M. varians, M. roseus* produces acid, aerobically, from β-D-(−)-fructose and D-(+)-glucose (though somewhat weaker acid) and often from D-(+)-xylose, reduces nitrates, and is oxidase-weak or -negative. Unlike other species, *M. roseus* usually produces acid from L-(+)-rhamnose and, unlike *M. varians,* it often produces acid from D-sorbitol. *M. roseus* is very susceptible to penicillin G and methicillin.

M. kristinae can be easily distinguished from other species on the basis of several distinctive characters. Colonies are small, convex to umbonate, and develop a pale cream to pale orange pigment. The pigment becomes intensified in the center of the colony upon prolonged incubation. Colonies usually have a granular texture and are very difficult to disperse in aqueous solutions. This species can grow slightly under anaerobic conditions in thioglycolate, produces relatively strong acid, aerobically, from D (+)-glucose, β-D-(−)-fructose, and D-(+)-mannose and usually moderate acid from glycerol and D-sorbitol, and produces relatively large amounts of acetoin. It does not usually grow on Simmons citrate agar and is very resistant to lysozyme.

M. nishinomiyaensis can be recognized by its small bright orange colonies that are convex to slightly umbonate in profile. Some strains produce an orange exopigment. Many strains produce weak to moderate acid, aerobically, from D-(+)-glucose and D-(+)-galactose, but fail to produce acid from β-D (−)-fructose, L-rhamnose, maltose, glycerol, or D-sorbitol. This species does not grow on Simmons citrate agar. It usually demonstrates oxidase activity and many strains reduce nitrates.

M. sedentarius produces cream white to deep buttercup yellow colonies that grow rather slowly compared to other species. Their profile is convex to pulvinate and they generally lack lustre. Many strains produce a brownish, water-soluble exopigment. All strains of this species that have been reported on to date are resistant to methicillin (MIC $\geq$ 100 μg/ml) and penicillin G (MIC $\geq$ 12.5 μg/ml) and, in practice, these antibiotic patterns can serve as useful taxonomic characters to identify this species. Like *M. luteus* and *M. lylae,* this species attacks only a very few carbohydrates. A few strains produce acid, aerobically, from D-(+)-glucose and α-lactose. It usually does not reduce nitrates or demonstrate oxidase activity and does not grow on Simmons citrate agar or produce acetoin. Most strains demonstrate arginine dihydrolase activity, a feature which also sets it apart from other species.

M. agilis has been only rarely isolated from human skin. It produces slow growing vivid red or red-orange colonies that have a relatively low convex profile. Growth is enhanced somewhat at cooler room temperatures. Unlike *M. roseus*, this species does not produce acid, aerobically, from L-rhamnose or D-xylose, is usually oxidase-positive, and hydrolyzes aesculin.

References

1. Auletta AE, and Kennedy ER (1966) Deoxyribonucleic acid base composition of some members of the Micrococcaceae. J Bacteriol 92:28–34
2. Baird-Parker AC (1965a) The classification of staphylococci and micrococci from world-wide sources. J Gen Microbiol 38:363–387
3. Baird-Parker AC (1965b) Staphylococci and their classification. Ann NY Acad Sci 128:4–25
4. Baird-Parker AC (1974a) The basis for the present classification of staphylococci and micrococci. Ann NY Acad Sci 236:7–14
5. Baird-Parker AC (1974b) Family I Micrococcaceae Pribram 1929. In: *Bergey's Manual of Determinative Bacteriology* Eighth Edition RE Buchanan, NE Gibbons (eds) pp 478–490. Williams and Wilkins, Baltimore
6. Bohacek J, Kocur M, and Martinec T (1965) Deoxyribonucleic acid base composition and taxonomy of the genus *Micrococcus*. Publ Fac Sci Univ J E Purkyne, Brno K35 pp 318–322
7. Bohacek J, Kocur M, and Martinec T (1970) DNA base composition of some Micrococcaceae. *Microbios* 6:85–91
8. Carr DL, and Kloos WE (1977) Temporal study of the staphylococci and micrococci of normal infant skin. *Appl Environ Microbiol* 34:673–680
9. Curry JC, and Borovian GE (1976) Selective medium for distinguishing micrococci from staphylococci in the clinical laboratory. J Clin Microbiol 4:455–457
10. Devriese LA, Hajek V, Oeding P, Meyer, SA, and Schleifer KH (1978) *Staphylococcus hyicus* (Sompolinsky 1953) comb nov and *Staphylococcus hyicus* subsp *chromogenes* subsp nov. Int J Syst Bacteriol 28:482–490
11. Durham DR, and Kloos WE (1978) A comparative study of the total cellular fatty acids of *Staphylococcus* species of human origin. *Int J Syst* Bacteriol 28:223–228
12. Emmett M, and Kloos WE (1975) Amino acid requirements of staphylococci isolated from human skin. Can J Microbiol 21:729–733
13. Emmett M, and Kloos WE (1979) The nature of arginine auxotrophy in cutaneous populations of staphylococci. J Gen Microbiol 110:305–314
14. Evans CA, Mattern KL, and Hallam SL, (1978) Isolation and identification of *Peptococcus saccharolyticus* from human skin. J Clin Microbiol 7:261–264
15. Evans JB (1976) Anaerobic growth of *Staphylococcus* species from human skin: effects of uracil and pyruvate. Int J Syst Bacteriol 26:17–21
16. Evans JB, and Kloos WE (1972) Use of shake cultures in a semisolid thioglycolate medium for differentiating staphylococci from micrococci. Appl Microbiol 23:326–331
17. Glass M (1973) Sarcina species on the skin of the human forearm. Transact St John's Hosp Dermatol Soc. 59:56–60
18. Gotz F, and Schleifer KH (1976) Comparative biochemistry of lactate dehydrogenase from staphylococci. In: Staphylococci and Staphylococcal Diseases. Jeljaszewicz J. (ed) pp 245–252, Gustav Fischer Verlag, Stuttgart
19. Hajek V (1976) *Staphylococcus intermedius*, a new species isolated from animals. Int J Syst Bacteriol 26:401–408
20. Kilpper R, and Schleifer KH (1978) *Peptococcus saccharolyticus:* an anaerobic staphylococcus? Abstr XII Int Congr Microbiol:86 (Abstr)

21. Kloos WE (1969) Transformation of *Micrococcus lysodeikticus* by various members of the family Micrococcaceae. J Gen Microbiol 59:247–255
22. Kloos, WE, (1980) Natural populations of the genus *Staphylococcus*. Ann, Rev Microbiol 34:559–592
23. Kloos WE, and Musselwhite MS (1975) Distribution and persistence of *Staphylococcus* and *Micrococcus* species and other aerobic bacteria on human skin. Appl Microbiol 30:381–395
24. Kloos WE, and Schleifer KH (1975a) Simplified scheme for routine identification of human *Staphylococcus* species. J Clin Microbiol 1:82–88
25. Kloos WE, and Schleifer KH (1975b) Isolation and characterization of staphylococci from human skin. II. Descriptions of four new species: *Staphylococcus warneri, Staphylococcus capitis, Staphylococcus hominis,* and *Staphylococcus simulans.* Int J Syst Bacteriol 25:62–79
26. Kloos WE, and Schleifer KH (1981) The genus *Staphylococcus* In: The Prokaryotes. MP Starr, H Stolp, HG Trüper, A Balows HG Schlegel (eds), Berlin, Heidelberg, New York: Springer-Verlag
27. Kloos WE, and Smith PB (1980) The staphylococci. In: Manual of Clinical Microbiology, (3rd ed) EH Lennette, A Balows, WJ Hausler, JP Truant (eds) Washington, D.C.: American Society for Microbiology, pp 83–87
28. Kloos WE, and Wolfshohl JF (1979) Evidence for deoxyribonucleotide sequence divergence between staphylococci living on human and other primate skin. Curr Microbiol 3:167–172
29. Kloos WE, and Wolfshohl JF, (1980) DNA divergence between human and nonhuman primate staphylococci. *Abstr Ann Meet Amer Soc Microbiol* 1980:94 (Abstr)
30. Kloos WE, Tornabene TG, and Schleifer KH (1974) Isolation and characterization of micrococci from human skin, including two new species: *Micrococcus lylae* and *Micrococcus kristinae.* Int J Syst Bacteriol 24:79–101
31. Kloos WE, Schleifer KH, and Noble WC (1976a) Estimation of character parameters in coagulase-negative *Staphylococcus* species. In: *Staphylococci and Staphylococcal Diseases.* Jeljaszewicz J (ed), pp 23–41, Gustav Fischer Verlag, Stuttgart
32. Kloos WE, Schleifer KH, and Smith RF (1976b) Characterization of *Staphylococcus sciuri* sp nov and its subspecies. Int J Syst Bacteriol 26:22–37
33. Kocur M, and Bohacek J (1974) DNA base composition and the classification of non-pigmented micrococci. Microbios 10A:31–38
34. Kocur M, and Martinec T (1972) Taxonomic status of *Micrococcus varians* Migula 1900 and designation of the neotype strain. Int J Syst Bacteriol 22:228–232
35. Kocur M, and Pacova Z (1970) The taxonomic status of *Micrococcus roseus* Flugge 1886. Int J Syst Bacteriol 20:233–240
36. Kocur M, and Schleifer KH (1975) Taxonomic status of *Micrococcus agilis* Ali-Cohen 1889. Int J Syst Bacteriol 25:294–297
37. Kocur M, Pacova Z, and Martinec T (1972) Taxonomic status of *Micrococcus luteus* (Schroeter 1872) Cohn 1872, and designation of the neotype strain. Int J Syst Bacteriol 22:218–23
38. Kocur M, Schleifer KH, and Kloos WE (1975) Taxonomic status of *Micrococcus nishinomiyaensis* Oda 1935. Int J Syst Bacteriol 25:290–293
39. Meyer SA, and Schleifer KH (1978) Deoxyribonucleic acid reassociation in the classification of coagulase-positive staphylococci. Arch Microbiol 117:183–188
40. Noble WC, and Somerville DA (1974) Microbiology of Human Skin. Saunders, Philadelphia
41. Ogasawara-Fujita N, and Saraguchi K (1976) Classification of micrococci on the basis of deoxyribonucleic acid homology. J Gen Microbiol 94:97–106
42. Phillips WE, King RE, and Kloos WE (1980) Isolation of *Staphylococcus hyicus* subsp *hyicus* from a pig with septic polyarthritis. Am J Vet Res 41:274–278

43. Rosenblum ED, and Tyrone S (1979) Deoxyribonucleic acid homologies among staphylococci: coagulase-positive reference strains. Curr Microbiol 2:171–174
44. Rosypal S, Rosypalova A, and Horejs J (1966). The classification of micrococci and staphylococci based on their DNA base composition and Adansonian analysis. J Gen Microbiol 44:281–292
45. Rupprecht M, and Schleifer KH (1977) Comparative immunological study of catalases in the genus *Micrococcus* Arch Microbiol 114:61–66
46. Schleifer KH, and Kloos WE (1975a) A simple test system for the separation of staphylococci from micrococci. J Clin Microbiol 1:337
47. Schleifer KH, and Kloos WE (1975b) Isolation and characterization of staphylococci from human skin I. Amended descriptions of *Staphylococcus epidermidis* and Staphylococcus saprophyticus and descriptions of three new species: *Staphylococcus cohnii, Staphylococcus haemolyticus,* and *Staphylococcus xylosus.* Int J Syst Bacteriol 25:50–61
48. Schleifer KH, Kloos WE, and Moore A (1972) Taxonomic status of Micrococcus luteus (Schroeter 1872) Cohn 1872: correlation between peptidoglycan type and genetic compatibility. Int J Syst Bacteriol 22:224–227
49. Schleifer KH, Schumacher-Perdreau F, Gotz, F, and Popp B (1976) Chemical and biochemical studies for the differentiation of coagulase-positive staphylococci. Arch Microbiol 110:263–270
50. Schleifer KH, Meyer SA, and Rupprecht M (1979) Relatedness among coagulase-negative staphylococci: deoxyribonucleic acid reassociation and comparative immunological studies. Arch Microbiol 122:93–101
51. Schleifer KH, Heise W, and Meyer SA (1979) Deoxyribonucleic acid hybridization studies among some micrococci. FEMS Microbiol Lett 6:33–36
52. Seidl PH, and Schleifer KH (1978) Rapid test for the serological separation of staphylococci from micrococci. Appl Environ Microbiol 35:479–482
53. Stackebrandt E, and Woese CR (1979) A phylogenetic dissection of the family Micrococcaceae. Curr Microbiol 2:317–322
54. Subcommittee on Taxonomy of Staphylococci and Micrococci (1965) Recommendations. Int Bull Bacteriol Nomen Tax 15:107–110
55. Subcommittee on the Taxonomy of Staphylococci and Micrococci (1976) Appendix 1. Identification of staphylococci. Int. J. Syst. Bacteriol. 26:333–334
56. Tornabene TG, Morrison SJ, and Kloos WE (1970) Aliphatic hydrocarbon contents of various members of the family Micrococcaceae. Lipids 5:929–937
57. Zimmerman RJ (1976) Comparative zone electrophoresis of catalase of *Staphylococcus* species isolated from mammalian skin. Can J Microbiol 22:1691–1698
58. Zimmerman RJ, and Kloos WE (1976) Comparative zone electrophoresis of esterases of *Staphylococcus* species isolated from mammalian skin. Can J Microbiol 22:771–779

Chapter 2
Coagulase-negative Staphylococci: Classification and Problems

R.R. MARPLES

The difficulties of classifying coagulase-negative staphylococci became evident to workers investigating the flora of the skin before most medical microbiologists were aware that a problem existed. The different staphylococci on the skin and the probable significance of this in ecological terms was realized,[12,7] but the classifications available were of no practical use. Standard procedure was to identify all coagulase-negative staphylococci as *Staphylococcus albus* and to recognize coagulase-positive strains as the pathogen *S. aureus*.

What was to become widely accepted classification was proposed by Baird-Parker[1,2] based on a few relatively simple biochemical tests, including the production of acid aerobically from arabinose, lactose, maltose, and mannitol and the production of coagulase, phosphatase, and acetoin. In this scheme the primary division between *Staphylococcus* and *Micrococcus* was made on the ability to ferment glucose under anaerobic conditions. The genus *Staphylococcus* was divided into six subgroups numbered SI–SVI, with subgroup SVI further subdivided into three types; and the genus *Micrococcus* was divided into eight subgroups, M1–M8. Later studies on the guanine-cytosine percentage composition of the DNA showed that at least the first four micrococcal subgroups were staphylococci, while M7 and M8 represented the genus *Micrococcus*. Baird-Parker amended his scheme[3] so that the original subgroup SI became *S. aureus;* SII–SVI became *S. epidermidis* biotype 1–4 with biotype 1 including SII and SV; and M1–4 became *S. saprophyticus* biotypes 1–4. This definition of *S. saprophyticus* includes a wider set of strains than later classifications allow.

The original classification has been widely used in studies of skin flora and infections and permitted the recognition of differing pathogenicity among the subgroups. Subgroup II became associated with infected surgical prostheses[6,24] and with diseased skin[17,13] while subgroup M3 was associated with domiciliary urinary infection in young women,[15] permitting the description of this newly differentiated disease.[20] Studies of the normal skin flora also showed differences in proportions of the subgroups at different sites[13,18] and in different environments.

We found it necessary to extend the classification of Baird-Parker to include two additional groups, SY and MX, to allow for strains that produce acid from mannitol and produce phosphatase and acetoin. We also subdivide M3 in the same way that Baird-Parker subdivided SVI. Pelzer et al.[21] found similar strains and added six further subgroups to the basic scheme.

Recently Kloos, Schleifer, and co-workers have described alternative classifications for both micrococci[9] and staphylococci[8] based on similar but not identical biochemical tests carried out by methods that differ from those of Baird-Parker. Their simplified scheme for staphylococci[8] forms a dichotomous key with alternative pathways leading to 11 subdivisions, most of which are named, including *S. aureus, S. epidermidis*, and *S. saprophyticus*. The definitions of *S. aureus* and *S. epidermidis* are compatible with the scheme of Baird-Parker, but the definition of *S. saprophyticus* is narrower.

This scheme has been applied to infections[19] and to normal human skin cocci.[10,11] Navamar et al.[16] have found the subdivision of Baird-Parker subgroup M3 by the classification of Kloos and Schleifer useful. For clinical work a scheme to identify *S. aureus, S. epidermidis*, and *S. saprophyticus* while detecting the other species has been advanced as a compromise.[25]

We studied 300 representative strains of catalase-positive gram-positive cocci by the methods of Baird-Parker and of Kloos and Schleifer. The collection included 96 strains from culture collections, 58 from animal sources, 16 from urinary infections, and 125 from human skin. For five strains the source was unknown. We then identified each strain by the two classifications.

Five biotypes gave very good correlation: *S. aureus* and *S. intermedius* equaled Baird-Parker SI; *S. epidermidis* equaled SII; *S. saprophyticus* equaled lactose positive M3; *S. cohnii* equaled MX; and *Micrococcus roseus* equaled M8. A minor redefinition of SIII to allow for maltose-positive strains made this subgroup equal *S. simulans;* lower reliance on the OF test incorporated SVI(3) and M3(3) as *S. capitis* and *S. xylosus* included both M5 and M6 strains.

Correlation between the remaining three species of Kloos and Schleifer—*S. haemolyticus, S. warneri*, and *S. hominis*—with the remaining staphylococcal biotypes of Baird-Parker was poor. Biotype M7 of Baird-Parker included the remaining *Micrococcus* species, though some strains of *M. kristinae* and *M. varians* were misclassified into staphylococcal biotypes. Further work is needed to improve the identification schemes for these strains to assess the significance of the different weights given to acid production from lactose and mannitol in the two classifications.

While *S. saprophyticus* is associated with urinary infection, the commonest biotype found in serious infections caused by coagulase-negative staphylococci is *S. epidermidis* (SII). Of 223 strains associated with nonurinary infection received by us in 1977, 181 were identified as this biotype. For epidemiological studies, methods of subdividing this biotype are required. Phage-typing schemes have been developed;[5] additional biochemical tests can be used.[6] Antibiotic susceptibility testing and, in some situations, studies of the host ranges of the phages carried by the field strains[23] may be helpful.

As an example of the application of these techniques to a clinical problem and to show how important the skin is in infections due to coagulase-negative staphylococci, let us take the investigation of infection in a cardiac surgery unit. We had the opportunity of retrospectively studying strains from a cardiac unit in which such infections were clinically significant; of studying strains carried on the skin of staff members[14] and of prospectively studying the acquisition of strains in the noses of the patients.

Over a 14-month period 382 operations, including 220 implantations of prosthetic valves, were performed. Infections occurred in 65 patients. These were mainly simple wound infections, but included 15 cases of septicemia (all after prosthetic implants) and 3 cases of osteomyelitis of the sternum. Coagulase-negative staphylococci were the most frequently isolated organisms, being recovered from the blood in pure culture in 12 of the patients with septicemia and in 40 episodes of infection in 35 patients. Single instances of septicemia were caused by *Pseudomonas maltophilia, Serratia rubidea,* and *Hemophilus influenzae.* Enterobacteria, usually in mixed culture, were recovered from 12 wounds; in 5 cases these were *Proteus,* in 3, *Escherichia* and unidentified enterobacteria, and 1 wound yielded a *Klebsiella.* Streptococci, all of group D, were isolated from 6 wounds.

Because of the recognition that 4 patients were simultaneously infected with a strain of *S. epidermidis* that was susceptible to a single phage, nose and skin swabs were taken from members of staff and some environmental samples were taken. In all, 511 isolates (285 from patients, including 111 preoperative swabs) were identified. Of these 409 were identified as *S. epidermidis* biotype 1. Only after employing all the techniques including "reverse" typing, direct phage typing, additional biochemical and antibiotic susceptibility tests was it possible to allocate 67 isolates to the epidemic strain and distinguish these from 21 isolates that made up another strain found in several members of staff. The epidemic strain was isolated from four septic wounds, the blood in ten episodes of septicemia, three reoperation specimens, the skin of one patient, and once from a series of six samples from the skin of a member of staff. The epidemic extended over 9 months, relapses occurred up to 6 months postoperatively, and five of the nine patients died.

Another strain was associated with five septic wounds and one replacement valve, while strains from four other septicemias that resulted in three deaths and 11 other wound infections could all be distinguished. Strains from infections were frequently resistant to many antibiotics, but this was not true of strains collected from patients before operation. Multiple resistant strains were rare in swabs from theater staff and from wound washouts, but were common on the skin of nurses in the intensive care unit (ICU). It appeared likely that the source of strains causing infection was the ICU. A prospective study was undertaken to confirm this supposition. For technical reasons this was restricted to nose swabs, taken from the patients pre- and

Table 2.1. Percentage of cardiac surgical patients' nose swabs giving the stated result at different times

	Preoperatively %	Postoperatively %	In ICU %
No growth	14	21	21
"Commensals"	14	3	0
Enterobacteria	0	4	4
S. aureus	29	24	9
Coagulase-negative staphylococci	54	59	87

postoperatively and on leaving the ICU, clinically required samples and a final survey of the strains carried by the staff. The results are summarized in Tables 2.1 and 2.2. The prophylactic antibiotics had the effect of reducing the carriage rate of *S. aureus* from 29% to 9% without permitting overgrowth of enterobacteria, but recovery of coagulase-negative staphylococci rose from 54% to 87%. Most (86%) were *S. epidermidis* biotype 1, and resistance to antibiotics increased markedly. In strains from the preoperative nose swabs, 68% were either sensitive to the 11 antibiotics tested or showed only one or two resistance mechanisms while only 20% of the strains recovered in the ICU and 26% of the strains from wounds and blood cultures were as sensitive as this. When all the strains from the staff were combined, only 20% of the strains from ICU staff showed two or fewer resistance mechanisms; the comparable figure for strains from theater staff was 83%. Medical staff strains were intermediate, with 40% of the strains in this group.

It was clear from the results of phage typing that the increase in resistance with time was due to the acquisition of new strains rather than selection of resistannt cells in a sensitive population. By including "reverse" typing it was possible to show that indistinguishable strains were present in both staff and patients, though the design of the study prevented a clear demonstration

Table 2.2. Antibiotic resistance mechanisms in strains from patients and staff

	No. of strains	Percentage of strains showing different numbers of resistance mechanisms						
		0	1	2	3	4	5	6
Patients								
Preoperative	48	8	27	33	4	10	8	8
Postoperative	36	6	25	14	22	19	8	6
ICU	30	0	10	10	23	7	43	7
Wounds, B/C[a]	50	0	18	8	14	24	30	6
Staff								
ICU	118	0	8	12	23	38	14	4
Theater	96	6	53	24	8	4	4	0
Medical	59	2	19	19	14	10	37	0

[a] Blood cultures

of the direction of transfer. The sequence of events appeared to be that the patient was admitted carrying antibiotic-sensitive coagulase-negative staphylococci that disappeared, as did *S. aureus,* as the prophylactic antibiotic regimen took effect. The patient was then recolonized by antibiotic-resistant strains present in the nose and on the skin of the staff of the ICU. Minor wound infections then led to endocarditis and, in a small proportion of cases, septicemia. If this scenario is accepted, it will imply changes in the manner in which aseptic and antiseptic procedures are carried out in the intensive care unit.

However, pathogenicity in cardiac patients is not restricted to *S. epidermidis.* We have been sent isolates from the blood and surgical specimens of an undoubted *Micrococcus* from six patients over a 3-year period. The six patients had all had multiple cardiac operations, and four had implanted prostheses. The organisms were remarkably similar. They produced no acid from "sugars," did not produce acetoin or reduce nitrate, but produced phosphatase and oxidase and hydrolyzed "Tweens." Unusual reactions were the hydrolysis of arginine, liquefaction of gelatin, and a positive CAMP test. All strains were resistant to methicillin and lysozyme. The species most similar was *M. lylae* but the type strain of this species did not hydrolyze arginine. Again the source seems to be human skin, as two strains with very similar characters were isolated from the skin of 18 normal volunteers, 13 of these carrying methicillin-resistant micrococci.

References

1. Baird-Parker AC (1963) A classification of micrococci and staphylococci based on physiological and biochemical tests. J Gen Microbiol 30:409–427
2. Baird-Parker AC (1965) The classification of staphylococci and micrococci from world-wide sources. J Gen Microbiol 38:363–387
3. Baird-Parker AC (1974) The basis for the present classification of staphylococci and micrococci. Ann NY Acad Sci 236:7–13
4. Corse J, Williams REO (1968) Antibiotic resistance of coagulase-negative staphylococci and micrococci. J Clin Path 21:722–727
5. Dean BA, Williams REO, Hall F, Corse J (1973) Phage typing of coagulase-negative staphylococci and micrococci. J Hyg (Camb) 71:261–270
6. Holt R (1969) The Classification of staphylococci from colonised ventriculo-atrial shunts. J Clin Path 22:475–482
7. Kligman AM (1965) The bacteriology of normal skin. In: Maibach HI, Hildick-Smith G (eds) Skin bacteria and their role in infection. McGraw-Hill, New York
8. Kloos WE, Schleifer KH (1975) Simplified scheme for routine identification of human *Staphylococcus* species. J Clin Microbiol 1:82–88
9. Kloos WE, Tornabene TG, Schleifer KH (1974) Isolation and characterization of micrococci from human skin including two new species: *Micrococcus lylae* and *Micrococcus kristinae.* Int J Syst Bacteriol 24:79–101
10. Kloos WE, Schleifer KH, Noble WC (1976a) Estimation of character parameters in coagulase-negative Staphylococcus species. Zbl Bakt Hyg I Abt Supplement 5:23–41
11. Kloos WE, Musselwhite MS, Zimmerman RJ (1976b) A comparison of the distribution of *Staphylococcus* species on human and animal skin. Zbl Bakt Hyg I Abt Suppl 5:967–973

12. Marples MJ (1965) The ecology of the human skin. Charles C Thomas, Springfield

13. Marples RR (1974) The microflora of the face and acne lesions. J Invest Dermatol 62:326–331

14. Marples RR, Hone R, Notley CM, Richardson JF, Crees-Morris JA (1978) Investigation of coagulase-negative staphylococci from infections in surgical patients. Zbl Bakt Hyg I Abt Orig A 241:140–156

15. Mitchell RG (1968) Classification of *Staphylococcus albus* strains isolated from the urinary tract. J Clin Path 21:93–96

16. Namavar F, de Graaf J, de With C, Maclaren DM (1978) Novobiocin resistance and virulence of *Staphylococcus saprophyticus* isolated from urine and skin. J Med Microbiol 11:243–248

17. Noble WC (1969) Distribution of the *Micrococcaceas*. Br J Dermatol 81 Supplement 1:27–32

18. Noble WC, Somerville DA (1974) Microbiology of human skin. Saunders, London

19. Nord CE, Holta-Oie S, Ljungh A, Wadstrom T (1976) Characterization of coagulase-negative staphylococcal species from human infections. Zbl Bakt I Abt Supplement 5:105–111

20. Pead L, Crump J, Maskell R (1977) Staphylococci as urinary pathogens. J Clin Path 30:427–431

21. Pelzer K, Pulverer G, Jeljaszewicz J, Pillich J (1973) Modification of Baird-Parker's classification system of *Staphylococcus albus*. Med Microbiol Immunol 158:249–257

22. Pulverer G, Pillich J, Klein A (1975) New bacteriophages of *Staphylococcus epidermidis*. J Infect Dis 132:524–531

23. de Saxe MJ, Notley CM (1978) Experiences with the typing of coagulase-negative staphylococci and micrococci. Zbl Bakt Hyg I Abt Orig A 241:46–59

24. Speller DCE, Mitchell RG (1973) Coagulase-negative staphylococci causing endocarditis after cardiac surgery. J Clin Path 26:517–522

25. Subcommittee on the taxonomy of staphylococci and micrococci (1976) Identification of staphylococci. Int J Syst Bacterior 26:332–334

Chapter 3
The Current Status of Aerobic Cutaneous Coryneform Bacteria

DAVID G. PITCHER AND PETER J. H. JACKMAN

Coryneform bacteria—gram-positive, nonsporing, pleomorphic rods—are found in almost every environment. The aerobic coryneforms of skin are commonly known in the medical professions as the "diphtheroids," a name coined to indicate a close relationship with the diphtheria bacillus. The more general term "coryneform" is more appropriate to these commensal skin bacteria, since many resemble *C. diphtheriae* in little but morphology. Although the majority of aerobic coryneforms can be assigned to the genus *Corynebacterium*, representatives of other coryneform genera are encountered, some of which may be residents. Others are derived from the environment and are termed transient or nomadic.[55]

The role of coryneforms on healthy skin and in diseases such as erythrasma and tinea will be discussed. An essential prerequisite is knowledge of current coryneform taxonomy.

Coryneform Taxonomy

Probably the most valuable method of assessing the genus affiliation of coryneforms has been that of cell-wall analysis. Three components must be considered: the diamino acids, the sugars, and the long-chain lipids (mycolic acids). These components may be easily and rapidly identified by paper and thin-layer chromatography of whole-cell hydrolysates.[33]

The genus *Corynebacterium* is characterised by its cell-wall sugars and diamino acids[11] and the presence of corynomycolic acids of chain length approx. 20–38 carbon atoms. *Nocardia* and *Mycobacterium* differ in possessing nocardomycolic acids (C_{50}) and mycolic acids (C_{80}), respectively.[28] *Brevibacterium* spp. possess fewer sugars and no mycolic acids.[19] Other cell-wall types do not contain mycolic acids and resemble *Arthrobacter globiformis;* others are undescribed species. Fatty acid patterns also separate *Corynebacterium* spp. from coryneforms related to *Brevibacterium* and *Arthrobacter.*[41]

Computer methods can be used to analyse the data from classical biochemical, morphological, and physiological tests. A numerical taxonomy of 134 strains with 97 biochemical tests[35] demonstrated two major groups within the *Corynebacterium* spp.: a group loosely clustered with reference strains of *C. pseudodiphtheriticum* that did not oxidize glucose and a group loosely clustered with reference strains of *C. minutissimum* and *C. xerosis* that oxidized glucose. In general, however, the skin strains seem distinct from reference strains from other habitats.

In a study of medically important *Corynebacterium* spp.[17] considering metabolic end products detected by gas–liquid chromatography (GLC), three broad groups were revealed: *C. diphtheriae* and some closely related species; a group including *C. minutissimum*, *C. renale*, and some strains of *C. xerosis*; and a group (comprising *C. pseudodiphtheriticum*, *C. aquaticum*, *Arthrobacter* spp. and *Brevibacterium* spp.) forming extremely few end products.

Chemical approaches include a detailed analysis of mycolic acids from skin *Corynebacterium* spp. by GLC and GLC mass spectrometry,[9] which indicate that the pattern and individual features of these components may be useful in intragenus classification.

Current research includes the taxonomic study of high-resolution whole-cell protein patterns obtained by gradient polyacrylamide gel electrophoresis. These patterns are individual fingerprints for strains and show potential use as epidemiological markers as well as in classification.

At present, most medically oriented studies of the aerobic cutaneous coryneforms limit the characterisation of isolates to strains that produce a coral-red fluorescence on special media or whose growth is stimulated by the presence of oleate (lipophiles), supplied as Tween 80, or those which possess esterases capable of breaking down Tween 80 (lipolytic). In some studies, coryneforms are referred to as being small or large colony. The latter are, in general, synonymous with the nonlipophilic coryneforms.

The Role and Distribution of Coryneforms in the Skin Flora

It has been suggested that coryneforms have a protective role on skin in that they are antagonistic towards gram-negative bacteria. Although there is in vitro evidence for this idea,[42] McBride et al.[26] have shown that patients who have repeated skin infections have a greater incidence of gram-negative bacteria but no obvious imbalance in the proportions of resident bacteria on most sites. However, in the axilla and toe webs of these patients a marked increase in coryneforms was noted.

These effects appear to negate a theory of coryneform protection. However, lipophilic coryneforms can be eliminated from the toe webs by the exaggerated use of antimicrobial soap leading to an overgrowth of gram-negative rods, maceration, and hyperkeratosis.[4] This could be interpreted as evidence for a protective role for the lipophilic coryneforms, but not the nonlipophiles.

The anterior nares have a more homogenous population of coryneforms than any other site. Several authors have noted the high incidence of strains possessing similar properties, particularly nitrate-reductase activity.[14, 18, 53]

The nares are the principle carriage sites for *S. aureus*; Pryjma and Heczko[36] noted that coryneforms in the nasal vestibule were of two types—those sensitive to the antagonistic properties of *S. aureus* and those which were resistant. Only the latter could coexist with *S. aureus* at this site.

Lipophilic Coryneforms

Noticeable differences in the ecology of the coryneforms emerge when the lipophilia is considered, though some obligate lipophiles may lose their requirement for Tween 80 after several subcultures or long-term storage, suggesting that the property may not be a stable one.[47]

Lipophilic coryneforms vary in their skin site distribution. On the scalp, though contributing only 10% of the aerobic flora, they were the second most prevalent organism detected by McGinley et al.[27] and occurred on 50% of scalps sampled.

In the region of the eye (cul-de-sac) McBride[25] found that coryneforms were the second most commonly isolated organism, comprising 8.8% of positive cultures (3.3% being lipophilic). Aly and Maibach[2] and Somerville[52] found an abundance of lipophilic coryneforms on the unoccluded forearm. A later study[4a] described the forearms as devoid of lipophilic coryneforms prior to occlusion, but they became the second most dominent component after 5 days of occlusion. According to Somerville[53] and Aly and Maibach,[1] lipophilic coryneforms are the most common types in the axilla, groin, and toe webs. After the experimental occlusion of glabrous skin, lipophiles may be the prevalent coryneforms where together with gram-negative rods, they appear to increase at the expense of the cocci.[4a,5]

The flora of the feet may be strongly influenced by the microclimate created by footwear. The lipophilic coryneforms increase, but nonlipophilic (large-colony) coryneforms show a rather erratic pattern.[4a,12,57]

Aly and Maibach[1] found a similar incidence of coryneforms in axilla, groin, and toe web. The only large difference in populations was in the axilla, where counts were over four times as high for nonlipophilic as for lipophilic coryneforms. Smith[48] found only small numbers of nonlipophilic coryneforms in proportion to the total aerobic coryneform flora. These conflicting reports probably indicate the great variation that may occur between groups of individuals chosen for surveys and the conditions under which these are carried out.

Nonlipophilic Coryneforms

On the evidence of cell-wall analysis, the property of lipophilia would appear to be confined to true *Corynebacterium* spp. of the skin though not all corynebacteria are lipophilic. Among the nonlipophilic coryneforms of skin, however, occur *Brevibacterium* spp. and coryneforms with cell-wall patterns uncharacteristic of *Corynebacterium*, some of which may be resident while others are undoubtedly derived from the environment. In conditions of poor hygiene, humidity, and occlusion these may increase.

The brevibacteria that occur on skin are similar to the type species *B. linens* which is found in dairy products. They differ principally in the lack of orange pigment, higher salt tolerance, and growth temperature of the cutaneous strains.

Sharpe et al.[43] discovered brevibacteria with these properties in dairy products and postulated that they were derived from human skin. Sharpe et al.[44] compared these strains to isolates from human skin and to *B. linens*—results indicated a close relationship. The principle feature of these coryneforms is their production of methane-thiol (CH_3SH) from L-methionine, a property said to contribute to the flavor and aroma of some cheeses.[24]

In addition, skin brevibacteria produce potent proteolytic enzymes and unlike most corynebacteria are lysozyme sensitive but penicillin resistant. Probably the most rapidly growing of skin coryneforms, they appear to favor the intertriginous regions, particularly the toe webs.[34]

Other nonlipophiles occur which cannot be assigned to *Corynebacterium* or *Brevibacterium*. Among these is a biotype producing propionic acid as its major metabolite of glucose. This would be indicative of a relationship with the *Propionibacterium acnes* group, but it is aerobic and has a different cell-wall polysaccharide. One property that may be of value in distinguishing this coryneform group is its reduction of nitrite. This characteristic appears uncommon in corynebacteria and brevibacteria, (Pitcher, unpublished material), though Smith et al.,[53] isolating aerobic coryneforms from normal and burned skin, oral cavities, rectal swabs, and the urine of burned children, found the incidence of nitrite-reducing coryneforms ranged from 27% (urine) to 66% (burned skin), all of which were nonlipoplilic. However, there is no evidence that these strains are of the same group as the propionic-acid producers. Pitcher,[33] on the cell-wall evidence, found only 2% of coryneforms to be of this type. They, like the brevibacteria, were more commonly isolated from the toewebs.

Lipolytic Coryneforms

The properties of lipophilia and lipolysis do not always coincide in aerobic coryneforms.

The substrates on which lipolytic coryneforms act appear to be limited. Smith and Willett (1968) found that nine out of ten strains were active only against Tween 80 and one strain against tributyrin, whereas staphylococci expressed a wide range of lipolytic activity against glycerol esters and natural oils though not against Tween 80.

The Occurrence of Coryneforms in Disease

Erythrasma

Erythrasma is recognized by scaling and a coral-red ultraviolet fluorescence of skin folds. *Corynebacterium minutissimum* has been implicated. It is defined as a coryneform producing coral-red fluorescence under ultraviolet light on suitably enriched media.[38] However, due to excess production of porphyrins, this property is widely distributed among bacteria,[20] including *C. xerosis* and *Micrococcus* spp.[52] In addition, Somerville[51] found that 2% of coryneform isolates were catalase-negative yet could also produce the coral-red fluorescence. These are probably derived from the environment, since skin corynebacteria and brevibacteria are catalase-positive whereas many soil and airborne coryneforms are negative.

The incidence of erythrasma in normal young adults was found by Somerville[51] to be 19%, but in a survey of mental patients the incidence was 43%.[56] There appeared to be an increase in the incidence of erythrasma with age, and it could be related to the hygienic status of the patients. Those who were the most mentally disturbed and received intensive nursing care exhibited little erythrasma. The isolation of fluorescent coryneforms from lesions could not alone explain the cause of erythrasma because they were isolated from only 35% of the fluorescent lesions and also from 25% of sites showing no fluorescence.

Sometimes fluorescent micrococci were isolated as the only fluorescent bacteria in a lesion. In terms of the number of bacteria on the lesions, Somerville[52] found that the mean count differences between sites showed high counts in area of scaling with fluorescence, less in scaling, nonfluorescent areas, and lower counts on normal areas. These counts were significant for both fluorescent and nonfluorescent coryneforms but not significant for fluorescent and nonfluorescent cocci. The nonfluorescent coryneforms outnumbered the fluorescent at all three areas.

Diabetes has been cited as a predisposing condition for erythrasma.[29,54] Factors such as sebogenesis and skin lysozyme deficiency may be linked to diabetes and have been suggested as contributory to the high incidence of erythrasma and other skin infections in diabetics.[8,37] However, corynebacteria and *S. aureus* are usually resistant to lysozyme so this factor may not contribute directly to an overabundance of these bacteria though brevibacteria, *S. epidermidis*, micrococci, and gram-negative bacteria may benefit from a reduction in lysozyme.

Keeping the area clean and dry is probably most effective in preventing erythrasma.

Tinea

Another condition in which coryneforms may play an important role is tinea. The area affected, usually the toe webs, is initially characterized by dry scaling, but not fluorescence, and the presence of dermatophyte fungi. However, dermatophytoses with erythrasma can occur.[13,40,56]

Leyden and Kligman[21] proposed that tinea is only initiated by the acquisition of a dermatophyte infection, resulting in a dry scaling. This is followed by dense colonization with resident microflora causing maceration, pruritis, and malodor. Tight shoes and hot weather were seen to exacerbate the condition, and these sequelae were attributed to overhydration. Antifungal treatments were found to have a low cure rate, but drying with aluminium compounds effectively inhibited bacteria and produced a high cure rate.

Goto[16] detected a massive increase in nonlipophilic, large-colony coryneforms from 28% of the flora in the toe web where only scaling occurred and fungi were present to 78% of the flora in soggy, macerated toe web where no fungi were isolated; the cocci increased from 30% to 50%; and lipophilic coryneforms from 47% to 50%. These findings are supported by Leyden and Kligman,[23] who found that lipophilic coryneforms predominated at the site of infection during the phase of simple scaling (dermatophytosis simplex). As the condition proceeded to a more serious lesion with itching, malodor, and maceration, nonlipophilic coryneforms increased enormously, and in severe cases (dermatophytosis complex) the flora was almost totally nonlipophilic, large-colony coryneforms. Coincident with this progression of symptoms was a general decrease in the recovery rate of fungi. Experimental occlusion confirmed these conclusions.

Experimental dermatophyte infection under occlusion was also carried out by Bibel and LeBrun[5] in an effort to elucidate changes in the resident bacterial flora. The coryneforms that they found to be increased during fungal infection are described as nonfluorescent, lipolytic, and lipophilic. The differences between infected and control sites were not shown to be significant, which does not bear out the findings of Leyden and Kligman.[23] We are not told whether the control area (volar forearm) was also occluded. However, there was a significant rise in the numbers of penicillin-resistant bacteria, both cocci and coryneforms, on the experimentally dermatophyte-infected forearms. Fundamental differences in the physico-chemical conditions between forearm and toe space may be responsible for these discrepancies.

With these results in mind, we may speculate on the nature of the nonlipophilic, large-colony coryneforms. The brevibacteria seem well suited to this role, being malodorous, proteolytic, and resistant to penicillin-like antibiotics. Dermatophyte fungi frequently produce such antibiotics, which could explain the selection of a resistant flora.[59]

Trichomycosis

The condition known as trichomycosis also appears to be due to an imbalance in the coryneform population at a site. It affects axillary or pubic hair. The coryneforms colonize the hair shaft forming concretions consisting entirely of bacterial cells. Trichomycosis affecting the pubic hair has received far less attention than the axillary form.[58a]

The phenomenon is frequently associated with poor hygiene and may go unnoticed since it causes little discomfort. Staining of the clothes and an offensive odor may direct attention to it.

Originally a causative organism was described as *C. tenuis*.[16a] Later work has shown that many different biotypes can be responsible for the condition[39] and no consistency in the type of coryneform could be found. McBride et al.[25] were able to artificially divide 19 strains from affected hairs into three groups. Group 1 strains were fluorescent, and Group 3 were obligate lipophiles. Group 2 was miscellaneous. All these bacteria were keratinolytic. However, Freeman et al.,[15] attempting in vitro colonization of hair with trichomycosis isolates, found that few strains were able to attack it. Keratinolysis was not confirmed by Orfanos et al.[31] using scanning electron microscopy; coryneforms were only seen to colonize the hair between the superficial horny layers. No significant relationship between trichomycosis and erythrasma could be found by Savin et al.[39] though fluorescent coryneforms were common in both conditions.

Coryneforms in Other Diseases of the Skin

Psoriatic plaques support an increased bacterial flora over that of uninvolved skin. Singh and Rao[46] did not divide coryneforms into groups and concluded that they remained in the same proportion of the flora on the plaques and uninvolved skin. However, Aly et al.[1] found the mean counts and incidence of lipophilic coryneforms on plaques to be significantly lower than on uninvolved skin. The incidence of the nonlipophilic strains was, however, slightly increased on the plaques.

In the lesions of atopic eczema, there was a lower incidence and mean count of coryneforms than on normal skin but only nonlipophiles could be isolated from the lesions.[4]

Coryneforms have also been investigated as a possible cause of diaper dermatitis. *Brevibacterium ammoniagenes,* a potent producer of ammonia from urea, was implicated by Cooke;[10a] however, Leyden et al.[22] found no difference between ureolytic coryneforms from diapers from the affected children and normal children. Of 18 strains of large-colony coryneforms, five were found to be potent producers of ammonia but their colonial morphology did not resemble that of *B. ammoniagenes*. Coryneforms like this organism were only rarely isolated. They[22] suggested that a more likely instigator of the condition is *Candida albicans.*

Pitted keratolysis is a condition in which crateriform pits occur in the plantar surface of the foot. Coryneform-like organisms have been seen in biopsies, but are frequently accompanied by filamentous bacteria.[58] There is, at present, no evidence to support a particular species as the etiological agent.[30]

Conclusion

Aerobic coryneform bacteria play a prominent role in the maintenance of a healthy normal flora on the skin.

Chemical or physical changes at the skin site may disturb the balance and promote excessive growth of the coryneforms and these, particularly the nonlipophilic group which includes the brevibacteria, may produce unpleasant effects.

Further studies of the ecological role of coryneforms on the skin must be preceded by an adequate classification system, otherwise there can be no correlation between conditions extant at a skin site with the properties of the bacterial occupying that niche.

References

1. Aly R, Maibach HI (1976) Effect of antimicrobial soap containing chlorhexidine on the microbial flora of skin. Appl Environ Microbiol 31:931–935
2. R, Maibach HI (1977) Aerobic microbial flora of intertrigininous skin. Appl Environ Microbiol 33:97–100.
3. Aly R, Maibach HI, Mandel A (1976) Bacterial flora in Psoriasis. Br J Dermatol 95:603–606
4. Aly R, Maibach HI, Shinefield HR (1977) Microbial flora of atopic dermatitis. Arch Dermatol 113:780–782
4a. Aly R, Shirley C, Cunico B, Maibach HI (1978) Effect of prolonged occlusion on the microbial flora, Ph, Co_2 and transepidermal water loss on human skin. J Invest Dermatol 71:378–381
5. Bibel DJ, LeBrun JR (1975a) Changes in cutaneous flora after wet occlusion. Can J Microbiol 21:496–500
6. Bibel DJ, LeBrun JR (1975b) Effect of experimental dermatophyte infection on the cutaneous flora. J Invest Dermatol 64:119–123
7. Bibel DJ, Lovell DJ (1976) Skin flora maps: a tool in the study of cutaneous ecology J Invest Dermatol 67:265–269
8. Binazzi M, Boncio L, Marconi P, Pitzurra M (1978) Serum and skin lysozyme activity in non-diabetic and diabetic subjects. Arch Dermatol Res 262:239–245
9. Corina DL, Sesardic D (1980) Profile analysis of the total mycolic acids from skin corynebacteria and from named *Corynebacterium* strains by gas-liquid chromatography and gas-liquid chromatography mass/spectrometry. J Gen Microbiol 116:61–68
10. Collins MD, Goodfellow J, Minnikin DE (1979) Isoprenoid quinones in the classification of Coryneform and related bacteria. J Gen Microbiol 110:127–136
10a. Cooke JV (1921) Etiology and treatment of ammonia dermatitis of gluteal region of infants. Am J Dis Child 22:481–490
11. Cummins CS, Lelliot RA, Rogosa M (1974) Genus *Corynebacterium*. Lehmann and Neumann. In: Buchanan RE, Gibbons NE (eds) Bergey's manual of determinative bacteriology, 8th edn. Williams and Wilkins Co, Baltimore, p 102
12. Duncan WC, McBride ME, Knox JM (1969) Bacterial flora: The role of environmental factors. J Invest Dermatol 52:479–484
13. English MP, Turvey J (1968) Studies in the epidemiology of Tinea pedis. IX. Tinea pedis and erythrasma in new patients at a chiropody clinic. Br Med J IV:228–232

14. Evans NM (1968) The classification of aerobic diphtheroids from human skin. Br J Dermatol 80:81–83
15. Freeman GR, McBride ME, Knox JM (1969) Pathogenesis of Trichomycosis axillaris. Arch Dermatol 100:90–95
16. Goto M (1970) Ecological study of athlete's foot. Jpn J Dermatol 80:130–132
16a. Grissey J, Rebell GC, and Laskas JJ (1952) Studies on causative organism of trichomycosis axillaries. J Invest Dermatol 19:187–197
17. Hine JE, Hill LR, Lapage SP (1978) Identification of medically important *Corynebacterium* spp by means of metabolic end-products detected bh gas-liquid chromatography. J Appl Bacteriol 45:v
18. Kasprowicz A, Heczko PB, Kucharezyk J (1974) A proposed classification of skin Corynebacteria. Med Dosw Mikrobiol 26:267–273
19. Keddie RM, Cure GL (1977) The cell wall composition and distribution of free mycolic acids in named strains of Coryneform bacteria and in isolates from natural sources. J Appl Bact 42:229–252
20. Lascelles J (1961) Synthesis of tetrapyrolles by microorganisms. J Biol Chem 41:417–441
21. Leyden JJ, Kligman AM (1975) Aluminium chloride in the treatment of symptomatic athlete's foot. Arch Dermatol 111:1004–1010
22. Leyden JJ, Kligman AM (1978a) The role of microorganisms in diaper dermatitis. Arch Dermatol 114:56–59
23. Leyden JJ, Kligman AM (1978) Interdigital athletes foot. Arch Dermatol 114:1466–1472
24. Manning DJ (1974) Sulphur compounds in relation to cheddar cheese flavour. J Dairy Res 41:81–87
25. McBride ME (1978) Evaluation of the microbial flora of the eye during wear of soft contact lenses. Appl Environ Microbiol 37:233–236
26. McBride ME, Duncan WC, Knox JM (1977) Cutaneous microflora of patients with repeated skin infections. J Cutan Pathol 4:14–22
27. McGinley KJ, Leyden JJ, Marples RR, Kligman AM (1975) Quantitative microbiology of the scalp in non-dandruff, dandruff and seborrhoeic dermatitis. J Invest Dermatol 64:401–405
28. Minnikin DE, Alshamaony L, Goodfellow M (1975) Differentiation of *Mycobacterium, Nocardia* and related taxa by thin layer chromatographic analysis of wholecell methanolysates. J Gen Microbiol 88:200–204
29. Montes, LF, Dobson H, Dodge BG, Knowles WR (1969) Erythrasma and diabetes mellitus. Arch Dermatol 99:674–680
30. Noble WC, Somerville DA (1974) Microbiology of human skin. Saunders, London
31. Orfanos E. Schloesser E, Mahrle G (1971) Hair destroying growth of *Corynebacterium tenuis* in the so-called Trichomycosis axillaris. Arch Dermatol 103:632–639
32. Pitcher DC (1976) Arabinose with *LL*-diaminopimelic acid in the cell wall of an aerobic Coryneform organism isolated from human skin. J Gen Microbiol 94:225–227
33. Pitcher DG (1977) Rapid identification of cell wall components as a guide to the classification of aerobic Coryneform bacteria from human skin. J Med Microbiol 10:439–445
34. Pitcher DG (1978) Aerobic cutaneous Coryneforms: recent taxonomic findings. Br J Dermatol 98:363–370
35. Pitcher DG, Noble WC (1978) Aerobic Diphtheroids of human skin. In: Bousfield J, Callely AG (eds) Special Publications of the Society for General Microbiology I. Coryneform Bacteria. Academic Press, London, pp 265–287
36. Pryjma J, Heczko PB, Bobr J (1971) Studies on the carrier state of

Staphylococcus aureus and diphtheroids of the nasal vestibule. Epidemiol Rev 25:197–202

37. Rebora A, De Marchi R (1975) Cutaneous, lipophilic Diphtheroids in diabetes mellitus. Giorn e min Derm 110:587–588

38. Sarkany I, Taplin D, Blank H (1961) The etiology and treatment of erythrasma. J Invest Dermatol 37:283–290

39. Savin JA, Somerville DA, Noble WC (1970) The bacterial flora of Trichomycosis axillaris. J Med Microbiol 3:352–356

40. Schlapper OLA, Roseblum GA, Rowden G, Phillips TM (1979) Concomitant erythrasma and dermatophytosis of the groin. Br J Dermatol 100:147–150

41. Sesardic D, Linnecar DFC, Noble WC (1978) Lipid composition as a taxonomic tool in the study of Coryneforms from human skin (abstr). Soc Gen Microbiol 6:31

42. Selwyn S, Ellis H (1972) Skin bacteria and skin disinfection reconsidered. Br Med J i:136

43. Sharpe ME, Law BA, Phillips BA (1976) Coryneform bacteria producing methanethiol. J Gen Microbiol 94:430–435

44. Sharpe ME, Law BA, Phillips BA, Pitcher DG (1977) Methanethiol production by Coryneform bacteria: Strains from dairy and human skin sources and *Brevibacterium linens*. J Gen Microbiol 101:345–349

45. Shehadeh NH, Kligman AM (1963) The effect of topical antibacterial agents on the bacterial flora of the axilla. J Invest Dermatol 40:61–71

46. Singh G, Rao DJ (1978) Bacteriology of psoriatic plaques. Dermatologica 157:21–27

47. Smith RF (1969) Characterization of the human cutaneous lipophilic Diphtheroids. J Gen Microbiol 55:433–443

48. Smith RF (1970) Fatty acid requirements of human cutaneous lipophilic Corynebacteria. J Gen Microbiol 60:259–263

49. Smith RF, Willett NP (1968) Lipolytic activity of human cutaneous bacteria. J Gen Microbiol 52:441–445

50. Smith RF, Blasi D, Dayton SL (1973) Isolation and characterization of Corynebacteria from burned children. Appl Microbiol 26:554–559

51. Somerville DA (1970) Erythrasma in normal young adults. J Med Microbiol 3:57–64

52. Somerville DA (1972) A quantitative study of erythrasma lesions. Br J Dermatol 87:130–137

53. Somerville DA (1973) A taxonomic scheme for aerobic Diphtheroids from human skin. J Med Microbiol 6:215–224

54. Somerville DA, Lancaster-Smith M (1973) The aerobic cutaneous microflora of diabetic subjects. Br J Dermatol 89:395–399

55. Somerville-Millar DA, Noble WC (1974) Resident and transient bacteria of the skin. J Cutan Pathol 1:260–264

56. Somerville DA, Seville RH, Noble WC, Savin JA (1970) Erythrasma in a hospital for the mentally subnormal. Br J Dermatol 82:355–359

57. Tachibana DK (1976) Microbiology of the foot. A Rev Microbiol 30:351–375

58. Taplin D, Zaias N (1968) The etiology of pitted keratolysis. XIII Int Cong Dermatol 1:593

58a. White, SW and Smith, J (1979) Trichomycosis Pubis. Arch Dermatol 115:444–445

59. Youssef N, Wyborn CHE, Holt G, Noble WC, Clayton YM (1978) Antibiotic production by dermatophyte fungi. J Gen Microbiol 101:105–111

Chapter 4
Factors Controlling Skin Bacterial Flora

RAZA ALY AND HOWARD MAIBACH

Several factors control skin colonization of bacteria, including pathogens, on human skin. When artificially applied on human skin, many pathogenic microorganisms (such as β-hemolytic streptococci and *Staphylococcus aureus*) do not survive for more than a few hours. Yet, on some, these organisms survive, multiply, and produce disease. Cutaneous bacterial flora do not numerically reflect the types of organisms in ambient air, suggesting that skin controls the survival of bacteria contacting it. In our view, the skin environment is unfriendly to most microbial species but sufficient nutrients are available to allow a certain number of bacteria to survive and multiply.

The ability to control exogenous bacterial life has been related to humidity, skin surface lipids, microbial antogonisms, bacterial adherence, desquamation, pH, toxic products, and secretory antibody.

Humidity

Superficially applied microorganisms survive much longer on wet skin.[10,26] The resident flora is denser in moist intertriginous regions. Most of the water on skin is derived from eccrine sweat and transepidermal water loss. In the axillary region there is also an intermittent contribution from apocrine sweat. The effect of hydration on the bacterial flora of the skin has been studied by Marples[21] and Aly et al.[8] Hydration not only elevates the microbial density, but also alters the relative ratios. We measured the effect of complete occlusion on the skin on aerobic microbial flora, pH, transepidermal water loss, and CO_2 emission rate.

Bacteria Under Occlusion

The arms were wrapped with plastic film Saran Wrap for 5 days (Fig. 4.1). Each day a succeeding portion of the wrap was removed to take measurements and microbial samples. Bacterial samples were removed by the detergent scrub method.[37]

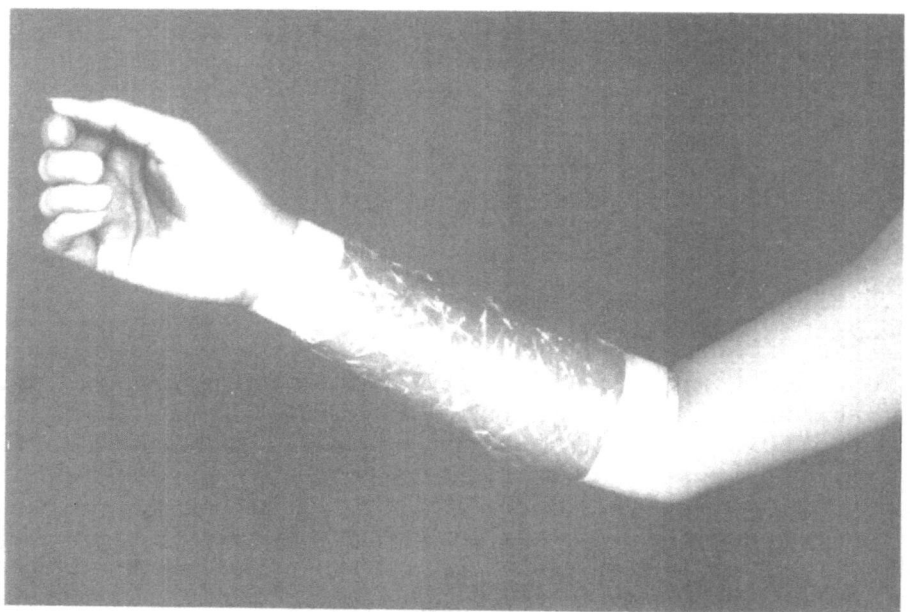

Fig. 4.1. The arm was occluded with a plastic film (Saran Wrap) for 5 days. This film was tightly secured at the wrist and just below the elbow with paper adhesive tape (Micropore).

The total average geometric microbial counts before occlusion were 1.8×10^2 organisms/cm²; after 24 h of occlusion this increased to 1.4×10^6/cm². Maxium counts (9.8×10^7/cm²) were noted on Day 4 of occlusion.

The types of microorganisms comprising the total counts changed (Fig. 4.2). The coagulase-negative staphylococci counts peaked on Day 4. The maximum counts for micrococci on Day 3 were 9.9×10^4/cm²; this reduced dramatically on Days 4 and 5. Lipophilic coryneforms were not detected before occlusion and peaked on Day 5 (1.5×10^5/cm²). They remained the second highest flora on Days 4 and 5. Lipophilic coryneform prefer a higher pH and moister environment than micrococci.[12] There was no definite trend for coryneforms—these fluctuated day to day. Gram-negative rods emerged after occlusion but never became the dominant flora. In Marples' study,[21] gram-negative rods increased to 10% of the total flora after 4 days of occlusion. In our study gram negatives were less than 1% of the total flora.

Skin pH Under Occlusion

The skin pH remains fairly constant in the normal adult population; most skin bacteria can grow under all pH conditions found on the skin. A slight change in pH might cause an ecological advantage in growth rate for a par-

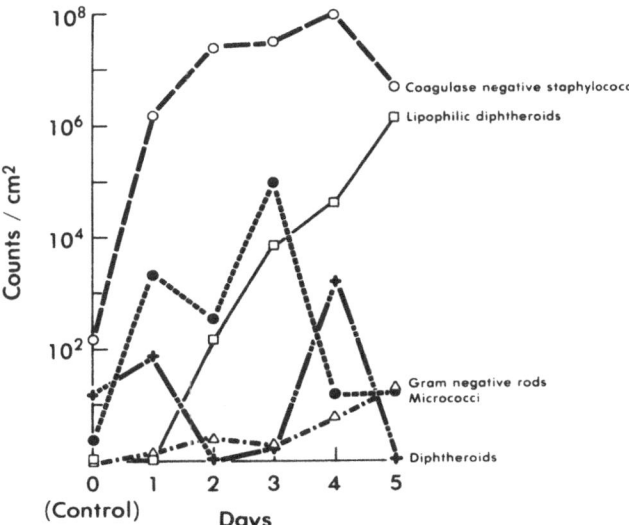

Fig. 4.2. Geometric microbial counts/cm² under the occlusion. The counts before occlusion were considered as control.

ticular organism with regard to its optimum hydrogen ion concentration. A pH meter was used to measure the skin's pH before and after occlusion (Fig. 4.3). The pH of the skin gradually shifted from the acidic range (4.38) before occlusion, to neutral (7.05) on Day 4 of occlusion. The pH of the skin may have reached an equilibrium with the interstitial fluid, as the final pH was

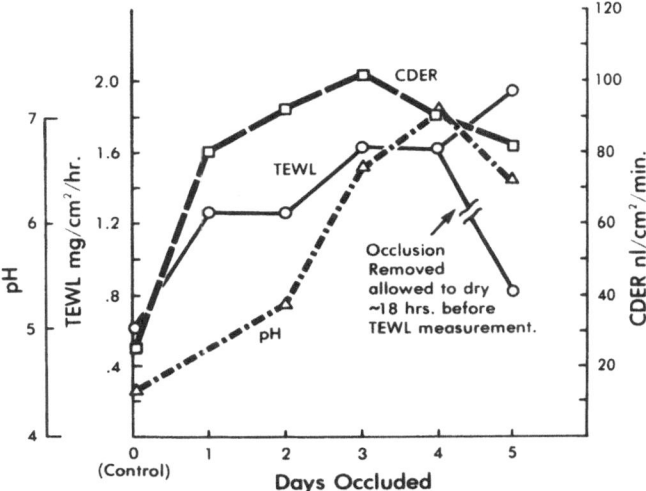

Fig. 4.3. Effect of occlusion on transepidermal water loss, carbon dioxide emission rate, and pH of skin. The unoccluded skin served as control.

close to the body pH. In general, skin pH is higher in humid intertriginous areas.

Transepidermal Water Loss (TEWL) Under Occlusion

Transepidermal water loss was measured with an electrolytic water analyzer (Fig. 4-3). TEWL reflects water diffusing through the moist membranes of the analyzer and water being lost as part of the hydration process. Increased TEWL reflects increase in water content of skin surface, provided no damage has occurred. TEWL was 0.56 mg/cm²/hr before occlusion and increased to 1.87/mg/cm²/hr on Day 5. With occlusion the skin was essentially saturated within 24 hours. TEWL was measured 18 hours after removal of occlusion to determine if there was any structural damage to the skin. These values, although higher than for previously occluded skin, were not significantly different from the control.

Carbon Dioxide Emission Rate (CDER) Under Occlusion

The CDER was measured with an infrared analyzer (Fig. 4-3). The CDER emission rate increased from less than 0.1% to over 5%. Changes in concentrations have effects on the type of microbial flora if two types are grown in competition.

We have shown that occlusion not only hydrates the skin but also alters pH and CO_2 emission rate. Other factors—such as bacterial metabolites, bacterial antogonisms, skin surface lipids, salt concentration and nutrient—may have been changed upon occlusion, thus exerting accumulative pressure on the skin flora.

Skin Surface Lipids

Eccrine sweat contains factors that inhibit the growth of *Microsporum gypsium* and *Epidermatophyton inguinale*.[22] This fungicidal property was attributed to acetic, propionic, caprylic and lactic acids, leading to the early development of fatty acid topical therapy of superficial cutaneous fungal infection. Burtenshaw[13] prepared crude ether extracts of skin; their antimicrobial effect against common bacterial pathogens was demonstrated. *Streptococcus pyogenes*, viridan streptococci, *Corynebacterium diphtheriae*, and certain strains of *Staphylococcus aureus* were sensitive to these extracts. Ricketts[27] extracted lipids from the forearms of volunteers with acetone and showed that unsaturated fatty acids rapidly kill a variety of microorganisms in vitro.

We substantiated earlier studies and demonstrated that when *S. aureus*, *S. pyogenes*, or *Candida albicans* were deposited on normal skin and

Table 4.1. Survival of *Staphylococcus aureus, Streptococcus pyogenes* and *Candida albicans* on forearms washed with acetone and unwashed (control) (Aly et al. 1972)

Subjects	Ratio between the two arms		
	S. aureus	*S. pyogenes*	*C. albicans*
Average of 10–15	1:80	1:102	1:37
subjects	P = 0.008	P = 0.08	P = 0.002

[a]The total bacterial counts on the acetone washed arms were divided by the total counts on the control arm and were expressed as a ratio.

occluded for 5 hours, their numbers were reduced significantly. In contrast, forearms washed with acetone allowed greater persistence of artificially applied bacteria.[4] The ratio of microbial survival between washed and unwashed forearms is shown in Table 4-1. Gram-negative rods (*Escherichia coli* and *Pseudomonas aeruginosa*) were not sensitive to the antimicrobial action of skin surface lipids.

Antimicrobial activity of the skin was restored when skin surface lipids were replaced on the skin (Table 4-2). Acetone-treated sites in which the extracts were not replaced had higher bacterial counts (ratio 1:68) than the sites in which the extracts were replaced on the sites (ratio 1:2). In an additional control study acetone was allowed to evaporate from the skin inside the delineated area. These acetone extracts were not removed. No significant difference was seen when *S. aureus* counts between acetone-treated arms and control (untreated) were compared. This suggests that it is not the solvent per se that increases the growth of microorganisms in the above study; it is more likely that specific antimicrobial lipids were removed.

The antimicrobial activity of skin surface lipids has been demonstrated in vitro. These organisms sensitive to skin surface lipids in vivo were also susceptible in vitro. *S. pyogenes* were most sensitive to lipids, followed by *C. albicans* and *S. aureus* (Table 4-3).

For some skin bacteria skin surface lipids may have a stimulatory rather than inhibitory role. Puhvel and Reisner[25] found that although saturated fatty acid (C_8, C_{16}) inhibited the growth in vitro of two strains of *propionibacterium acnes*, oleic acid (unsaturated C_8) promoted their growth. Oleic acid promotes growth of cutaneous aerobic lipophilic diphtheroids in vitro.[24] Pillsbury and Rebell[23] showed that many fatty acids present in

Table 4.2. A comparison of the average effect in 10 subjects of (1) removal of acetone extract of skin, (2) removal and replacement of acetone extract, and (3) treatment and evaporation of acetone from the skin on artificially applied *S. aureus* (Aly et al. 1972)

Untreated arm vs. treated arm		
(1)	(2)	(3)
1:68	1:2	1:1

Table 4.3. Antimicrobial effect of acetone skin extracts on *S. aureus, C. albicans* and *S. pyogenes* (Aly et al. 1974)

S. aureus	Percent recovery of organisms C. albicans	S. pyogenes	Control[b]
63	41	30	100

[a]The colony counts from tubes containing skin extracts were divided by the counts from tubes with acetone and were expressed as the percentage of each organism recovered.
[b]Acetone only

sebaceous secretions, especially caproic and caprylic acids, were lethal in vitro for several organisms.

It has been suggested that the more varied bacterial flora of infants and children, as compared with those of adults, could be due to the relatively low level of free fatty acids on the skin of the children.[31] Streptococci that are very sensitive to antimicrobial effect of fatty acids are rarely found on adult skin but occur on infants' skin.

Microbial Antagonism

Microbial antagonism regulating the resident flora on the skin surface is undoubtedly mechanistically complex. Some mechanisms may involve more or less direct interactions between various types of indigenous microbes. Interferring organisms elaborate antimicrobial substances that may inhibit the growth of transient microbes. The role of bacteriocins in the microbial ecology of skin is not fully investigated. Bacteriocins are produced by many gram-negative and gram-positive bacteria. Those produced by gram-positive bacteria are antibacterial only against strains of the same or closely related species, while bacteriocin produced by gram-negative bacteria have a wider range or activity.

Bacteriocins are a chemically heterogenous group of substances ranging from polypeptides and simple proteins to proteins associated with carbohydrates and lipids and small phagelike particles.

In the early antibiotic era many workers[15,17,34] searched for potentially useful antibiotic producers among bacterial species indigenous to man. The first application of bacterial interrelationships in maintaining the bacterial status quo was by Sieberth, et al.[31] They were unable to detect *Proteus* in the colon of turkey poults until the density of *E. coli* and enterococci strains elaborating bacteriocidin against *Proteus* were greatly reduced by oral oxytetracycline. Organisms from the intestinal tract of turkey poults were demonstrated to be antagonistic to *Proteus* on solid media.

Alpha hemolytic streptococci inhibit growth of bacteria in the oropharynx.[32] We have observed that certain strains of viridans streptococci isolated from the nose of normal subjects inhibit *S. aureus* and *S. pyogenes* when grown together on trypticase soy agar (to be published). Prior intra-allontoic infection of chicken embryos with viridans streptococci protected them from the virulent strain of *S. aureus.*[5]

More recently the phenomenon of bacterial interference has been utilized to control cross-infection of the skin and nasal mucosa by staphylococci. *S. aureus* 502A obtained from a nurse and infants in the hospital nurseries who were free from staphylococcal disease was capable of colonizing nasal mucosa and umbilicus when artificially applied. *S. aureus* 502A drastically reduced the incidence of infection of these body sites by virulent strains of *S. aureus*. This aspect of bacterial interference has been utilized as a therapeutic tool in patients with recurrent furunculosis.[29] No antibacterial substance seems to be produced by this strain[9]; therefore, its mode of interference is not related to the production of bacteriocin or other inhibitor. Viable *S. aureus* 502 is required to demonstrate bacterial interference phenomenon.[6] Patients with extensive burns, with neoplastic conditions, who are chronically ill or immunosuppressed would be potential beneficiaries from such ecological approaches. In these patients administration of antibiotics often leads to the overgrowth of more dangerous microbial flora.

The opposite effect—satellitism, i.e., growth enhancement of one organisms by another—is a little understood phenomenon of the microbial ecology of skin. It is not known whether the bacteria involved in this phenomenon exert their effects by production of essential growth factors or destruction of toxic substances. The role of satellitism in the ecology of skin requires systematic study.

Bacterial Adherence

Adherence of microorganisms to skin surface is another factor determining colonization. Not all bacteria adhere to skin even when competition from other bacteria are eliminated.[7] The binding is a specific event brought about by the molecular interaction of the adhesion and the receptor. Such specificity is of ecological advantage because it increases the probability that the bacteria become located on a surface suitable for colonization. The epithelial cell-bacterial interaction requires the presence of binding substances on the surface of the bacteria. Several studies indicate that Pili in gram-negative bacteria and fimbriae in gram-positive bacteria are involved in binding the organisms to host surfaces.[11,14,33]

Table 4.4. Adherence of bacteria to skin (forearm) epithelial cells

Organisms	Average bacterial counts[a]
Staphylococcus aureus	36
Staphylococcus epidermidis	42
Streptococcus pyogenes	42
Viridan streptococci	12
Pseudomonas aeruginosa	98
Klebsiella pneumoniae	22
Aerobic diphtheroids	95
Candida albicans	11

[a]The average of 50 epithelial cells.

We have shown a selective ability of bacterial adherence to human skin epithelial cells. Significant adherence occurred with *P. aeruginosa*, *S. epidermidis*, *S. aureus*, *S. pyogenes*, and diphtheroids but not with viridan streptococci and *Candida albicans*. The feeble attachment of viridan streptococci was of considerable interest and may correlate with the fact that their presence on the skin is seldom noted. *P. aeruginosa* and *S. pyogenes* showed good attachment despite the fact that these are not seen in healthy adult populations. It seems that other host and ecological factors may be operative in vivo and may have been altered in the in vitro model.

The capacity of a given microorganism to colonize a particular epithelial surface is proportional to the ability of the organism to adhere to that surface. Group A streptococci isolated from skin adhere in greater number to human epithelial cells than to cells obtained from buccal mucosa, whereas streptococci isolated from the throat tend to adhere better to buccal epithelial cells.[2]

Desquamation

The ecological aspect of epithelial cells' desquamation has not been generally appreciated. This factor of host defense mechanism is effective in limiting bacterial population. Regions with heavy microbial burden, such as the intestinal canal, have a higher turnover of epithelial cells. In germ-free animals where the turnover rate is much slower, the desquamation rate is less.[1] It seems that indigenous flora will reattach upon desquamation, while feebly adhering bacteria will eventually be eliminated. Therapists have used agents (salicyclic acid) that promote the casting off of horny scales with this effect in mind.

Secretory Antibodies

A wide spectrum of antibacterial antibodies occur in external secretions of the normal host. It is conceivable that these antibodies play an important role in the regulation of normal flora and pathogenic bacteria. The role of secretory antibody against bacterial infection has been studied. The predominant class of antibody generally has been IgA and occasionally IgG, IgM, and IgE.[28] The mechanism by which secretory IgA mediates its antibacterial activity on mucous surfaces remain a matter of speculation. Theoretically, antibodies in secretions may react with bacteria while they are multiplying on the mucosal surfaces; hence, when the organisms become dislodged, they would have a reduced ability to reattach.

IgA antibodies found in parotid saliva can inhibit the attachment of oral streptococci to epithelial cells.[36] It is expected that this property would inhibit streptococcal colonization without the involvement of phagocytes or complement system. Such mechanisms may explain, at least in part, how IgA antibodies provide protection on mucosal surfaces.

In naturally acquired immunity against pathogenic bacteria, the role of secretory antibodies is important in controlling the bacterial colonization. Many natural infections are followed by a state of immunity in which the host is less likely to be colonized by similar types of pathogens.

Conclusion

We have only begun to scratch the surface in understanding the factors controlling skin bacterial flora. It is reasonable to believe that many factors affect the composition of the resident flora and influence host resistance or susceptibility to pathogenic organisms. The ecological principles that we have discussed may be simple in concept but their underlying molecular basis is more complex. Several factors independently or in combination to control skin biota. One cannot single out one individual factor as the one solely responsible in directing the survival of organisms on the skin. Some of these host or bacterial factors may be more effective acting in combination rather than alone. We have already learned that occlusion of the skin results not only in hydration, but also alters several ecological factors. We also know that not all bacteria are susceptible to the same treatment. Gram-negative bacteria are more sensitive to dessication than to skin lipids. Gram-positive bacteria, particularly S. pyogenes and S. aureus, are susceptible to skin lipids. On the other side of the spectrum, some diphtheroids require lipids for their growth. It is a complex issue; there is much to be learned.

References

1. Abrams GD, Bauere H, Sprinz H (1963) Influence of the normal flora on mucosal morphology and cellular renewal in the ileum. Lab Invest 12:355–364
2. Alkan M, Ofek I, Beachey EH (1977) Adherence of pharyngeal and skin strains of group A streptococci to human skin and oral epithelial cells. Infect Immun 18:555–557
3. Aly R, Maibach HI, Strauss WG, Shinefield HR (1972) Survival of pathogenic microorganisms on human skin. J Invest Dermatol 58:205–210
4. Aly R, Maibach HI, Shinefield HR (1975) Correlation of human in vivo and in vitro cutaneous antimicrobial factors. J Infect Dis 131:579–583
5. Aly R, Maibach HI, Shinefield HR (1974) Protection of chick embryos by viridan streptococci against lethal effect of S. aureus. Infect Immun 9:559–563
6. Aly R, Maibach HI, Shinefield HR (1974) Bacterial interference among strains of S. aureus in man. J Infect Dis 129:720–724
7. Aly R, Shinefield HR, Maibach HI, Strauss WG (1977) Bacterial adherence to nasal mucosal cells. Infect Immun 17:546–549
8. Aly R, Shirley C, Cunico B, Maibach HI (1978) Effect of prolonged occlusion on the microbial flora, pH, CO_2 and transepidermal water loss. J Invest Dermatol 71:378–381

9. Anthony BF, Wannamaker BFA (1967) Bacterial interference in experimental burns. J Exp Med 125:319–336

10. Arnold L (1942) Relationship between certain physico-chemical changes in the cornified layer and the endogenous bacterial flora of the skin. J Invest Dermatol 5:207–223

11. Beachey EH, Ofek I (1976) Epithelial cell binding of group A Streptococci by Lipoteichoic acid on fimbriae denuded of M protein. J. Exp Med 143:759–771

12. Blank IH, Dawes RK (1950) The water content of the stratum corneum. IV. The importance of water in promoting bacteria multiplication on cornified epithelium. J Invest Dermatol 31:141–145

13. Burtenshaw JM (1942) The mechanisms of disinfection of the human skin and its appendages. J Hyg 42:184–210

14. Costerton JW, Geesey GG, Cheng K-J (1978) How bacteria stick. Sci Am 238:86–95

15. Florey HW (1946) The use of microorganisms for therapeutic purposes. Yale J Biol Med 19:101–118

16. Gibbons RJ, Van Houte J (1975) Bacterial adherence in oral microbial ecology. Ann Rev Microbiol 29:19–44

17. Halbert SP, Swick LS (1950) In vivo antibiotic production by E. Coli. J Immunol 65:675–686

18. King RD, Dillavou CL, Greenberg JH, Jeppsen JC, Jaegar JS (1976) Identification of CO_2 as a dermatophyte inhibitory factor produced by Candida albicans. Can J Microbial 22:1720–1727

19. Michalek SM, McGhee JR, Mestecky J, Arnold RR, Bozzo L (1976) Ingestion of Streptococcus mutans induces secretory immunoglobulin A and caries immunity. Science 192:1238–1240

20. Montgomery PC, Cohn J, Lally ET (1974) In: Mestecky J, Lawton AR (eds) The immunoglobulin A system, Plenum, New York, pp. 453–462

21. Marples RR (1963) The effect of hydration on the bacterial flora of the skin. In: Maibach HI, Hildick-Smith G (eds) Skin bacteria and their role in infection. McGraw-Hill Book Co., New York, p 33

22. Peck S, Rosenfeld H, Leifer W, Bierman W (1939) The role of sweat as a fungicide. Arch Derm Syph 39:126–148

23. Pillsbury DM, Rebell G (1952) The bacterial flora of the skin. Factors influencing the growth of resident and transient organisms. J Invest Dermatol 18:173–186

24. Pollack MR, Wainwright SD, Mansion EED (1949) The presence of oleic acid requiring diphteroids on human skin. J Pathol Bacteriol 61:274–276

25. Puhvel M, Reisner RM (1970) Effect of fatty acids on the growth of Corynebacterium acnes in vitro. J Invest Dermatol 54:48–52

26. Rebell G, Pillsbury DM (1950) Factors affecting the rapid disappearance of bacteria placed on the normal skin. J Invest Dermatol 14:247–263

27. Ricketts CR, Squire JR, Topley E (1951) Human skin lipids with particular reference to the self sterilizing power of skin. Clin Sci 10:89–111

28. Rossen RD, Wolff SM, Butler WT (1967) The antibody response in nasal washings and serum to S. typhosa endotoxin administered intravenously. J Immunol 99:246–254

29. Shinefield H, Ribble JC, Boris M, Aly R, Maibach HI (1974) Bacterial interference between strains of S. aureus. Ann NY Acad Sci 236:445–455

30. Sieberth JM, McGinnis J, Skinner CE (1952) The effect of terramycin on the antagonism of certain bacteria against species of Proteus. J Bacteriol 64:163–169

31. Somerville, DA (1969) The normal flora of the skin in different age groups. Br J Dermatol 81:248–253

32. Sprunt K, Leidy GA, Redman W (1971) Prevention of bacterial overgrowth. J Infect Dis 123:1–10
33. Swanson J (1973) Studies on gonococci infection. IV. Pili: Their role in attachment of gonococci to tissue culture cells. J Exp Med 137:571–589
34. Thompson R, Shibuya M (1946) The inhibitory action of saliva on the diphtheria bucillus. J Bacteriol 51:671–684
35. Wannamaker LW (1970) Difference between streptococcal infections of the throat and of the skin. N Engl J Med 282:23–30, 75–85
36. Williams RC, Gibbons RJ (1972) Inhibition of bacterial adherence by secretory immunoglobulin A: a mechanism of antigen disposal. Science 177:697–699
37. Williamson P, Kligman A (1965) A new method for the quantitative investigation of cutaneous bacteria. J Invest Dermatol 45:498–500

Chapter 5

Microbiology of Specialized Sites in Relation to Infection

W. C. NOBLE

The specialized sites of human skin are the scalp, axillae, perineal area, and toe webs. Much of the published work on these sites has been performed because of commercial interest rather than for an intrinsic interest in the area or its relation to infection. A full description of the general microbiology of these sites is published.[16] This discussion will concern itself with specialized sites in relation to infection; because of obvious similarities, only one aspect of infection will be discussed in relation to each site.

The Scalp

The principle microorganisms in the scalp are the yeasts *Pityrosporum;* these have been well studied in relation to dandruff and are rarely involved in other infections. McGinley et al.[13] reported that the normal, nondandruff scalp carries 5×10^5 *Pityrosporum* per cm², 3×10^5 *P. acnes* complex/cm², and 2×10^5/cm² aerobic bacteria. No studies of *P. acnes* dispersal from the scalp have been conducted; it would be difficult to identify those from scalp as distinct from facial or thoracic organisms, but *P. acnes* does occur in surgical wound infection, especially when a prosthesis such as a hip joint is involved.

Among the aerobic bacteria on the scalp are gram-negative bacilli and *Staphylococcus aureus.* The hair acts as a sampling device for environmental bacteria and has been described as a potential source of dispersed organisms.[21,15,4] The beard has also been reported as a potential hazard[3,10]; it may be significant that editorial comments outnumber experimental studies two to one.

It has been difficult to demonstrate that organisms on the hair are other than a potential hazard.[5,6] Only Dineen and Drusin[7] have associated contaminated hair with surgical wound infection; in their study there was evidence of frank scalp infection that may have been more important than carriage. Huismans-Evers[10] has shown that contaminated scalp and beard hair can lead to dispersal of *S. aureus* and thus may pose a potential problem in operating rooms. That potential has not been fully explored.

The Axilla

The axilla has been studied mainly as an area that requires the application of a deodorant or antiperspirant. We do not yet know the importance of the various coryneforms inhabiting the axilla and contributing to odor; the studies of Pitcher and Jackman (see Chapter 3) throw some light on this subject.

Recently there has been an increased interest in the organisms known as *Acinetobacter*. Originally described as *Mima* or *Herellea*, recent taxonomic advances have shown them to be more closely related than was formerly believed: *Mima polymorpha* is now *Acinetobacter calcoaceticus var lwoffi* while *Herellea vaginicola* is now *A. calcoaceticus var anitratus*. Taplin, Rebell, and Zaias[22] (Table 5.1) reported these organisms as inhabitants of the axilla, groin, and toewebs.

There has been a substantial review of the role of these organisms in infection from CDC, Atlanta, by Retailliau et al.,[17] who report a summer maximum of infection over each of the last five years. Where laboratories have reported varieties separately, only *A. calcoaceticus var anitratus* contributes to this maximum. Males are more frequently infected than females; indeed males account for 57% of infection with *Acinetobacter* although they form only 43% of the population infected with other organisms—the young adult male is especially at risk. In a study of infection with *A. calcoaceticus var anitratus* alone, Glew et al.[8] found that the very old and very young plus those who had some impairment of immunity or continuing surgical interference were most often infected. Nine of their 53 patients had septicemia; all had indwelling intravenous catheters at the time of infection.

Kloos and Musselwhite[11] previously reported a summer maximum in skin carriage of this organism though, curiously, they found few *Acinetobacter* in the axilla—perhaps a reflection of deodorant use. They suggest the higher levels might be due to increased sweating in the hotter, more humid summer months. Males sweat more than females, and this may account for the excess of infected males, while the general increase in sweating may account for the summer increase. Alkhoja and Darrell[1] have suggested skin as a source of *Acinetobacter* in blood.

Although these organisms are only rare pathogens, appearing in fewer than 1% of infections reported to CDC, it seems that they present a clear example of infection with a normal skin inhabitant, the headquarters of which are the axilla and groin. The precise relationship between skin populations and infection remains to be fully explored.

Table 5.1. Recovery of *Acinetobacter calcoaceticus* from skin

| | Taplin, Rebell and Zaias (1963) | | Somerville and Noble (1970) | |
	v. lwoffi	*v. anitratus*	*v. lwoffi*	*v. anitratus*
Axilla	1	10	21	14
Groin	7	17	21	13
Toe webs	6	20	19	8
Forearm	7	15	ND	ND

Perineal Area

One particular aspect of infection associated with the perineal area provides a good example of our lack of detailed knowledge of the skin flora, even in relation to a well-accepted infection: urinary tract infection in young women caused by *Staphylococcus saprophyticus*. In the Baird-Parker[2] classification of Micrococcaceae, the urinary pathogen together with other strains isolated from skin appeared as *S. saprophyticus* biotype 3. Revisions of the taxonomy by Kloos and Schleifer[12,18] have shown that urinary tract strains are *S. saprophyticus*; the skin strains are classified as *S. cohnii* or *S. warneri*[14,9] see Table 5.2.

In a recent doctoral thesis from the University of Amsterdam, Namavar[14] points out that although *S. saprophyticus* can be found sparsely on human skin, it does not seem sufficiently common for us to attribute urinary tract infection merely to its presence. Sellin et al.[19] remarked that this organism is rarely found in the introitus or urethra of normal young women or in the prepuce or urethra of normal young men, suggesting that this is not a sexually transmitted disease, nor does a fecal source seem likely.

It seems most probable that *S. saprophyticus* (sensu Kloos and Schleifer), which acts as a urinary tract pathogen, is also an inhabitant of human skin. Yet it is a relatively rare one, appearing on normal skin with about the same frequency as *S. aureus*. However, the principle carrier sites of *S. aureus* are well known while those of *S. saprophyticus* remain to be discovered. Presumably, like *S. aureus*, it requires some particular conditions to initiate disease. Mechanical interference alone probably does not suffice, since coccal urinary tract infection in hospital is caused by *S. epidermidis*.

The Toe Webs

Much of what has been written above could also be applied to the toe webs, which are most frequently studied in relation to tinea pedis. However the toe webs form an excellent reservoir of miscellaneous gram-negative rods. In a recent study by Pitcher, Noble, and Seville (unpublished work), a much greater variety of gram-negative rods was encountered in the toewebs than in the axilla or perineum (Table 5.3). Gram-negative rods now form the

Table 5.2. Coagulase-negative cocci from urinary tract infection classified as *S. saprophyticus* biotype 3 sensu Baird-Parker (1974)

	Identification sensu Kloos and Schleifer (1975) Percentage distribution			
	S. saprophyticus	*S. warneri*	*S. cohnii*	Others
Namavar et al.[14] 77 isolates	77	13	8	2
Hovelius and Mardh[g] 50 isolates	98	0	0	2

Table 5.3. Gram-negative bacilli isolated from the toe webs of a sample of 43 persons

Nomenclature according to API 20E	Subjects
Moraxella lwoffi	4
Enterobacter cloacae	3
Alcaligenes	3
Aeromonas hydrophila	2
Proteus mirabilis	2
Acinetobacter calcoaceticus	1
Pseudomonas aeruginosa	1
Pseudomonas maltophila	1
Escherichia coli	1
Citrobacter freundii	1
Others[a]	9

[a]Not identified by API 20E alone.

most frequent infecting organism in hospitals and, although it is fashionable, and often successful to examine environmental sources for gram-negative bacilli found to be infecting patients, it may be worth remembering, when all else fails, that the patient may carry his own "environmental" bacteria with him.

References

1. Alkhoja, MS, Darrell JH (1979) The skin as a source of *Acinetobacter* and *Moraxella* occurring in blood cultures. J Clin Pathol 32:497–499
2. Baird-Parker AC (1974) The basis for the present classification of staphylococci and micrococci. Ann NY Acad Sci 236:7–14
3. Barbeito MS, Matthews CT, Taylor LA (1967) Microbiological laboratory hazard of bearded men. Appl Microbiol 15:899–906
4. Black T (1966) The bacterial flora of the skin and its relation to post-operative wound infection. Trans Soc Occup Med 16:18–23
5. Black WA, Bannerman CM, Black DA (1974) Carriage of potentially pathogenic bacteria in the hair. Br J Surg 61:735–738
6. Cozanitis DA, Mäkelä P, Grant J (1977) Microorganisms in the hair of staff and patients in an intensive care unit. Anaesthestist 26:578–580
7. Dineen P, Drusin L (1973) Epidemics of post-operative wound infections associated with hair carriers. Lancet II:1157–1159
8. Glew RH, Moellering RC, Kuny LJ (1977) Infection with *Acinetobacter calcoaceticus* (*Herellea vaginicola*): clinical and laboratory studies. Medicine (Baltimore) 56:79–97
9. Hovelius B, Mardh PA (1977) On the diagnosis of coagulase-negative staphylococci with emphasis on *Staphylococcus saprophyticus*. Acta Pathol Microbiol Scand [B] 85:427–434
10. Huismans-Evers AGM (1978) Results of routine tests for the detection of dispersers of *Staphylococcus aureus*. Arch Chir Neerl 30:141–150
11. Kloos WE, Musselwhite MS (1975) Distribution and persistence of *Staphylococcus* and *Micrococcus* species and other aerobic bacteria on human skin. Appl Microbiol 30:381–395
12. Kloos WE, Schleifer KH (1975) Isolation and characterization of staphylococci from human skin. II Description of four new species: *Staphylococcus warneri,*

Staphylococcus capitis, *Staphylococcus hominis* and *Staphylococcus simulans*. Int J System Bacteriol 25:62–79
13. McGinley KJ, Leyden JJ, Marples RR, Kligman AM (1975) Quantitative microbiology of the scalp in non-dandruff, dandruff and seborrhoeic dermatitis. J Invest Dermatol 64:401–405
14. Namavar F, deGraaff J, MacLaren DM (1978) Taxonomy of coagulase negative staphylococci: a comparison of two widely used classification schemes. Antonie Van Leeuwenhoek J Microbiol Serol 44:425–434
15. Noble WC (1966) *Staphylococcus aureus* on the hair. J Clin Pathol 19:570–572
16. Noble WC (1981) Microbiology of human skin. Lloyd-Luke Medical Books Ltd, London
17. Retailliau HF, Hightower AW, Dixon RE, Allen JR (1979) *Acinetobacter calcoaceticus:* a nosocomial pathogen with an unusual seasonal pattern. J Infect Dis 139:371–375
18. Schleifer KH, Kloos WE (1975) Isolation and characterization of staphylococci from human skin. I Amended description of *Staphylococcus epidermidis* and *Staphylococcus saprophyticus* and descriptions of three new species *Staphylococcus cohnii*, *Staphylococcus haemolyticus* and *Staphylococcus xylosus*. Int J System Bacteriol 25:50–61
19. Sellin M, Cooke DI, Gillespie WA, Sylvester DGH, Anderson JD (1975) Micrococcal urinary tract infection in young women Lancet II:570–572.
20. Somerville DA, Noble WC (1970) A note on the Gram negative bacilli of human skin. Europ J Clin Biol Res 15:669–671
21. Summers MM, Lynch PF, Black T (1965) Hair as a reservoir of staphylococci. J Clin Pathol 18:13–15
22. Taplin D, Rebell G Zaias N (1963) The human skin as a source of Mima-Herellea infections. JAMA 186:952–953

Chapter 6
Newer Methods of Quantifying Skin Bacteria

R.R. MARPLES

Different methods of quantifying skin bacteria as previously discussed[2] included surface sampling by swabbing and contact plates, stripping off the surface layers with cellophane tape, air sampling to assess dispersion (particularly of *Staphylococcus aureus*,[21] and sampling by means of a wash fluid either in successive basins, as in the Price technique,[18a] or in a localized area defined by a glass cylinder.[24]

Williamson and Kligman[24] introduced the use of a nonionic detergent wash fluid (0.1% Triton X-100 in phosphate buffer; pH 7.9) to elute the organisms from the skin and to disperse clumps of organisms. Because skin organisms survive for some time without multiplying in this fluid, volumetric dilution and plating of the sample is possible. Originally the skin was rubbed with a Teflon scrubber for 1 min in 1 ml of wash fluid held in a glass cylinder enclosing an area of 3.8 cm^2, and the first two washes were pooled. Incorporation plates were prepared and incubated for 48 h at 37°C.

All aspects of the Williamson and Kligman technique have been modified. We changed from the use of incorporation plates to the more convenient drop inoculation procedure[13] and, more recently, have sampled 5.5 cm^2 inside a stainless steel cylinder.[22] The technique has been employed to enumerate the desquamating cells of the epidermis[9] and to determine their dimensions.[18]

Selwyn and Ellis reduced the procedure to a single wash of 30 s and compared the numbers of colony-forming units obtained to those liberated from full-thickness skin biopsies agitated in the same detergent wash fluid.[20] They found that the detergent wash technique and a detergent-moistened swab technique were more efficient than tape stripping or direct contact plate methods but gave only about a tenth of the number of colony-forming units extractable from biopsy material.

Biopsy methods have been applied to the quantification of bacteria in pilosebaceous follicles in normal skin.[19] The normal flora of such follicles differed from the surface flora in that it contained only 16% of the number of aerobic bacteria but more than ten times the number of anaerobic bacteria found by surface sampling. The lack of cocci in many normal follicles is unlike the findings in individual lesions in acne[11,14] where, in addition, some selection of coccal strains was apparent.

Improvement of surface sampling methods for areas where sebaceous follicles are frequent and for acne subjects has been attempted by modification of the surface biopsy method[10] in which a cyanoacrylate adhesive is applied to the skin and spread out with a glass disk. The outer layers of skin and follicles are removed and dispersed into detergent wash fluid for enumeration.[4] This blocks the surface flora to some extent but the method samples mainly small follicles.[17]

The detergent wash method can be mechanized. A reciprocating machine was developed at Letterman Army Institute of Research and used by Bibel and Lovell[1] to develop maps of the skin flora. The cylinder was modified to a stainless-steel square section and 25 strokes of a stainless-steel spatula loaded to 250 g were made in 1 ml of detergent wash fluid. The procedure was repeated, a rinse step inserted, and the three samples pooled. A grid was drawn on the skin so that no zone was greater than 10×10 cm; samples were taken from the center of each zone. A normal volunteer and a patient with atopic dermatitis were studied. The technique of mapping has been used to investigate the heterogeneity of the skin flora and the response of the flora on the backs of volunteers to a small occlusive dressing.[2]

Rotatory machines were introduced before the detergent wash method[16] and recently the wash technique has been compared with the use of a similar rotatory machine.[3] This removed two or three times as many organisms as the manual wash. Used correctly rotatory machines may be satisfactory, but frictional damage to the skin appears to be a problem.

We attempted to speed up the removal of bacteria from the skin by using ultrasonic impulses and compared the methods with the detergent wash method.[22] A metal probe (25 mm diam.) connected the generator with 1 ml wash fluid in a cylinder (26.5 mm diam.) held to the skin of the forearm in nine volunteers. A power setting of 30 w was too little while at 90 w there was damage to the skin. Serial 10-second washes using the ultrasonic generator at 50 w gave the same results as 30-second manual washes. In some experiments the number of corneocytes was lower in ultrasound washes than in manual washes, while the bacterial counts were identical, suggesting a superficial location for the aerobic flora.[9] A single attempt to take ultrasound washes from the forehead was made. The sensation was unpleasant and further experiments were not made.

The most obvious anomaly in the early work was the difference in numbers of colony-forming units detected by the detergent wash technique and those detected by contact plates. Holt,[5] comparing a velvet-pad technique with a modified detergent wash technique, stated that the velvet pad gave an estimate of the number of microcolonies while the detergent wash estimated the number of viable cells. On this assumption, microcolonies could be as large as 10^5 viable cells. He found that the apparent increase in the number of microcolonies found after bathing by contact sampling was accompanied by a reduction in viable cells as determined by the wash technique. Similar findings have been reported by others; microcolonies have been seen by use of scanning electron microcopy.[15] The accuracy of the estimate of the size of the microcolonies by comparing a contact plate count with a detergent wash count is dependent on the efficiency of the contact plate. It is unlikely that a flat surface or even a velvet pad makes contact with the entire skin surface. As the detergent wash method gives good dispersion of the bacterial cells,[5] the size of the average microcolony is likely to be overestimated.

If microcolonies are large and localized, their dispersion from the skin on desquamating cells will be infrequent. Many studies have shown that individuals differ markedly in the numbers of bacteria that they liberate into the air.[15] We recently attempted to examine the numbers and sizes of skin fragments and the proportion of cells carrying aerobic bacteria in six selected individuals.[8] The volunteers performed a standard exercise in a test chamber enclosing the lower half of the body. The chamber was inside a unidirectional flow clean-air room.[7] There was a thousandfold difference in the number of bacteria-carrying particles liberated on exercise between the highest and lowest disperser (Table 6-1). The proportion of particles carrying bacteria gave a 25-fold difference and the number of particles liberated gave a 43-fold difference between the highest and lowest and in the same sense. The corneocyte counts for each individual were within an order of magnitude; it seems likely that low dispersers must lose cells at other times than during the experiments. This anomaly deserves further investigation. The high disperser carried the largest number of organisms on the skin as determined by the detergent wash method. We examined the sizes of dry dispersed epithelial cells, particularly in the context of prevention of dispersal by different garments. Epithelial cells collected from the air in a dry condition measured 21×33 μm, while corneocytes measured 33×44 μm—indicating the large increase in size on hydration.

The test chamber can be used to determine the efficiencies of containment of bacteria-laden particles by different fabrics[7] and the correlation of these results with bench tests.[6] The effects of skin treatments on dispersal of organisms from the skin can also be studied.

Table 6.1. Dispersal of cells and bacteria. Numbers of skin fragments, numbers of microcolonies and percentage of fragments carrying bacteria liberated during $2^{1}/_{2}$ min of standard exercise in the test chamber (Mackintosh et al. 1978[8])

Subject	Skin fragments	Microcolonies	Proportion (%)
Male			
1	53 150	28 700	54
2	13 530	6 300	47
3	11 360	2 840	25
Female			
1	7 550	343	4.5
2	2 920	385	13
3	1 225	27	2.2

Table 6.2. Effect of washing procedures on the number of organisms transferred (Marples and Towers 1979[12])[a]

Time	Percentage reduction in number of organisms transferred after procedure of that transferred by the other hand before the procedure		
	Detergent in still water	Bar soap under running water	70% ethanol (still)
Immediate	95.7	98.4	99.99
15-min delay	98.4	95.4	100

[a]Both hands grasped inoculated cloth-covered bottles, then one hand (control) grasped a sterile cloth-covered bottle. The procedure was carried out and the other (test) hand grasped another recipient bottle.

A final method that can be employed in the study of transmission of organisms between individuals is concerned with contact transfer by the hands.[12] In this method, a cloth rectangle is attached to a bottle and contaminated with the test organism. The bottle is then grasped firmly. Another fabric-covered bottle is subsequently grasped and the number of test organisms on the recipient cloth enumerated. Using this model to study hand-washing procedures (Table 6-2) we found that simple washing of the hands reduced the number of organisms transferred by 95% and, in another experiment, even the application of small volumes of alcohol reduced the transfer by more than 90%.

The variety of methods employed in the investigation and quantification of skin bacteria suggest that no one method is outstanding. Different methods give results requiring careful interpretation if errors are to be avoided.

References

1. Bibel DJ, Lovell DJ (1976) Skin flora maps: a tool in the study of cutaneous ecology. J Invest Dermatol 66:265–269
2. Bibel DJ, Lovell DJ, Smiljanic RJ (1976) Effects of occlusion upon population dynamics of skin bacteria. Br J Dermatol 95:607–612
3. Fleurette J, Transy MJ (1978) Essai d'amélioration et de standardisation du prélevement des micro-organismes cutanés au moyen d'un appareil électrique. Revue de l'Institut Pasteur de Lyon 11:493–501
4. Holland KT, Roberts CD, Cunliffe WJ, Williams M (1974) A technique for sampling micro-organisms from the pilo-sebaceous ducts. J Appl Bacteriol 37:289–296
5. Holt RJ (1971) Aerobic bacterial counts on human skin after bathing. J Med Microbiol 4:319–327
6. Lidwell OM, Mackintosh CA (1978) The evaluation of fabrics in relation to their use as protective garments in nursing and surgery. I. Physical measurements and bench tests. J Hyg (Camb) 81:433–451
7. Lidwell OM, Mackintosh CA, Towers AG (1978) The evaluation of fabrics in relation to their use as protective garments in nursing and surgery. II. Dispersal of skin organisms in a test chamber. J Hyg (Camb) 81:453–469
8. Mackintosh CA, Lidwell OM, Towers AG, Marples RR (1978) The dimensions of skin fragments dispersed into the air during activity. J Hyg (Camb) 81:471–479

9. McGinley KJ, Marples RR, Plewig G (1969) A method for visualizing and quantitating the desquamating portion of the human stratum corneum. J Invest Dermatol 53:107–111

10. Marks R, Dawber RPR (1971) Skin surface biopsy: an improved technique for the estimation of the horny layer. Br J Dermatol 84:117–123

11. Marples RR (1974) The microflora of the face and acne lesions. J Invest Dermatol 62:326–331

12. Marples RR, Towers AG (1979) A laboratory model for the investigation of contact transfer of micro-organisms. J Hyg (Camb) 82:237–248

13. Marples RR, Fulton JE, Leyden J, McGinley KJ (1969) Effect of antibiotics on the nasal flora in acne patients Arch Dermatol 99:647–651

14. Marples RR, Leyden JJ, Stewart RN, Mills OH, Kligman AM (1974) The skin microflora in acne vulgaris. J Invest Dermatol 62:37–41.

15. Noble WC, Somerville DA (1974) Microbiology of human skin. Saunders, London

16. Pillsbury DM, Kligman AM (1954) Some current problems in cutaneous bacteriology. In: Modern trends in dermatology. Butterworth, London

17. Plewig G, Kligman AM (1975) Acne: morphogenesis and treatment. Springer-Verlag, Berlin

18. Plewig G, Marples RR (1970) Regional differences of cell size in human stratum corneum. Part I. J Invest Dermatol 54:13–18

18a. Price PB (1938) The bacteriology of the normal skin: a new quantitative test applied to a study of the bacterial flora and disinfection action of Mechanical Cleansing. J Infect Dis 63:301–318

19. Puhvel SM, Reisner RM, Arnirian DA (1975) Quantification of bacteria in isolated pilosebaceous follicles in normal skin. J Invest Dermatol 65:525–531

20. Selwyn S, Ellis H (1972) Skin Bacteria and skin disinfection reconsidered. Br Med J 1:136–140

21. Shooter RA (1965) The environment. In: Maibach HI, Hildick-Smith G (eds) Skin bacteria and their role in infection. McGraw-Hill, New York

22. Stringer MF, Marples RR (1976) Ultrasonic methods for sampling human skin micro-organisms. Br J Dermatol 94:551–555

23. Ulrich J (1965) Dynamics of bacterial skin populations. In: Maibach HI, Hildick-Smith, G (eds) Skin bacteria and their role in infection. McGraw-Hill, New York

24. Williamson P, Kligman AM (1965) A new method for the quantitative investigation of cutaneous bacteria. J Invest Dermatol 45:498–503

Chapter 7
Treatment of Nasal Carriers of Coagulase-positive Staphylococci

L. Joseph Wheat, Richard B. Kohler and Arthur White

Coagulase-positive staphylococci can frequently be isolated from man's environment. In most circumstances, the organisms isolated from the air, bed clothing, or floors in hospitals or around outpatients are derived from multiplication of the organisms in or on the human host. Several major primary reservoirs of staphylococci exist. Lesions, particularly infections of the skin, pneumonias, or other open infections disseminate large numbers of virulent staphylococci into the environment. In the absence of clinical disease, the major reservoir for coagulase-positive staphylococci is the anterior nares or perineal areas. From these sites, the organism is shed onto the skin and is often disseminated in large numbers from the skin into the air and onto floors and other sites.[22,30,32,35]

If one takes repeated cultures, 80% of adult subjects can be shown to be carriers of coagulase-positive staphylococci in the anterior nose (Table 7.1). In most individuals, recovery of staphylococci is transient; these individuals are infrequently sources of dissemination. Between 20%–40% of adults will be carrying coagulase-positive staphylococci consistently in the anterior nares, often for years. Usually, patients who are consistent carriers of staphylococci are carrying the organism in large numbers, with one to ten million organisms recovered per swab.[30,32,35]

Perineal carriers occur less frequently. Approximately 13% of patients admitted to a medical ward have *Staphylococcus aureus* in the perineum alone or combined with nasal or throat cultures.[2,8,30] Of these, only 3% had carriage of the organism of the perineum without carriage in the nose or throat. In most circumstances, neither perineal nor nasal carriage of coagulase-positive staphylococci is associated with disease at any given time. However, in a small proportion of patients, nasal carriage or perineal carriage of staphylococci has been associated with recurrent furuncles and other types of staphylococcal skin lesions. In addition, postoperative infections due to staphylococci are thought to result from either auto-infection by organisms from the patient's own nose or dissemination from organisms present in the nasal site of personnel.

Local treatment of nasal staphylococci reduces the frequency of recurrent staphylococcal skin lesions.[3,7,10,31]   Postoperative infections have been reduced by nasal application by some investigators but not by others.[4,5,9,24,33] If permanent eradication of the reservoirs were possible, a marked reduction in staphylococcal infections would be likely.   Over the last several years, hospitals have developed a high proportion of staphylococcal resistance to penicillin, tetracycline, and erythromycin.   When these drugs are used for the treatment of a variety of infections, they also result in elimination of susceptible staphylococci in the anterior nares but the organisms are often replaced by drug-resistant ones from the hospital environment.   The nasal carrier rates may not be suppressed and may be increased but the organisms which are carried in the nose are now multiple drug-resistant strains.   By using phage typing, it has been possible to show that organisms carried on admission that are drug susceptible are of a different phage type from drug-resistant organisms that patients acquired while on therapy.

   If antimicrobial drugs to which there are few or no resistant organisms available are administered, the frequency with which treated patients are

Table 7.1. Some nasal carrier rates for *Staphylococcus aureus* (Noble and Sommerville 1974)

Population	Number sampled	Carrier rate, %	Studies
Normal adults outside hospital	various	<10 to >40	Williams, 1963 a review (16 papers)
Population of S. Sudan	194	9	Harper, 1967
Coastal New Guinea	1588	13	Kariks & Battey, 1966
Papua/New Guinea	344	7	Rountree, 1956
Nigerian adults	310	46	Davis & Davis, 1965
Nigerian children	200	70	
Random samples of rural Netherlands	1021	29 (19–40)	Noble, Valkenburg & Wolters, 1967
Normal population U.K., adults	378	33	
Normal population U.K., children	382	49	Noble, 1969a
Controlled adult diabetics	98	32	Somerville & Lancaster-Smith, 1973
Diabetic children	157	76	
Insulin-requiring diabetic adults	144	53	Smith, O'Connor & Willis, 1966
Orally treated diabetic adults	150	35	
Nondiabetic children	166	44	
Normal adults	254	34	
Patients on admission to hospital	various	>20 to <50	Williams, 1963 a review (8 papers)
Surgical patients: on admission to hospital	2925	40	Noble et al., 1964
on discharge		43	
Dermatology patients on admission to skin hospital	1548	48	Wilson, White & Noble, 1971

L. Joseph Wheat, Richard B. Kohler and Arthur White

Table 7.2. Effect of systemic therapy on nasal carriage of coagulase-positive staphylococci

Drug (No. subjects)	Carrier rate			Comments	Reference
	Before	During	After		
Cephalexin (6)	100%	67%	67%	60-day follow-up	1
Erythromycin (27)	100%	33%	77%	29-day follow-up	42
Rosamycin (30)	100%	73%	76%		
Placebo (30)	100%	90%	93%		
Untreated	12%	23%	—	All treatments	36
Tetracycline	16%	54%	—	markedly increased	
Penicillin G	23%	23%	—	frequency of drug	
Multiple antibiotics	33%	46%	—	resistant staphylococci	
Cyclacillin					
Penicillinase producers	100%	100%	80%	60-day follow-up	20
Nonpenicillinase producers	100%	25%	68%		
Control	100%	88%	100%		
Erythromycin (6)	100%	0	100%	1-month follow-up	15
Placebo (5)	100%	100%	100%		
Penicillin V (16)	100%	60%	70%	2-month follow-up	40
Tetracycline (10)	100%	20%	80%	1-month follow-up	
Methicillin (22)	36%	0	0	11-day follow-up	39
Oxacillin (13)	100%	30%	—		38
Nafcillin					
I.M. (10)	100%	0	—		27
P.O. (15)	100%	70%	—		

α Phenoxypropyl penicillin					
Penicillinase producers (13)	100%	80%	—		25
Nonpenicillinase producers (12)	100%	12%	—		
Oxacillin					
Penicillinase producers (12)	100%	48%	—	Only 6 of 41 patients who were noncarriers at the end of therapy were noncarriers 30 days after therapy	26
Nonpenicillinase producers (11)	100%	36%	—		
Penicillin V					
Penicillinase producers (16)	100%	74%	—		
Nonpenicillinase producers (5)	100%	30%	—		
Propicillins					
Penicillinase producers (13)	100%	84%	—		
Nonpenicillinase producers (12)	100%	29%	—		
Fusidic acid (12)	100%	39%	41%	10-day follow-up	28
Tetracycline (25)	16%	100%	—	Only carriers of multiple-resistant phage Group III recorded	13
Penicillin (30)	18%	80%	—		
Tetracycline	16%	9%	—	7–9 days of therapy	14,16
Penicillin	23%	10%	—		
Untreated	20%	11%	—		

L. Joseph Wheat, Richard B. Kohler and Arthur White

Table 7.3. Effect of topical therapy on nasal carriage of coagulase-positive staphylococci

Drug (No. subjects)	Carrier rate			Comments	Reference
	Before	During	After		
Gramicidin + neomycin + polymixin (105)	100%	11%	28%	1-month follow-up	11
Controls (60)	100%	73%	—		
Lysostaphin (21)	100%	9%	63%	4-month follow-up	19
Gentamicin (21)	100%	14%	74%		
Controls (22)	100%	100%	91%		
Neomycin + chlorohexidine (12)	100%	25%	42%	3–5-day follow-up	41
Gentamicin (23)	100%	30%	43%		
Chlorhexidine (12)	100%	58%	75%		
Halguinol (12)	100%	75%	50%		
Vancomycin (12)	100%	58%	50%		
Controls (69)	100%	93%	87%		
Framycetin + gramicidin					
Personnel (30)	100%	10%	90%	28-day follow-up	30
Patients (20)	100%	40%	90%	10-day follow-up	
Gentamicin (20)	100%	0%	73%	7–10-day follow-up	37
Control (20)	100%	88%	100%		
Lysostaphin (32)	100%	20%	62%	61–90-day follow-up	18
Methicillin (22)	100%	00%	46%	30-day follow-up	32
Controls (7)	100%	100%	100%		

Treatment					
Chlorhexidine + neomycin + framycetin + gramicidin (266)	41%	27%	—		17
Bacitracin + neomycin (39)	100%	28%	—		34
None (37)	100%	78%	—		
Placebo (22)	100%	77%	—		
Neomycin (27)	100%	0%	77%	2–3-week follow-up	17
Penicillin G (41)	100%	23%	31%	2-week follow-up	6
Streptomycin (21)	100%	29%	51%	1-month follow-up	
Chloramphenicol (18)	100%	33%	74%	20-week follow-up	
Chlortetracycline (21)	100%	14%		No difference; in different ointment	
Oxytetracycline (20)	100%	0%			
Neomycin or kamamycin in sesame oil (30)	100%	33%	87%		21
Neomycin spray (10)	100%	40%	100%		
Framycetin (42)	100%	40%	68%		23
Oxacillin (27)	100%	12%	25%	7–10-day follow-up	29
Placebo (13)	100%	95%	—		

carriers of coagulase-positive staphylococci is markedly reduced. The numbers of staphylococci in those patients who remain carriers is also reduced. Unfortunately, after effective antimicrobial drugs are stopped there is usually rapid reacquisition of staphylococci in the nose so that the total carrier rate is only slightly reduced a month after discontinuation of antimicrobial agents. The effects of a number of systemic drugs are summarized in Table 7.2.

Topical treatment has been attempted with a wide variety of drugs including methicillin, neomycin, gentamicin, gramicidin, and lysostaphin (Table 7.3). There has been a reduction in the nasal carriage of organisms with most active compounds. A marked reduction in the frequency with which staphylococci can be recovered from various sites of the skin and from air samples obtained around the treated patients followed only local nasal treatment.[29,30,32] In patients who are perineal carriers of S. aureus, dissemination onto skin and into air has been markedly reduced by treatment with local hexachlorophene emulsion.[30] In all studies with adequate follow-up, recurrence of nasal staphylococci has occurred in 63%–100% of subjects after therapy was discontinued.

To date, there has been no single topical or systemic antimicrobial drug that has been demonstrated to be effective in permanently eradicating the nasal carrier site. Although recommendations for treatment of furuncles of a recurring nature have included topical application of a variety of ointments (including bacteriocin, neomycin, and gentamicin), none of these have been demonstrated to permanently eradicate the carrier site. Therefore, topical therapy should be continued for prolonged periods to continue suppression of nasal staphylococci.

There is suggestive published and unpublished evidence that rifampin may be as effective in eradicating staphylococci from anterior nares as it is effective in eradicating meningococci from carriers. But this drug has not been investigated in sufficently controlled manner to clearly demonstrate that it is effective in permanent eradication of staphylococci.

References

1. Aly R, Maibach HI, Strauss WG, Shinefield HR (1970) Effects of systemic antibiotic on nasal bacterial ecology in man. Appl Microbiol 20:240–244
2. Boe J, Solberg CO, Vogelsang TM, Wormnes A (1964) Perineal carriers of staphylococci. Br Med J II:280–281
3. Copeman PWM (1958) Treatment of recurrent styes. Lancet II:728–729
4. Gillespie WA, Alder VG, Ayliffe GAJ, Bradbeer JW, Wypkema W (1959) Staphylococcal cross-infection in surgery. Effects of some preventive measures. Lancet II:781–784
5. Gillespie WA, Alder VG, Ayliffe GAJ, Powell DEB, Wypkema W (1961) Control of staphylococcal cross-infection in surgical wards. A four-and-a-half-year study. Lancet I:1299–1303
6. Gould JC (1956) The effect of local antibiotic on nasal carriage of staphylococcal pyogenes. J Hyg (Camb) 53:379–385
7. Green KG (1961) The role of the carrier in staphylococcal disease. Lancet II:921–923

8. Hare R, Ridley M (1958) Further studies on the transmission of *Staph. aureus*. Br Med J I:69–73
9. Henderson RJ, Williams REO (1961) Nasal disinfection in prevention of postoperative staphylococcal infection of wounds. Br Med J II:330–333
10. Hobbs BC, Carruthers HL, Gough J (1947) Sycosis barbae. Serological types of *Staphylococcus pyogenes* in nose and skin and results of penicillin treatment. Lancet II:572–574
11. Hunter DT, Baker CE (1967) Control of staphylococcal carriers in three hospitals. Public Health Rep 82:329–333
12. Hussar AE (1962) Neomycin spray in the treatment of nasal carriers of staphylococcus. Clin Pharmacol Ther 3:441–446
13. Knight V, Holzer AR (1954) Studies on staphylococci from hospital patients. I. Predominance of strains of group III phage patterns which are resistant to multiple antibiotics. J Clin Invest 33:1190
14. Knight V, White A (1956) Drug resistant staphylococci. I. Their distribution in hospital patients. South Med J 49:1173
15. Knight V, White A, Martin MP (1958) The effect of antimicrobial drugs on the staphylococcal flora of hospital patients. Ann Intern Med 49:536–543
16. Knight V, White A, Foster F, Wenzel T (1956) Studies on staphylococci from hospital patients. II. Effect of antimicrobial therapy and hospitalization on carrier rates. Ann NY Acad Sci 65:206
17. Lindborm G, Laurell G (1967) Studies of the epidemiology of staphylococcal infections. Acta Pathol Microbiol Scand 69:237–245
18. Martin RR, White A (1967) The selective activity of lysostaphin in vivo. J Lab Clin Med 70:1–8
19. Martin RR, White A (1968) The reacquisition of staphylococci by treated carriers: A demonstration of bacterial interference. J Lab Clin Med 71:791–797
20. Martin RR, White A (1971) Quantitative nasal culture: a tool in antibiotic research. Appl Microbiol 22:397–400
21. Martin WJ, Nichols DR (1960) Henderson ED The problem of management of nasal carriers of staphylococci. Mayo Clin Proc 35:282–292
22. Noble WC, Somerville DA (1974) Microbiology of human skin. WB Saunders Company, Philadelphia
23. Porter IA, Miller IT, McNeill IF, Green CA (1963) Effect of topical framycetin on staphylococcal nasal carriage. Br Med J I:1515–1518
24. Rountree PM, Loewenthal J, Tedder E, Gye R (1962) Staphylococcal wound infection: the use of neomycin and chlorhexidine ("Naseptin") nasal cream in its control. Med J Aust II:367–370
25. Smith J, White A (1963) Antistaphylococcal activity of alpha-phenoxypropyl penicillin. Antimicrob Agents Chemother 1962 ed. JC Sylvester. pub. Am Society for Microbiology pp 734–738
26. Smith J, White A (1963) Activity of 3 penicillins against staphylococci. J Lab Clin Med 61:129–137
27. Smith J, White A (1964) Activity of sodium nafcillin [6-(2-Ethoxy-1-Napthamido-) penicillanic acid] against staphylococci in vivo. Antimicrob Agents Chemother 1963 ed JC Sylvester. pub. Am Society for Microbiology pp 354–361
28. Smith J, White A (1964) Development of resistance to fusidic acid during treatment of nasal carriers of staphylococci. Antimicrob Agents Chemother 1963 ed JC Sylvester. pub. Am Soc For Microbiology pp 155–159
29. Smith JW, White A (1964) Nasal reservoir as the source of extranasal staphylococci. Antimicrob Agents Chemother 1963 ed JC Sylvester. pub. Am Society for Microbiology pp 679–683
30. Solberg CO (1965) A study of carriers of *Staphylococcus aureus*: with special regard to quantitative bacterial estimates. Acta Med Scand 178:Supplement 436

31. Tulloch LG, Alder VG, Gillespie WA (1960) Treatment of chronic furunculosis. Br Med J II:354–356
32. Varga DT, White A (1961) Suppression of nasal, skin and aerial staphylococci by nasal application of methicillin. J Clin Invest 40:2209–2214
33. Weinstein HJ (1959) The relation between the nasal-staphylococcal-carrier state and the incidence of postoperative complications. N Engl J Med 260:1303–1308
34. Weinstein HJ (1959) Control of nasal-staphylococcal-carrier states. N Engl J Med 260:1308–1310
35. White A (1961) Relation between quantitative nasal cultures and dissemination of staphylococci. J Lab Clin Med 58:273–277
36. White A (1964) Increased infection rates in heavy nasal carriers of coagulase-positive staphylococci. Antimicrob Agents Chemother 1963 ed. JC Sylvester. pub. Am Soc. for Microbiology p 667–670
37. White A (1964) The use of gentamicin as a nasal ointment. Am J Med Sci 248:52–55
38. White AC, Smith J (1962) Antistaphylococcal activity of penicillin P-12. (Oxacillin). Am J Med Sci 241:202–208
39. White A, Varga DT (1961) Antistaphylococcal activity of sodium methicillin. Arch Intern Med 108:671–678
40. White A, Hemmerly T, Martin MP, Knight V (1959) Studies on the origin of drug-resistant staphylococci in a mental hospital. Am J Med 27:26–39
41. Williams JD, Waltho CA, Ayliffe GA, Lowbury EJL (1967) Trials of five antibacterial creams in the control of nasal carriage of Staphylococcus aureus. Lancet II:390–392
42. Wilson SZ, Martin RR, Putnam M, Greenberg SB, Wallace RJ Jr, Jemsek JG (1979) Quantitative nasal cultures from carriers of Staphylococcus aureus: effects of oral therapy with erythromycin, rosamicin, and placebo. Antimicrob Agents Chemother 15:379–383

Chapter 8
Practical Office Microbiology

JOSEPH GREENBERG

Most practicing physicians use a certain amount of office microbiology. Perhaps more can be accomplished in our office than we are presently attempting. We may be able to find answers to clinical questions in a faster, more convenient, and less expensive way than previously. Frequently physicians see patients with diseases suspected of having a microbiologic etiology. Often it is helpful to have a laboratory verify this etiology. Other times a patient may have a disease that a physician is sure is caused by a certain microorganism, but with penicillin resistance frequent and resistance to tetracycline and erythromycin increasing, it would be advantageous if we could test for these sensitivities either prior to or shortly after the onset of treatment. This information allows the physician to follow not only the most efficacious therapy, but also the most cost-effective one. Occasionally a physician knows a disease has a microbiologic cause but wants definite proof[1]. These cases can be evaluated in the correctly equipped office laboratory. The basics of office microbiology include a microscope, refrigerator, incubator with carbon dioxide ability, a vital stain (Gugol blue), KOH, slides, cover slips, media (culturettes, T-M media, bi-plates with crystal-violet blood agar and manitol salt egg-albumen) and silica gel swabs, and an autoclave to dispose of all contaminated material.

A refrigerator is necessary for media storage; this unit need not be large or expensive.

A microscope for looking at KOH and Tzanck preparations is a requirement. This should be a standard four-objective model, either monocular or binocular.

A small incubator can be purchased inexpensively. It should have a temperature gauge and controls and be able to hold 10% carbon dioxide atmosphere (this is achieved by placing an Alka-Seltzer tablet in a small cup of water inside the incubator). A vital stain is needed for doing Tzanck smears; this may help in determining the cellular morphology of vesicular-pustular diseases. Staining of vesicular fluid with this stain can lead to an immediate diagnosis of herpes group infection (Fig. 8.1); this is extremely helpful in areas where herpes group viruses are not cultured. For many reasons this

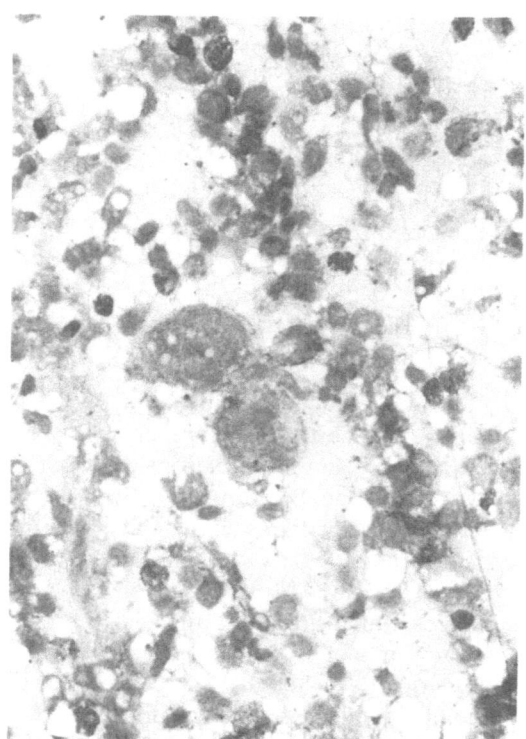

Fig. 8.1. Tzanek preparation demonstrating two multinucleated giant cells. Diagnostic of herpes group infection.

test is rarely used by most physicians. The reagents are bulky to store, messy to use, and too time consuming to be efficiently performed in the office. A vital stain primarily used in veterinary medicine seems to solve these objections. Gugol blue is easy to handle without causing staining of equipment or skin; with it a physician can take a sample from the patient and have it stained and under the microscope in about one minute.

Potassium hydroxide preparations are necessary for successful patient management. They are time consuming, especially for the person not utilizing them frequently. Thick scales are not rapidly penetrated by this reagent, and time is needed to adequately interpret the slide. The addition of 20% DMSO to the KOH speeds up the penetration and dissolution process and improves the clinical practicality of this test. Balloting the cover slip with a sharp object while observing the preparation at 100X magnification speeds the dissolution, and also lights up hyphal forms. This procedure makes the differentiation between cell walls and hyphae easier to see. If clinical suspicion is high and no hyphae are seen, allowing the slide to sit for an hour or so often makes the preparation easier to interpret. (Fig. 8.2).

For culture of microorganisms in the office, the outstanding advances

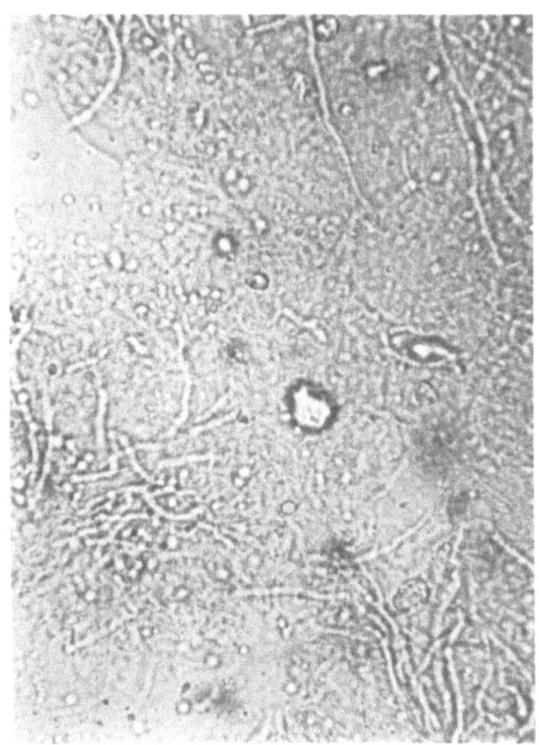

Fig. 8.2. Markedly positive KOH preparation. Allowing time for dissolution of cell walls makes for easier interpretation.

have been the development of defined media that isolate specific organisms. Through the use of special dyes or other markers they allow a minimally trained person to quickly learn to identify the organism in question. These media allow for specific culturing of dermatophytes, β-hemolytic streptococcus, *Staphytococcus aureus*, and *Neisseria gonococcus*. When other organisms are suspected the physicians must send a specimen to a commercial laboratory; for this he should have a good transfer medium.

The development of DTM (dermatophyte test media) made culturing of dermatophytes a simple procedure and allowed untrained individuals to differentiate dermatophytes from contaminants. A drawback of DTM, however, is that it does not grow deep fungi. This seems to be more of a theoretical than practical problem, though, as deep fungi are rarely seen in clinical practice in the United States.

Researchers at Letterman Army Hospital had a great deal of experience with a biplate containing two media. One side contained a crystal violet blood agar (CVBA) combination that isolated β-hemolytic streptococcus, while the other side contained a mannitol salt egg albumin media (MSEA) that isolated *S. aureus* (Fig. 8.3 a-d). A third chamber to do bacterial sensi-

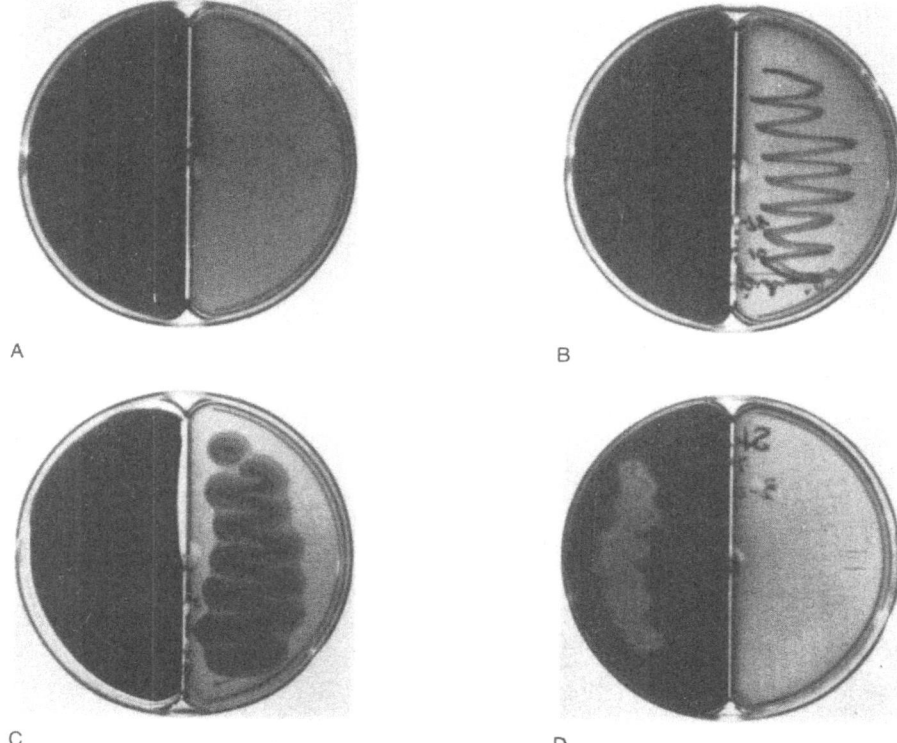

Fig. 8.3. Detection of β-hemolytic streptococci and *S. aureus on plates contain-ing* crystal violet blood agar late (CVBA) and mannitol salt egg agar (MSEA) (biplate). **A** Biplate—control (no growth). Left chamber contained CVBA; right chamber contained MSEA. **B** Demonstration of *S. epidermidis* on MSEA. **C** Dem-onstration of *S. aureus* on MSEA. Note clear zone. **D** Demonstration of β-hemolytic streptococcus on CVBA.

tivities can be added, which increases the practicality of this plate for clinical use.   Thayer-Martin medium is readily available and will grow *N. gonococcus*.   Since cultures at a commercial laboratory cost upward of $20—and sensitivities, about $15—an appreciable amount of money can be saved by doing office culturing and sensitivity testing.

At present a company is investigating the feasibility of a triplate contain-ing media for *S. aureus,* β-hemolytic streptococcus, and sensitivity testing. Hopefully this will be marketed.

An office can be equipped for the practice of clinical microbiology for be-tween $1 000–$2 000. With this equipment a physician can give his patient quicker answers to important questions. He can treat more specifically with less guesswork, and through these procedures the physician can save the pa-tient both time and money.

Chapter 9

Microbial Interactions and Antibiosis
SYDNEY SELWYN

In the various ecosystems of man, including the epidermis, the members of the commensal microflora interact dynamically among themselves and with any intruders. Paradoxically, most investigators have confined their attention to the study of isolated microorganisms under the artificial conditions of pure culture. During the past 110 years, scattered observations on microbial interactions have accumulated and have recently been amplified by more systematic studies.

Early Observations and Applications (1870–1960)

After centuries of empirical use in folk medicine, the antibacterial activity of fungi—notably *Penicillium* species—was established scientifically between 1870–1876.[37] This work by English pioneer bacteriologists, including Joseph Lister, is relevant not only as a general starting point for our subject, but also in connection with recent research on skin ecology.

Bacillus anthracis was the first skin pathogen investigated; in 1877 Pasteur and Joubert reported its inhibition by an airborne bacterium of unknown identity. They advocated the use of bacterial antagonism in the treatment of infection, and the principle was enthusiastically applied for years—with equivocal results—to treat a wide range of infections, including diphtheria and tuberculosis. In 1887, English physician Henry Tomkins reported the inhibition of anthrax bacilli by a skin micrococcus and described for the first time a simple form of mixed-growth curve study.[45] In 1889, using a solid medium system, Doehle published the earliest known illustration of this antagonistic interaction.[13]

"Antibiosis" in its modern sense was defined in 1899 by Marshall Ward: "where one of two associated organisms is injuring the other, as exemplified by many parasites...this state of affairs has been termed *Antibiosis*."[48]

Because of its therapeutic potential, the powerfully antagonistic effect of some strains of *Staphylococcus aureus* upon *Corynebacterium diphtheriae* was studied in detail during the 1920s and 1930s.[14,31] From 1945 to the mid-1950s, in the search for clinically useful antibiotics, coagulase-negative members of the *Micrococcaceae* received attention but the active products of the strains tested possessed a relatively narrow antibacterial spectrum.[18,24]

Colonization Priority and Bacterial Interference

Staphylococcus Aureus

With the development of bacteriophage typing in the 1950s it became clear that colonization of the skin of the newborn or of the anterior nares in all age groups by one strain of *S. aureus* tended to prevent subsequent colonization by other strains. This phenomenon, called "colonization priority," does not usually involve the production of antibiotic substances. Instead, population pressure resulting from competition for available nutrients is thought to be responsible.[50]

This simple form of interstrain competition is the basis of the commonest variety of *bacterial interference*; it was systematically studied and applied by Shinefield, Eichenwald, and their colleagues in the prevention of staphylococcal cross-infection during the serious epidemics of the 1960s.[10,43] The organism mainly used for artificial colonization was a group III nonantibiotic-producing strain of *S. aureus* (502A), of relatively low virulence. Within a day of birth it was applied to the infant's nostrils and umbilical stump as a saline dilution of a broth culture containing about 1×10^3 colony-forming units. No antibacterial preparation of the inoculation site was necessary under these conditions of minimal prior colonization. In older individuals, despite the use of larger bacterial inocula (e.g., 10^6), it was necessary to suppress the preexisting microflora of the anterior nares by topical and systemic antistaphylococcal agents such as oxacillin.[10] The nares of some persistent noncarriers could not be colonized due to unknown local factors—possibly the presence of strains of commensals with marked abilities to produce interference.

Strain 502A gave generally good protection against staphyloccal cross-infection in the newborn[23] and, to a more limited extent, in adults.[25] Some success was achieved in the control of recurrent furunculosis in adults.[25] A firm basis for the clinical use of bacterial interference was provided by studies on experimental burns in rabbits.[3] Unfortunately, occasional severe infections among patients treated with 502A demonstrated that it retained considerable pathogenicity.[9,21] This has been the main reason for a decline in enthusiasm for the use of bacterial interference involving *S. aureus*.

Skin Commensals

During the past decade, attention has turned—albeit on a small scale—to achieving bacterial interference by means of coagulase-negative staphylococci as a safer alternative to *S. aureus*. Numerous clinical observations emphasize the ecological importance of maintaining the normal microflora of the skin as a defense against colonization by potential pathogens. Thus suppression of the normal gram-positive flora by deodorants can lead to overgrowth by *Proteus* species.[42] Hexachlorophene has produced similar effects in neonates,[15] while both topical and systemic antibacterials in adults promote colonization and clinical infection by gram-negative bacilli and other organisms which normally cannot establish themselves on skin.[1,26,34,36]

In addition to the removal of simple nutritional competition, as has been discussed, suppression of the skin's commensals greatly reduces the production of a wide range of antimicrobial substances, notably fatty acids (which is, of course, the express purpose of deodorants). Similar factors associated with the normal intestinal flora have recently been investigated in detail; there has been general agreement that systemic antibacterial therapy should be designed to produce the minimum disturbance of the bowel's "colonization resistance." This is a powerful defense mechanism depending on the maintenance of anaerobic commensal bacteria in the colon.[37] Antibiotics such as cephalosporins, which do not reach high concentrations in the intestine, are preferable to broad-spectrum penicillins and tetracyclines. Drugs in the latter group reduce colonization resistance and facilitate the overgrowth of potential pathogens, notably gram-negative bacilli such as *Klebsiella* species. Close analogies with the skin are evident, despite the considerably greater population densities of microorganisms in the colon.

Although the normal microflora of the skin is obviously a valuable defense mechanism, its occasional failures are readily apparent in primary pyoderma and secondary infection of dermatoses, wounds, and burns. Studies on animals and in man, summarized below, show that commensal strains differ considerably in their abilities to interfere with colonization by pathogens. Pathogenic bacteria show enhanced growth potential under the exudative condition of dermatoses, burns, and other wounds.[12,39]

Both of the main components of the skin flora undoubtedly fulfill important defensive roles. The few published investigations on bacterial interference by commensals deal with coagulase-negative staphylococci. This group comprises *Staphylococcus* types 2–6 and *Micrococcus* types 1–7 of Baird-Parker's 1965 classification, and *Staphylococcus epidermidis, Staphylococcus saprophyticus,* and *Micrococcus* biotypes in his 1974 classification.[5] Reports on the effect of diphtheroids are awaited.

Four pioneer in vivo studies have involved the inoculation of chick embryos[29] and experimental animals with commensal staphylococci.[17,33,50]

In the 1950's Halbert, Swick, and their colleagues showed that a number of antibiotic-producing strains of *Micrococcus* obtained from the human conjunctiva protected against *Clostridium septicum* infection in mice and diphtheria in guinea pigs, when injected at the same time as the pathogens.[17] Protection against staphylococcal infection was obtained with non-antibiotic-producing strains from the skin in experimental burns in guinea pigs,[50] and in experimental abscesses in rabbits and mice.[33] In the latter study, *S. aureus* 502A proved less effective, and even produced infections itself. Chick embryos have been found useful as a test system; interference was observed against *Salmonella, Escherichia,* and *Proteus* species, and against *S. aureus* and pneumococci.[29]

Personal Investigations

Our own studies involved the use of a wide range of in vitro and in vivo systems. Screening for antagonistic abilities among commensal bacteria has been performed by direct and deferred competitive growth on solid media. Detailed quantitative studies on bacterial interference have been performed in liquid and solid in vitro systems, using batch cultures and chemostats.

Interference by diphtheroids outside the coryneform group was seen only once. This was in a long-term growth curve study, when *S. aureus* was inhibited after 140 of mixed growth (unpublished data). In contrast, a relatively large number of skin staphylococci exerted powerful antagonism against a wide range of commensals and pathogens. Although the effects were often evident both in liquid-batch culture and continuous (chemostat) culture,[27,28] inhibition of relatively resistant strains of *S. aureus* and other bacteria was clearly seen only on a solid test system, which provides a more realistic model of the epidermal surface.[30] The outcome of the interaction of any two bacteria was influenced by the relative and absolute inoculum size of each strain and whether an antibiotic was produced by either of the organisms (Fig. 9.1).

Similar results have been obtained on the intact skin of rabbits under occlusion.[41] Further studies on the semioccluded skin of human volunteers and on experimental burns of pigs have shown that selected strains of *S. epidermidis* possess an impressive ability to prevent infection by *S. aureus* (Selwyn, Sethna, and Winter, unpublished work).

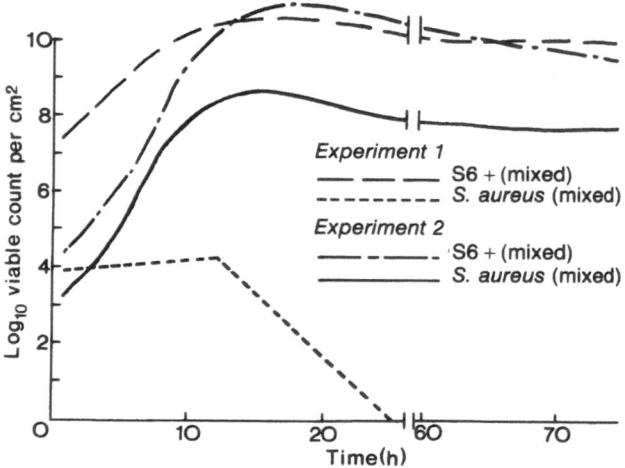

Fig. 9.1. Growth curves of mixed cultures of *S. epidermidis* strain S6[+] and *S. aureus* on nutrient agar at 37°C (experiments 1 and 2).

In a less direct approach to bacterial interference on human skin, a combined epidemiological and clinical study has provided further evidence of the protective role of the normal flora. The presence of inhibitory commensal staphylococci in the skin lesions of 263 dermatologic inpatients was associated with a significantly low incidence of colonization and infection by gram-positive pathogens at the time of admission and during hospital stay.[22]

Antibiotic Production by Skin Microorganisms

Members of each of the subgroups of skin commensals and strains of *S. aureus* and the dermatophyte fungi are capable of producing antibiotics. The antibacterial effect of whole skin—living or cadaver[11]—may be due partly to antibiotics produced by the skin microflora, although organic acids and other antibacterial components are probably more important factors.

Commensal Staphylococci

The resident skin flora of approximately 20% of healthy subjects includes one or more strain of antibiotic-producing commensal Staphylococcus (or Micrococcus).[22,40] The results of direct and deferred antagonism tests show that wide variation in the potency and the range of antibacterial activity of the inhibitory products (Figs. 9.2, 9.3). Deferred tests, when the producer strain is grown before the indicator strain, reveal greater activity than direct tests (Fig. 9.3) and are ecologically more relevant.

Antibiotic-producing strains, where present, were predominant on the healthy skin of only about 40% of individuals. In contrast, these bacteria were predominant in the skin lesions of almost 80% of dermatologic patients—notably in cases of eczema.[22] Evidently such strains possess minimal competitive advantages on normal, relatively dry intact skin; but in skin lesions—particularly of an exudative nature—inhibitory commensals benefit from the higher population densities and better conditions for the production and diffusion of antibiotics. Under these conditions, such bacteria appear to be an asset to their hosts. Nevertheless, 80% of individuals, while not possessing such defensive bacteria, remain generally free of skin infection.

A representative strain of *S. epidermidis* biotype 4 isolated from my own skin has been fully investigated. Referred to as "S6 +" (since it had been identified as *Staphylococcus* type 6 in the earlier classification), the organism antagonized all the representative strains of gram-positive bacteria tested. These included species of *Bacillus, Clostridium, Corynebacterium, Staphylococcus, Micrococcus,* and *Streptococcus,* and the principle species of the gram-negative genera: *Neisseria, Hemophilus, Pasteurella, Brucella, Bordetella, Mycoplasma,* and *Bacteroides.* There was no cross-resistance with the standard antibiotics (Fig. 9.3). Strains of the *Enterobacteriaceae* family and *Pseudomonas* genus were usually resistant in the tests, as were

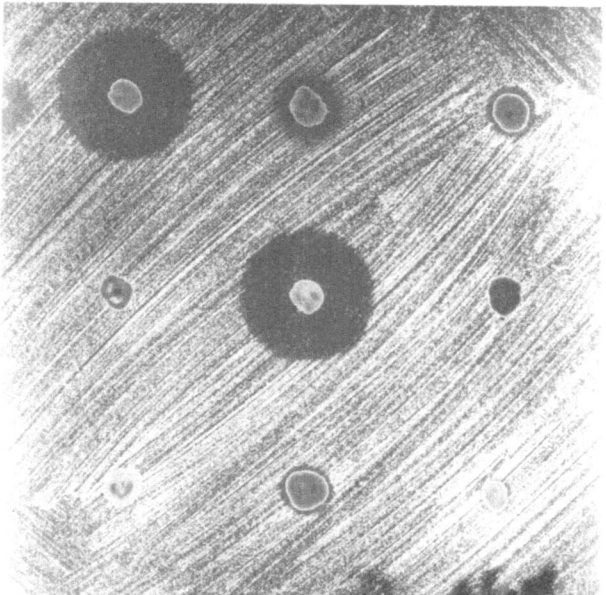

Fig. 9.2. Part of large agar plate (actual size) with *Micrococcus* type 7 background inoculated with nine other skin commensals. Various inhibition zones are present.

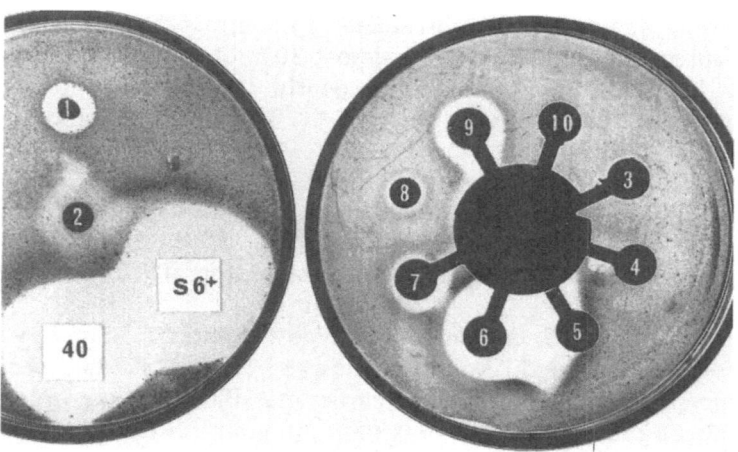

Fig. 9.3. Sensitivity of a prevalent *S. aureus* strain on 5% salt agar to direct antagonism (1 = S6⁺) and deferred antagonism (S6⁺ and 40) by commensals; resistance seen to methicillin (2), erythromycin (3), lincomycin (4), tetracycline (5), fusidic acid (6), neomycin (7), chloramphenicol (8), gentamicin (9), and benzylpenicillin (10). The discs contained 10 μg of each antibiotic except for benzylpenicillin (1.5 units), chloramphenicol (25 μg), and lincomycin (2 μg).

fungi; but the semipurified active extract also inhibited representative members of these groups.[41]

The antibiotic is produced early in the growth cycle and is a bactericidal polypeptide that appears to have a cyclical structure.[41] A second, less active antibiotic is produced simultaneously (unpublished results). Analogous but chemically distinct antibiotics ("epidermidins") have been described by Hsu and Wiseman.[22] In any investigations on such antibiotics, they must be carefully distinguished from bacteriophages, bacteriocines, lysozyme- and lysostaphinlike agents.

Diphtheroid Bacilli

The frequency of antibiotic production among coryneforms is approximately the same as that in commensal staphylococci, but antibacterial activity is narrower, usually being limited to members of the same genus.

In a recent study of *Propionibacterium acnes*, inhibitory strains appeared to possess no competitive advantage over others in acne lesions.[20] While the production by diphtheroids of antimicrobial substances, notably fatty acids, seems to be an important defense against skin infection, further epidemiological and experimental investigations are required to assess the significance of antibiotic production by these bacteria.

Staphylococcus Aureus

Bacterial antagonism is chiefly confined to *S. aureus* of bacteriophage group II (type 71 and related types). The antibiotics responsible form a homogeneous class active against a narrow range of gram-positive bacteria, including *Streptococcus pyogenes* and *Corynebacterium diphtheriae*.[6,32] The activity against diphtheria bacilli was exploited clinically over 40 years ago. Since the antibiotic-producing strains are those that are responsible for most cases of impetigo, there has been speculation about the possible part played by antibiotics in the epidemiology or pathogenesis of the disease. The question has not been resolved, except for the likelihood that bacterial interference by these strains of *S. aureus* limits the incidence of mixed infection with *S. pyogenes*.

Lysostaphin is an entirely different antibacterial substance produced by certain strains of *S. aureus*.[38] It destroys the integrity of cell wall peptidoglycan by cleaving the pentoglycine bridges. The cell wall-deficient cocci subsequently lyse. Since the effect is limited to *S. aureus*, it has been used as a differential test for this species. Lysostaphin has been successfully used in the treatment of experimental staphylococcal infection.[16]

Skin Fungi

The inhibitory effect of the commensal yeast *Pityrosporum ovale* on dermatophyte fungi[49] is probably due to the products of its lipolytic action rather than to a true antibiotic. In contrast, dermatophytes of the *Epidermophyton* and *Trichophyton* genera themselves can produce a wide range of powerful antibiotics, notably penicillins and streptomycinlike agents.[51] These can be detected in artificial culture media more readily than on stratum corneum cultures. In the latter system, antibiotic production has been studied by microbiological assay and indirectly by the morphological changes produced in micrococci, which were strikingly demonstrated under the scanning electron microscope.[8] In the latter investigation, *Trichophyton mentagrophytes* hyphae did not produce antibiotic uniformly, as judged by the effects on adjacent micrococci. However, not all the bacteria could be expected to be at a suitably active phase of growth for penicillins and related antibiotics to affect cell wall synthesis.

Production of antibiotics in vivo has been inferred by indirect effects, notably the increased incidence of antibiotic-resistant bacteria in the vicinity of tinea lesions.[7,44]

Several types of interaction between dermatophytes and skin bacteria have been observed. The most obvious is unilateral antagonism by the fungus against antibiotic-susceptible commensal bacteria. The fungus benefits by the removal of competition for nutrients; however, antibiotic-resistant strains of *S. aureus* or other bacteria benefit and possess a survival advantage.[44,47] Unilateral enhancement (synergy) of the fungus by the bacteria is evident from the directional growth of hyphae, presumably along a gradient of lipid or other nutrients derived from bacterial activities. Subsequent destruction of the bacteria by antibiotic may release further nutrients.[47] These effects could play a significant part in helping to establish mycoses. Conversely, some of the bacterial products possess antifungal activity,[2] thus creating the relatively uncommon interaction of mutual antagonism.

Synergistic Interactions

If small-colony forms of skin staphylococci or diphtheroids are used as indicator strains they will often show "satellitism" (or unilateral synergy) around colonies of other skin commensals (Fig. 9.4). In the first systematic study of this phenomenon, 12% of healthy individuals possessed such growth-enhancing commensals.[40] Although there is a general resemblance to the well-known satellitism of *Hemophilus influenzae*, which occurs around colonies of staphylococci, no part is played by either X or V factor (hematin or diphosphopyridine nucleotide-related coenzymes). Products of bacterial lipolysis are undoubtedly responsible for some of the observed satellitism, and the growth of lipophilic diphtheroids and *Pityrosporum* yeasts depends to a significant extent on this process; but growth curve and other studies reveal more complex aspects (Milyani and Selwyn, to be published). Some strains of staphylococci have their growth stimulated by *Candida albicans*,[46] and lipid factors do not seem to be involved. Conversely, *C. al-*

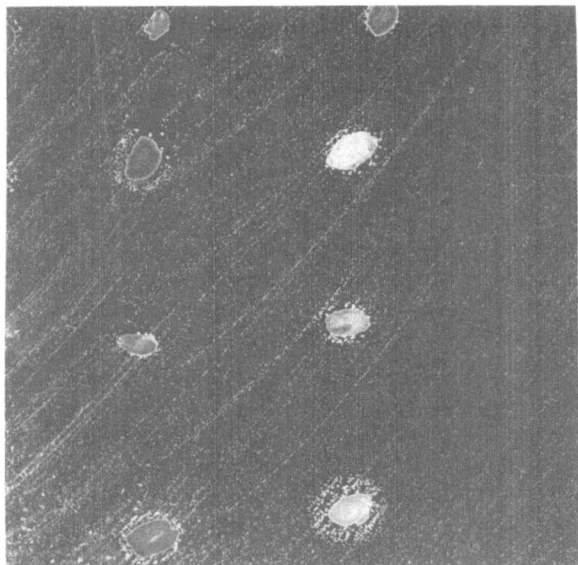

Fig. 9.4. Part of a large agar plate (actual size) with a background growth of a small-colony form of a coagulase-negative staphylococcus. Six of the strains of skin commensals enhance the background growth to varying degrees ("satellitism").

bicans inhibits the growth of some bacteria, including *Neisseria gonorrhoeae*.[19]

Clinically important examples of bacterial enhancement include the potentiation of anerobic infections by staphylococci (e.g., in symmetrical synergistic gangrene) and the protection during penicillin therapy of antibiotic susceptible bacteria by organisms that produce penicilliinase (β-lactamase). The potentiation of experimental *Proteus* infections by staphylococci[4] further emphasizes the wider pathogenetic and clinical implications of synergistic relationships.

After years of neglect the great importance of microbial interactions on and in the body is now abundantly evident. The fascinating interrelationships already elucidated on the skin should greatly stimulate further work on this most accessible of human ecosystems.

References

1. Aly R, Maibach HI, Strauss WG, Shinefield HR (1970) Effect of a systemic antibiotic on nasal bacterial ecology in man. Appl Microbiol 20:240–244
2. Aly R, Maibach HI, Shinefield HR, Strauss WG (1972) Survival of pathogenic microorganisms on skin. J Invest Dermatol 58:205–210
3. Anthony BF, Wannamaker LW (1967) Bacterial interference in experimental burns. J Exp Med 125:319–336
4. Arndt WF (1965) In: Synergistic infections in skin bacteria and their role in infection. Maibach HI, Hildick-Smith G (eds) McGraw-Hill Book Company, New York, pp 113–123

5. Baird-Parker AC (1974) The basis for the present classification of staphylococci and micrococci. Ann NY Acad Sci 236:7–13
6. Barrow GI (1963) Microbial antagonism by *Staphylococcus aureus*. J Gen Microbiol 31:471–481
7. Bibel DJ, Lebrun JR (1975) Effect of experimental dermatophyte infections on cutaneous flora. J Invest Dermatol 64:119–123
8. Bibel DJ, Smiljanic RJ (1979) Interactions of *Trichophyton mentagrophytes* and micrococci on skin culture. J Invest Dermatol 72:133–137
9. Blair EB, Tull AH (1969) Multiple infections among newborns resulting from colonization with *Staphylococcus aureus* 502A. Am J Clin Pathol 52:42–49
10. Boris M, Sellers TF, Eichenwald HF, Ribble JC, Shinefield HR (1964) Bacterial interference. Protection of adults against nasal *Staphylococcus aureus*. Infection after colonization with a heterologous *S. aureus* strain. Am J Dis Child 108:252–261
11. Castaigne F, Micheau P, Lacaze B (1973) Étude microbiologique et biochimique de difffcrentes greffes de peau (humaine et bovine). Ann Chir 27:203–208
12. Colebrook L, Lowbury EJL, Hurst L (1960) The growth and death of wound bacteria in serum, exudate and slough. J Hyg (Camb) 58:357–366
13. Doehle P (1889) *Beobachtungen über einen Antagonisten des Milzbrandes*. Habilitationsschrift, Kiel
14. Dujardin-Beaumetz E (1932) Action antibiotique d'une variété de staphylocoque à l'égard des bacilles gram-positifs et acido-résistants. CR Soc Biol Paris 110:1210–1213
15. Forfar JO, Gould JC, MacCabe AF (1968) Effect of hexachlorophane on incidence of staphylococcal and gram-negative infection in the newborn. Lancet 2:177–178
16. Goldberg LM, Defranco JM, Watanakunkorn C, Hamburger M (1968) Studies in experimental staphylococcal endocarditis in dogs. VI. Treatment with lysostaphin. Antimicrob Agents Chemother 1967:45–53.
17. Halbert SP, Kazar CS, Swick LS (1957) Mixed bacterial infections in relation to antibiotic activities. III. Staphylococcaldiphtheria infections of guinea pigs. AMA Arch Ophthalmol 57:716–723
18. Halbert SP, Locatcher-Khorazo D, Sonn-Kazar C Swick L (1957) Further studies on the incidence of antibiotic-producing micro-organisms of the ocular flora. AMA Arch Ophthalmol 58:66–76
19. Hipp SS, Lawton WD, Chen NC Gaafar HA (1974) Inhibition of *Neisseria gonorrhoeae* by a factor produced by *Candida albicans*. Appl Microbiol 27:192–196
20. Holland KT, Cunliffe WJ, Eady EA (1979) Intergeneric and intrageneric inhibition between strains of *Propionibacterium acnes* and Micrococcaceae, particularly *Staphylococcus epidermidis*, isolated from normal skin and acne lesions. J Med Microbiol 12:71–82
21. Houck PW, Nelson JD, Kay JL (1972) Fatal septicemia due to *Staphylococcus aureus* 502A: report of a case and review of the infectious complications of bacterial interference programs. Am J Dis Child 123:45–58
22. Hsu C-Y, Wiseman GM (1972) The nature of epidermidins, new antibiotics from staphylococci. Can J Microbiol 18:121–125
23. Light IJ, Walton RL, Sutherland JM, Shinefield HR, Brackvogel V (1967) Use of bacterial interference to control a staphylococcal nursery outbreak. Deliberate colonization of all infants with the 502A strain of *Staphylococcus aureus*. Am J Dis Child 113:291–300
24. Loeb LJ, Moyer A, Murray RGE (1950) An antibiotic produced by *Micrococcus epidermidis*. Can J Res 28:212–216
25. Maibach HI, Strauss WG, Shinefield HR (1969) Bacterial interference: relating to chronic furunculosis in man. Br J Dermatol 81 (Supplement 1):69–76

26. Marples RR, Kligman AM (1971) Ecological effects of oral antibiotics on the microflora of human skin. Arch Dermatol 103:148–153
27. Marsh PD, Selwyn S (1977a) Studies on antagonism between human skin bacteria. J Med Microbiol 10:161–169
28. Marsh PD, Selwyn S (1977b) Continuous culture studies on interactions among human skin-commensal bacteria. J Med Microbiol 10:261–265
29. McCabe WR (1967) Bacterial interference induced in embryonated eggs by staphylococci. J Clin Invest 46:453–462
30. Milyani RM Selwyn S (1978) Quantitative studies on competitive activities of skin bacterial growing on solid media. J Med Microbiol 11:379–386
31. Papacostas G Gaté J (1928) Les Associations Microbiennes, Leurs Applications Therapeutiques Gaston Doin et Cir, Paris
32. Parker MT, Simmons LE (1959) The inhibition of *Corynebacterium diphtheriae* and other Gram-positive organisms by *Staphylococcus aureus*. J Gen Microbiol 21:457–476
33. Pryjma J, Heczko PB, Wilburg J (1972) The effect of staphylococcal antagonism on development of infection in experimental animals. Exp Med Microbiol 24:86–90
34. Savin JA (1967) *Pseudomonas aeruginosa* infections in a skin ward. Trans St. John's Hosp Dermatol Soc 53:75–79
35. Selwyn S (1979b) Microbiological and clinical implications of the pharmacokinetics of cephalosporins. In: van der Waaij D, Verhoef J (eds) New criteria for antimicrobial therapy: maintenance of digestive tract colonization resistance. Excerpta Medica, Amsterdam, pp 147–156
36. Selwyn S (1975). Natural antibiosis among skin bacteria as a primary defence against infection. Br J Dermatol 93:487–493
37. Selwyn S (1979a) Pioneer work on the 'penicillin phenomenon,' 1870–1876. J Antimicrob Chemother 3:249–255
38. Selwyn S (1979c) The mechanisms and range of activity of penicillins and cephalosporins. In: The beta-lactam antibiotics: penicillins and cephalosporins in perspective, Chap 2. Hodder and Stoughton Educational, London:
39. Selwyn S, Chalmers D (1965) Dispersal of bacteria from skin lesions. A hospital hazard. Br J Dermatol 77:349–356
40. Selwyn S, Ellis H (1972) Skin bacteria and skin disinfection reconsidered. Br Med J 1:136–140
41. Selwyn S, Marsh PD, Sethna TN (1976) *In vitro* and *in vivo* studies on antibiotics from skin *Micrococcaceae*. In: Williams JD, Geddes AM (eds) Chemotherapy, Vol. Plenum Press, New York, pp 391–396 (Proceedings of 9th International Congress of Chemotherapy, 1975)
42. Shehadeh NH, Kligman AM (1963) The effect of topical antibacterial agents on the bacterial flora of the axilla. J Invest Dermatol 40:61–71
43. Shinefield HR, Ribble JC, Eichenwald HF, Boris M, Sutherland JM (1965) Bacterial interference. In: Maibach HI, Hildick-Smith G (eds) Skin bacteria and and their role in infection. McGraw-Hill Book Company, New York, pp 235–251
44. Smith JMB, Marples MJ (1965) Dermatophyte lesions in the hedgehog as a reservoir of penicillin resistant staphylococci. J Hyg (Camb) 63:293–303
45. Tomkins H (1887) Note on the cultivation of Bacillus anthracis. Br Med J 1:328–329
46. Virtanen O (1959) Observations on the symbiosis of some fungi and bacteria. Ann Med Exp Biol Fenn 29:352–358
47. Wallerstrom A (1968) Production of antibiotics by *Epidermophyton floccosum*. 2. Microflora in *Epidermophyton* infected skin and its resistance to antibiotics produced by the fungus. Acta Path Microbiol Scand 74:531–542
48. Ward M (1899) Symbiosis. Ann Bot 13:549–562

49. Weary PE (1968) *Pityrosporum ovale:* observation on some aspects of host-parasite interrelationship. Arch Dermatol 98:408–422
50. Wickman K (1970) Studies of bacterial interference in experimentally produced burns in guinea pigs. Acta Path Microbiol Scand [B] 78:15–28
51. Youssef N, Wyborn CHE, Holt G, Noble WC Clayton YM (1978) Antibiotic production by dermatophyte fungi. J Gen Microbiol 105:105–111

TOPICAL SKIN
ANTIBACTERIALS

Chapter 10
Clinical Trials of Topical Antimicrobials

ALFRED M. ALLEN

Clinical trials of topical antimicrobials have improved substantially over the past 20 years. Two decades ago, trials of these agents typically lacked "double blindedness" and randomized allocation of medication. Many did not even have concurrent controls; comparisons were based on investigators' impressions of what would have happened in the absence of treatment.

Great strides have been made since then. Today virtually every report contains the stock phrase "randomized, controlled trial," and there is a general appreciation of the necessity of this kind of investigation. But along with this progress has come a disturbing tendency to use this phrase as a guarantee of quality. No amount of retrospective data analysis can make up for such design flaws as inadequate numbers of patients and noncomparable groups of patients.

The destructiveness of these flaws is seldom apparent to people other than statisticians and epidemiologists, partly because of the technical complexity of the issues involved. But they can no longer be ignored by the medical community, as recently attested to by the editor of the *New England Journal of Medicine*.[11] These issues, as they apply to topical antimicrobials, are illustrated below by examples from published reports.

Comparing Competitive Treatments

Development of topical antimicrobials has long since reached the stage where more than one agent can be expected to prove highly effective as compared to placebos or preantibiotic forms of treatment. Clinical comparison of two or more active agents creates a need to conduct much larger trials than those which were suitable for demonstrating the efficacy of a new drug. Published reports indicate that this need is rarely met and, indeed, is seldom recognized.

Trials of competing topical antimicrobials typically have involved from 20 to 50 patients in each treatment group. This number of patients was generally adequate to demonstrate a statistically significant difference between a placebo or vehicle and an active drug. However, it has generated few results that indicate that one agent is significantly better than another. This comes as no surprise to statisticians, who recognize that the lack of significant results is as likely to be due to including too few patients as it is to trivial differences in efficacy.

The fact is that present-day trials are usually far too small to generate answers that are vital to the interests of both physicians and patients. This can be shown by applying basic statistical principles to published results.

Doctors from Lebanon reported a study in which there was no statistically significant difference between responses to gentamicin and placebo in the treatment of pyoderma.[14] Based on a trial involving a total of 46 patients, they concluded that topically applied antibiotics were ineffective in the treatment of pyodermas. No mention was made of the possible influence of the small number of patients studied on the outcome of the trial.

If the true long-run cure rate from placebo was 50% and that from gentamicin 75%, trials involving only 46 patients would yield a statistically significant result less than a third of the time. In other words, two out of three trials of this size would fail to detect a great difference in efficacy. To be virtually certain (i.e., 95% certain) that this trial would show a statistically significant difference, there must be 100 patients in each of the treatment groups.[10] In other words the Lebanese study would have to have been some five times as large as it was to provide good assurance that gentamicin would not benefit half again as many patients as placebo treatment.

In another trial gentamicin and neosporin ointments were compared in the treatment of pyoderma.[5] The authors stated that neosporin was more effective than gentamicin, but their data showed no statistically significant differences. The only defensible conclusion is that neosporin may be more effective than gentamicin, but this trial does not provide adequate evidence for it.

In the neosporin–gentamicin trial there were approximately 50 patients in each treatment group. The cure rates after a week of treatment were 45% and 33%, respectively. Using these figures as guidelines, assume that the real underlying one-week cure rates are 33% and 50%. With only 50 patients in each treatment group, a statistically significant result would occur in only one-third of trials comparing neosporin to gentamicin. Again, adequate evidence of a substantial difference in efficacy would probably not be obtained because the number of patients was too small.

The answer to the problem of inadequate evidence is simple: enroll more patients in trials, and start thinking in multiples of 100 rather than in multiples of 10 to arrive at the number of patients required. Precise figures can be derived in investigators are willing to state what differences between treatments are important to detect.[2] For example, if true differences in efficacy of 30% are considered too important to miss, investigators must be prepared to enroll at least 80 patients in each treatment group.[6] This would provide them with good assurance that statistically significant differences will be found when the long-run cure rates between the two treatments are

30% and 60%. Only 5% of trials would fail to show a significant difference when the true differences in efficacy are this large.

The number of patients required is dependent upon what the trial is intended to demonstrate.[2] If this can be specified in numerical terms, the number of patients needed in each group can be determined by consulting published tables.[2,6,10]

The information in sample size tables will no doubt come as a shock to many, especially those who have gone through the ordeal of conducting a trial themselves. The news that 500 patients rather than 50 are required will doubtless be unwelcome. However, the unpalatability of dealing with vastly more patients to complete a trial does not negate the validity of the argument. Once the relationship between sample size and statistical significance is recognized, the issue is no longer scientific but merely a problem in logistics.

Equivalent Efficacy

The concept of equivalent efficacy has existed in clinical medicine for a long time, but the expression itself is new and is not in common usage.[3] The term refers to the situation where one form of treatment is considered or proven to be no more effective than another for a given condition. The value of this information to a patient or physician is that it allows choices of therapy to be based on considerations other than efficacy; for example, cost, convenience, and personal preference.

There can also be considerable commercial value to a claim of equivalent efficacy. A drug company may have reason to promote a product with the clear implication that it is as effective as a competitor's, particularly when the competitor is making substantial inroads into the market with claims that its product is superior.

This situation can be clearly seen in a recent, highly publicized advertising campaign involving over-the-counter preparations for the treatment of athlete's foot. To counter claims for the superiority of tolnaftate cream, a company that sells undecylenic acid ointment (Desenex) has sponsored numerous nationwide television and radio commercial announcements that say in part: "If you suffer with athlete's foot...new scientific evidence that short of a prescription nothing is better than Desenex. Nothing." Obviously, there is in this statement an inherent claim of equivalent efficacy.

At least three segments of society—physicians, patients, and drug companies—have a stake in the issue of equivalent efficacy. For this reason it is worthwhile to correct some widely prevalent misconceptions about the nature of equivalent efficacy and how it can be scientifically demonstrated.

The Desenex advertisements provide a good illustration of the problems that beset anybody who is involved in deciding whether one treatment is as good as another. The company's claim, as indicated in its commercials, rests on the results of a study performed on students at the University of New

Mexico. A published report on the study[13] state that "there was no statistically significant difference between the results from the two antifungal agents" (undecylenic acid ointment and tolnaftate cream). The implied claim of equivalent efficacy therefore appears to be based on the absence of a statistically significant outcome in a single clinical trial.

Unlike achieving statistical significance, the absence of statistical significance has meaning only in relation to the size of the samples being compared. In the New Mexico study, there were just 30 patients in each treatment group, a minuscule number when a question of equivalent efficacy is involved. Table 10.1 shows that with this number of patients, a statistically significant difference is unlikely to be demonstrated even when there are marked differences in true, long-run efficacy rates. Consequently, stating that there was no statistically significant difference does not say much about similarities or differences in efficacy.

Had a much larger number of patients been studied, say 300 instead of 30 in each treatment group, a statistically insignificant result would have meant a great deal more. Table 10.1 shows that with 300 patients statistically significant differences would likely be demonstrated even when true efficacy rates are rather similar. Under these circumstances, an absence of statistical significance would indicate that true differences in efficacy are small and perhaps of little clinical consequence.

In practical usage, equivalent efficacy is only a relative term.[3] It is statistically impossible to prove exact equivalency, so recourse must be had to demonstrating that any differences are too small to be important. This requires a clinical judgment as to what differences in efficacy are so small that, for practical purposes, there is equivalency.

Large numbers of patients must be studied in order to demonstrate equivalent efficacy. This is illustrated in the following example.

Suppose that in a trial of two athlete's foot remedies a true difference in ef-

Table 10.1. Probabilities of demonstrating a statistically significant difference between two treatments (P < .05)

No. patients in each group	True efficacy rates (%) Drug A	Drug B	Probabilities (%)
30	25	50	41
	50	67	19
	67	75	6
50	25	50	68
	50	67	34
	67	75	10
100	25	50	95
	50	67	64
	67	75	19
300	25	50	99
	50	67	97
	67	75	63

Table 10.2. Probabilities that the true differences between two treatments are no greater than 10%

No patients in each group	True efficacy rate (%) of more effective drug	Probabilities (%)
100	25	36
	50	~20
	75	~30
300	25	82
	50	63
	75	70
500	25	96
	50	86
	75	91

[a]Assuming that no statistically significant differences are found in the trials.

ficacy of 10% or less would be considered unimportant. With this as the criterion of equivalent efficacy, the trial is conducted with 100 patients in each treatment group and the results are not statistically significant. Table 10.2 shows that this provides little assurance of equivalent efficacy, even though there were more than three times as many patients as were involved in the New Mexico study. In contrast, had there been 300 or 500 patients in each group, a statistically insignificant difference in results would have provided a much higher degree of assurance that the two remedies were equivalent in efficacy (Table 10.2).

Such large numbers of patients are virtually unheard of in present-day trials. This lays bare the inconclusiveness of the evidence when the results are reported to be not statistically significant.[3] It also shows the futility of continuing to conduct trials that are too small to provide information on which to base medically sound decisions.[2]

The size of trials required to demonstrate or refute a claim of equivalent efficacy can be determined by consulting statisticians or published tables.[2] The unaccustomedly large number of patients required will greatly increase the expense and logistic complexities of the trials, but these problems should not be insurmountable when dealing with such common conditions as dermatophytosis and pyoderma.

Noncomparable Treatment Groups

One of the fundamental tenets of a clinical trial is that like is being compared to like. Ideally, the groups of patients being compared should be alike in all relevant respects except the treatments under evaluation. Trial design and analysis techniques are available to assist in meeting this ideal.[1,3] Unfortunately, their use appears to have been the exception rather than the rule in trials of therapeutic agents for skin infections. This pattern can be reversed with no increase in expense in conducting future trials.

Patients with skin infections differ in a number of respects that can influence the outcome of a trial. Among these are the patient's age and sex, the infecting organism, and the location and severity of the infection. The clinical presentation can also vary in other ways, for example, acute vs. chronic dermatophytosis and primary vs. secondary bacterial infections.

Nearly all trials deal with patients that are heterogeneous in these respects. Most reports faithfully catalog the percentage distribution of these characteristics in the study population as a whole, but few present evidence that this information is taken into account in analyzing and interpreting results.

Of the four published reports on trials of tolnaftate vs. undecylenic acid for treatment of dermatophytosis,[8,9,12,13] just one[13] indicated that the treatment groups were comparable with respect to infecting fungi. Only one[8] compared treatment groups according to severity of lesions at the start of the trial. Three of the trials[8,9,12] dealt with infections at various sites (i.e., tinea pedis, tinea cruris, other tineas); only two of the reports[9,12] presented results according to location of the lesions. The sex of the patients was mentioned in only two[8,13] of the four reports; no breakdown by treatment groups was given.

Similar omissions are characteristic of reports on topical therapy of pyoderma.[4,5,7,14] Age might be expected to play an important role, especially when the range is from 4 months to 65 years,[14] but no mention is made of the age distribution in the treatment groups.

A possible reason for these omissions is the undue faith put in randomization as a means of equalizing treatment groups.[7] Randomization has a sole purpose—prevention of bias in allocating treatments. Using simple randomization alone, equal distribution of various categories of patients can be expected only when the numbers involved are far beyond those ordinarily employed in clinical trials.

Another possibility is the difficulty in dealing with several variables at the same time, either in assignment of patients to treatment groups or in analyzing results. The usual approach is to avoid the issue in the design of the trial. Later, partial compensation is made in the analysis by stratifying the data according to one characteristic, say location of lesions. This can result in an apparently well-balanced comparison (Table 10.3) when there are approximately equal numbers of patients in each treatment group. On the other

Table 10.3. Balanced comparisons of tolnaftate vs. undecylenic acid treatments for tinea pedis

	Tolnaftate		Undecylenic acid	
	No. Patients	No. good results (%)	No. Patients	No. good results (%)
Lubowe & Wexler (9)	14	13 (93)	12	8 (67)
Roberts & Champion (12)	14	10 (71)	15	11 (73)
Tschen et al. (13)	30	21 (70)	30	27 (90)

Table 10.4.   Unbalanced comparisons of tolnaftate vs. undecylenic acid treatments
for dermatophytoses other than tinea pedis

	Tolnaftate		Undecylenic acid	
	No. Patients	No. good results (%)	No. Patients	No. good results (%)
Lubowe & Wexler (9)				
T. cruris	11	9 (82)	7	2 (29)
Other	0	— —	3	1 (33)
Roberts & Champion (12)				
T. cruris	8	6 (75)	4	4 (100)
Other	4	3 (75)	9	4 (44)

hand, an overtly unbalanced comparison (Table 10.4) can occur when patients with lesions at a particular location are either under-represented or not represented at all in one of the groups. Unbalanced comparisons are not inherently undesirable, but they can lead to erroneous conclusions when summary comparisons are made on the overall results.[3]

There is an elegantly simple solution to these problems. It is a technique known as the *randomized block* trial design.[1] Use of this technique allows investigators to match subgroups of patients without fear of biasing the assignment of treatment. Subgroups in dermatophytosis studies might consist of patients who are alike in infecting organism and location and severity of lesions. In pyoderma studies, subgroups could be alike in age, infecting organism, and clinical variety of infection. Even if there were small numbers in a particular subgroup, the numbers would be equally divided between the treatments and thus make an equal contribution to the overall results. Moreover, though the numbers might be too small for separate statistical analysis, there would be an equal amount of evidence for each treatment.

Table 10.5 shows how a randomized block design benefit the presentation and interpretation of the results of a trial comparing two topical antifungal agents. Note that there are equal comparisons within each disease severity/lesion location subgroup. For realism, the overall results and the total number of patients with lesions at each location are the same as in a

Table 10.5.   Effect of using randomized block design to balance trial results with respect to location and severity of infection

Type of disease	Tolnaftate		Undecylenic acid	
	No. Patients	No. good results (%)	No. Patients	No. good results (%)
T. corporis/cruris				
Severe	6	6 (100)	5	0 (0)
Not severe	24	16 (67)	24	16 (67)
T. pedis				
Severe	3	3 (100)	3	0 (0)
Not severe	14	10 (71)	13	5 (38)
Overall results	47	35 (74)	45	21 (47)

published report.[8] The rest of the data are of necessity hypothetical since the authors provided no similar breakdown of results. A more complete application of the randomized block design could have included age, sex, and infecting species as well.

Utopia or Reality?

The issues discussed in this chapter are unavoidable in presentday life, but this does not make the answers any more palatable. The idea that 1000 patients may be required to demonstrate a point will undoubtedly seem utterly utopian to those who are accustomed to clinical trials less than a tenth this size. The idea may be even more difficult to accept when the diseases in question are so trivial, and the remedies so commonplace that they can be found on the counter of any drugstore. Though these feelings may influence the rate at which change is adopted, they will not change the rules for acquiring scientifically valid evidence.

Is change coming? Will clinical investigators and their supporters be forced to empanel vastly larger numbers of patients and to employ more intricate trial designs?

The answers are only partly conjectural. Journal editors, including those of the *New England Journal of Medicine* and the *Journal of the American Medical Association*, have already gone on record in this regard.[11] They have taken steps to improve the rigor of the statistical reviews of manuscripts submitted for publication. Most importantly, they have begun to become much more critical about these issues.

One of the reasons that journal editors can reasonably expect such high standards in trials is that some investigators have shown by example that they can be met.[11] This sets a precedent that is hard to ignore.

Compared to other classes of drugs, topical antimicrobials pose few problems for implementation of these improvements. Ethical constraints that loom large in life-threatening diseases or potentially hazardous treatments are virtually nonexistent. Patients with skin infections are abundant; over 8 million people in the United States alone have tinea pedis.[13] And the courses of treatment are conveniently short, only a month or so at most.

The time has come for full understanding of these issues by clinical investigators and their supporters. The issues are not arcane, nor are they particularly new. All have been explained in detail with relevant dermatologic examples.[1,2,3] This should make clear both the need and the means to achieve the goal of obtaining adequate information on which to base rational therapeutic decisions.

References

1. Allen AM (1978) Clinical trials in dermatology: Part 1, Experimental design. Int J Dermatol 17:42–51
2. Allen AM (1978) Clinical trials in dermatology: Part 2, Numbers of patients required. Int J Dermatol 17:194–203

3. Allen AM Clinical trials in dermatology: Part 4, Analysis and interpretation. Int J Dermatol (to be published)
4. Dillon HC (1970) The treatment of streptococcal skin infections. J Pediat 76:676–684
5. Esterley NB, Markowitz M The treatment of pyoderma in children. JAMA 212:1667–1670
6. Fleiss JC (1973) Statistical methods for rates and proportions. John Wiley & Sons, New York, pp 176–194
7. Hughes WT, Wan RT (1967) Impetigo contagiosa: etiology, complications, and comparison of therapeutic effectiveness of erythromycin and antibiotic ointment. Am J Dis Child 113:449–453
8. Kuflik EG, Howie CR, Nelson PI (1971) An evaluation of topical antifungal therapy in servicemen. Cutis 7:287–290
9. Lubowe II, Wexler L (1965) Comparative study of tolnaftate solution and compound undecylenic acid ointment N.F. in the treatment of epidermophytoses. Curr Ther Res 7:401–405
10. Mainland D (1963) Elementary Medical Statistics. 2nd ed. WB Saunders Co., Philadelphia. pp 368–370
11. Rennie D (1978) Editorial: Vive La Difffcrence (P < 0.05). N Engl J Med 299:828–829
12. Roberts SOB, Champion RH (1968) Treatment of ringworm infections with tolnaftate ointment. Practitioner 199:797–798
13. Tschen EH, Becker LE, Ulrich JA, Hoge WH, Smith EB (1979) Comparison of over-the-counter agents for tinea pedis. Cutis 23:696–698
14. Zaynoun ST, Matla MT, Uwayda MM, Kurban AK (1974) Topical antibiotics in pyodermas. Br J Dermatol 90:331–334

Chapter 11
Preoperative Shower Bath with 4% Chlorhexidine Detergent Solution: Reduction of *Staphylococcus Aureus* in Skin Carriers and Practical Application

STAFFAN SEEBERG, ANDERS LINDBERG

AND BO R. BERGMAN

Our interest in the skin bacteria of surgical patients originated during a study of orthopedic postoperative wound infections at Sahlgrens Hospital and at our laboratory, Göteborg, Sweden. The dominant bacterium in early infections following operations of the hip was *Staphylococcus aureus*, as was the case in a study from Lund, Sweden, by Ericsson et al.[4] In environmental studies, bacteria from the orthopedic patient's specific environment—from the operating room or recovery unit—could not be traced in patients with *S. aureus* infections. However, in many cases *S. aureus* with the same phage type was isolated before operation from the patient's nose and/or skin.[7] It was concluded that in all probability the patients were their own sources of infection.

Several authors demonstrated a relationship between the patient's skin and nose staphylococci and contamination of the surgical wound[13] or subsequent wound infection.[8,16,17] The patient's entire skin, particularly the groin, acts as a bacterial reservoir.[2,5,6,7]

We concluded that it might be desirable to reduce the number of skin bac-

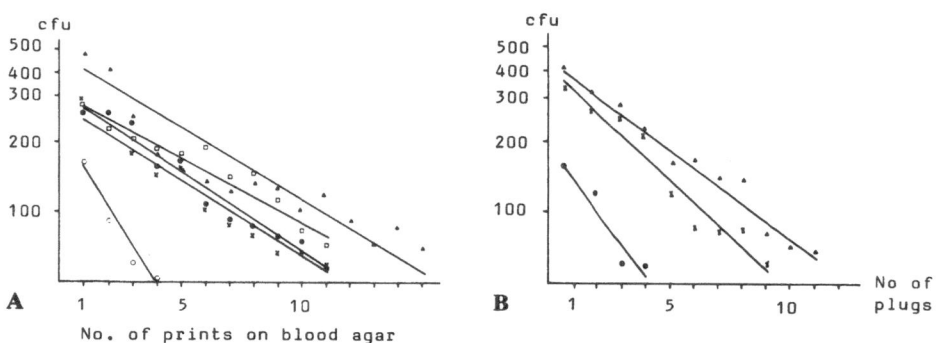

Fig. 11.1. Test of bacterial sampling method (6 cm² cardboard plug): **A** Colony counts after repeated prints with one plug onto blood-agar plates. **B** Colony counts after serial sampling from a defined skin area with plugs, each printed once onto blood-agar.

teria, especially *S. aureus*, in certain surgical patients, and that total body disinfection might have advantages over local disinfection of the operation field only by preventing bacterial contamination from adjacent areas of the skin and delaying recolonization of the skin during the first few days after the operation.

As demonstrated by Lowbury and Lilly and others,[3,9,10,11,12,14] chlorhexidine solutions are effective in reducing the bacterial flora of the skin, including *S. aureus*.

The present work develops a simple program for preoperative treatment of skin to reduce the risk of bacterial transfer from the patient's own skin to his wound.

Chlorhexidine Treatment of S. Aureus Skin Carriers

As a first step we investigated effect of 4% chlorhexidine gluconate detergent solution (Hibiscrub) on the *S. aureus* of five staphylococcal skin carriers. Five male volunteers, aged between 19 and 62 years, their skin heavily colonized with *S. aureus*, applied the chlorhexidine solution to their wet skin in three consecutive 5-min periods. Each application was followed by rinsing under a shower, drying, and bacterial sampling.

Sampling was done with a 6 cm² cardboard plug, moistened and pressed to the skin with a rotating movement to achieve a desquamating effect. The plug was used for repeated printing onto blood-agar medium containing recommended neutralizing substances for chlorhexidine (1% Tween 80, 0.5% lecithin, and 1% natriumthiosulfate).

This sampling method showed that with repeated prints with one plug, the plug could be emptied of its colony-forming units (cfu); there was a linear decrease in the number of cfu (Fig. 11.1a,b). A linear decrease in the number of colony-forming units was found when a series of plugs was pressed onto an exactly defined area of skin and printed once on blood-agar (Fig.11.1b), indicating that the skin area was depleted of its bacteria. This variant of replica printing permitted a quantitative estimation of those staphylococci transferable by contact and of possible clinical significance. Parallel use of Petri dishes containing blood-agar for direct contact sampling showed a good correlation between these two contact methods.

Figure 11.2 shows the mean values for the samples from the upper lip, axilla, right and left groin.

A reduction of the staphylococcal cfu was found after the first chlorhexidine treatment; after three washings, few or no *S. aureus* were isolated from the groin and axilla. A small number of staphylococci were isolated from the lip, probably because the nose consituted a reservoir. The duration of the effect was studied: all volunteers were followed for 3 days; two of them, for 9 days after the chlorhexidine treatment (Fig. 11.3). The *S. aureus* cfu level was still low compared to the original values. This is probably due to the residual antibacterial action on the skin of the 4% chlorhexidine preparation.

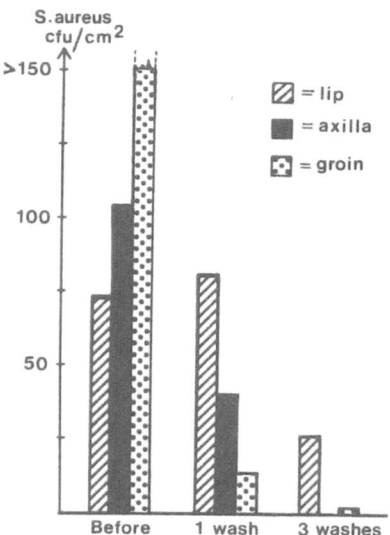

Fig. 11.2. **Mean** *S. aureus* colony counts from lip, axilla and groin of five skin carriers before and after total body washing with 4% chlorhexidine detergent solution.

Program for Preoperative Shower Bath With Chlorhexidine

The next step was to develop a model for total body washing of patients having elective orthopedic surgery. The patients started the skin disinfection program in their homes. We wished to ascertain whether our patients would cooperate in such a program. This was documented with bacterial sampling.

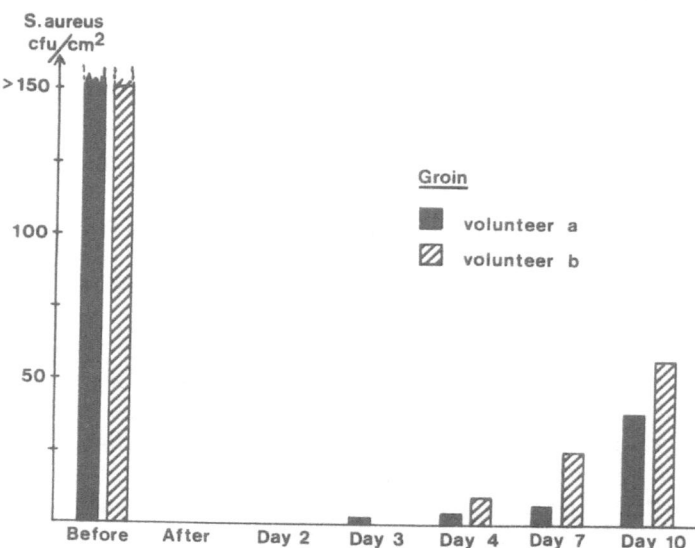

Fig. 11.3. *S. aureus* colony counts from groin of two skin carriers before and up to 10 days after shower bath with 4% chlorhexidine detergent solution.

Twenty-eight orthopedic patients of both sexes, aged between 19 and 78 years, participated. Bacterial sampling was performed with contact plates containing blood-agar (as described by Raahave[15]), this method being more suitable for large-scale sampling than the method described above. Samples were taken from the sternum, the right and left groin, and the operation site. Nasal swabs were taken.

The patients were called to the hospital and given information about the purpose of skin disinfection and washing routines 1 week before the operation. The first bacterial sampling was performed at this visit; the patient was given the chlorhexidine preparation and brief printed information about the washing procedure. The patients were instructed to wash their entire bodies, including the scalp, with the chlorhexidine preparation the day before admission. The washing procedure entailed two consecutive thorough applications of the chlorhexidine preparation and rinsing under a shower. The patients were encouraged to dress in a clean change of clothes after total body washing and instructed to use the chlorhexidine preparation for any subsequent washing.

The washing procedure was repeated immediately after admission to hospital, this time under supervision of the nursing staff.

On the following day (the day of operation) the patients were prepared in the ususal way; the operation site was washed with the chlorhexidine preparation once again, and bacterial samples were taken as previously described. At this last sampling the patient's chest and groin had thus been washed twice by the patient himself, and the operation site had been washed three times, the last by the nursing staff. The samples gave an illustration of the patient's skin flora immediately before the final skin disinfection of the operation site with chlorhexidine in alcohol.

The results from the bacterial cultures from the patients' skin show that a well-motivated patient usually complies with directions. This is most clearly demonstrated by the samples taken from the patients' groin (Fig. 11.4a,b), where the total aerobic bacterial count on the contact plates was often high before washing. The figure shows the results of sampling before and after the skin disinfection program. The number of cfu in samples taken after treatment from all but three patients was low, in most cases less than 100 cfu per plate, regardless of the pretreatment values.

The same tendency was found in the samples taken from the chest.

The operation site, having been treated an extra time with chlorhexidine, showed virtually no bacterial growth at the second sampling (Fig. 11.4b).

S. aureus was isolated from the skin in eight patients, but only three of them had high amounts; *S. aureus* in these patients constituting up to 30% of the total colony count. In five of the eight, *S. aureus* was not isolated after the skin disinfection program; in three, one of whom had high amounts in pretreatment samples, single colonies were still isolated in samples from chest and groin. All patients with *S. aureus* on the skin had it in the nose; in all cases the staphylococci from the skin and nose were of the same phage type.

No complications in the form of skin irritation, etc. were observed.

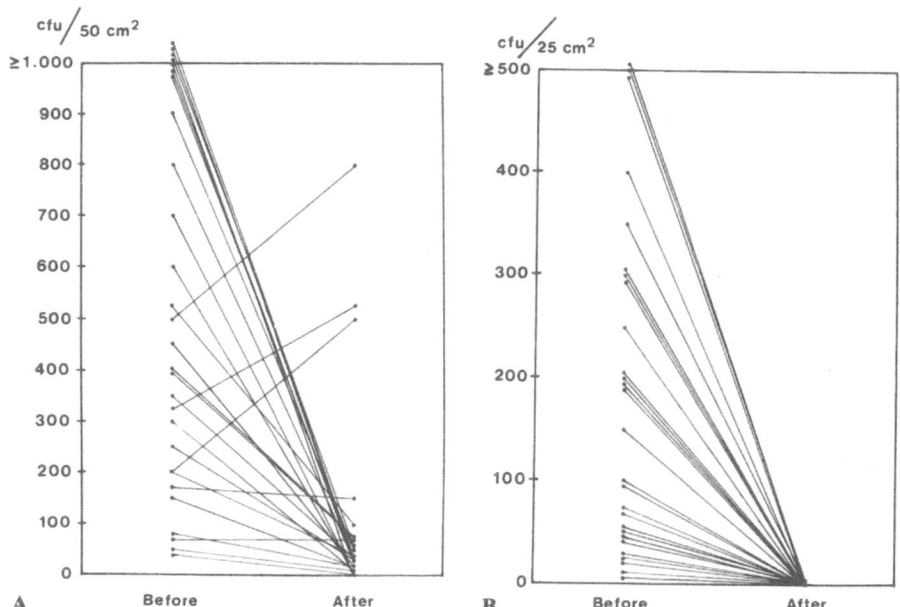

Fig. 11.4. Aerobic colony counts on contact-plates from 28 orthopedic patients before and after the preoperative skin-disinfection program. **A** Samples from right + left groins. **B** Samples from operation site.

It is important to have some form of supervision of the patients during the second total body washing at the hospital because some of the patients may have physical handicaps. In this study we had three patients who had high numbers of bacteria in posttreatment samples from the chest or groin or both. Two of these had physical handicaps (severe arthrosis), the third had a language problem. In the program presented, the patients started the chlorhexidine treatment at home and in most cases did not require assistance of the hospital personnel for the second treatment. The extra work load for the staff was minimal.

Whether or not the procedure of total body washing with chlorhexidine can reduce the already low incidence of postoperative wound infections in orthopedic surgery remains to be shown. But for those patients who are at a high risk as regards postoperative wound infections—such as patients undergoing total arthroplasty of the knee or hip joint—we have adopted the present program.

References

1. Backelin B, Bergman BR, Brandberg A The infection panorama in orthopaedic surgery. To be published
2. Bethune DW, Blowers R, Parker M, Pask EA (1965) Dispersal of *Staphylococcus aureus* by patients and surgical staff. Lancet 27:480–483

3. Davies J, Babb JR, Ayliffe GAJ, Ellis SH (1977) The effect on the skin flora of bathing with antiseptic solutions. J Antimicrob Chemother 3:473–481

4. Ericson C, Lidgren L, Lindberg L (1973) Cloxacillin in the prophylaxis of postoperative infections of the hip. J Bone Joint Surg (Am)55:808–813

5. Hare R, Ridley M (1958) Further studies on the tranmission of *Staph. aureus*. Br Med J 11:69–73

6. Hill J, Howell A, Blowers R (1974) Effect of clothing on dispersal of *Staphylococcus aureus* by males and females. Lancet 9:1131–1133

7. Hoborn J (1977) Mensch, Bekleidung und Reinraumtechnik. Medita 7 Nr. 9:3–8

8. Jarvis AW, Wigley RD (1961) Staphylococcal surgical wound infections, with particular reference to skin flora, and a comparison of two pre-operative skin preparations. Med J Aust 30:542–545

9. Lilly HA, Lowbury EJL (1971) Disinfection of the skin: an assessment of some new preparations. Br Med J 18:674–676

10. Lindberg A (1977) Staphylococcus aureus i hudfloran. En metod för semikvantitativ bedömning - effekt av tvättning med 4 proc klorhexidin. Lakartideningen 12:1185–1186

11. Lowbury EJL, Lilly HA (1960) Disinfection of the hands of surgeons and nurses. Br Med J 14:1445–1450

12. Lowbury EJL, Lilly HA (1973) Use of 4% chlorhexidine solution (Hibiscrub) and other methods of skin disinfection. Br Med J 3:510–515

13. McLaughlan J, Logie JRC, Smylie HG, Smith G (1976) The role of clean air in wound infection acquired during operation. Surg Gynecol Obstet 143:6–8.

14. Nielsen ML, Raahave D, Stage JG, Justesen T (1975) Anaerobic and aerobic skin bacteria before and after skin-disinfection with chlorhexidine: an experimental study in volunteers. J Clin Pathol 28:793–797

15. Raahave D (1973) Agar contact plates in evaluation of skin-disinfection. Dan Med Bull 20:204–208

16. Rountree PM, Harrington M, Loewenthal J, Gye R (1960). Staphyloccoccal wound infection in a surgical unit. Lancet II:7140–7145

17. Weinstein HJ (1959) The Relation between the nasal-staphylococcal-carrier state and the incidence of postoperative complications. N Engl J Med 260:1303–1308

Chapter 12

Preoperative Whole Body Disinfection by Shower Bath with Chlorhexidine Soap: Effect on Transmission of Bacteria from Skin Flora

ÅKE BRANDBERG AND INGRID ANDERSSON

Total body washing of patients is part of the standard routine in most hospitals. Acceptable cleaning or "social hygiene" can be effected through many methods; the effect on microorganisms as a prophylaxis against hospital infections can be dubious. Meers et al.[12] found a 17-fold increase in release of particles carrying viable bacteria after washing with soap. Lowbury and Lilly[11] investigated the reduction of the superficial bacterial flora of the skin by chlorhexidine solutions and found a marked reduction. Total body washing of staphylococcal skin carriers with 4% chlorhexidine gluconate soap solution (Hibiscrub) resulted in a marked reduction of the staphylococci in the skin flora, the reduction lasting over 1 week.[10]

Bruun showed a decrease from 8% to 2% of the incidence of postoperative infections caused by *S. aureus* by reducing the number of patients preoperatively colonized with *S. aureus* in the skin from 50% to 20%. Thomsen[13] studied sources and routes of postoperative wound infections in general surgery and judged few to be transmitted during the operation. In 75% of the patients the infection was transmitted at the ward. This was confirmed by Jepsen.[9] Bröte and Niléhn[3] emphasized the complex and heterogeneous causalities, such as resident skin flora, preoperative colonization, preoperative contamination of the wound by *S. aureus* from the operating staff and surroundings (demonstrated by Howe and Marston,[8] Burke,[4] and Gierhake[7]), heterografts, scars, and a decreased resistance in compromised host-patient. They found postoperative contamination of wounds of minor interest as a cause of infection. *S. aureus* found preoperatively in the resident skin flora and in cultures from the wound at the end of the operation were strongly correlated to postoperative infection.

Early postoperative infections in Sweden in orthopedic, plastic, and vascular surgery are almost always caused by *S. aureus*.[1,3,6] The aim of the present study was to illustrate the transmission of aerobic bacteria from skin by direct contact before and after a shower bath.

Materials and Methods

The illustrations are based on results obtained from samples taken from 42 different loci of the body surface of 16 volunteers, (11 males and 5 females) before and after a shower bath. Shower bathing in hot water was performed with nonmedicated soap and, as a comparison, with 4% chlorhexidine gluconate soap solution (Hibiscrub).

The degree of direct bacterial contact contamination was judged from the number of colony-forming units (cfu) obtained on contact hygiene agar plates (25 cm^2) after a 48-h aerobic incubation at 37°C.

The following blood-agar medium was employed: 1 000 ml of placental infusion, 10 g of neutralized bacteriological Peptone, Oxoid L 34, 3 g of NaCl, 0.5 g of glucose, 1 g of Na$_2$HPO$_4$.H$_2$O, 12 g of agar no. 1, Oxoid L 11, 50 ml of defibrinated horse blood. No suitable inactivator for chlorhexidine is available.[2]

The washings were performed according to standard instructions: the volunteers were told to wash their entire bodies, including the scalp, with the chlorhexidine solution. The washing procedure entailed two consecutive thorough applications of the soap solution and rinsing under a hot shower. No samples were taken before the skin was dry.

The figures were drawn by translating the number of cfu/plate into shadings of corresponding darkness. A key for translation of gradation of darkness to numbers of cfu/plate is presented in Fig. 12.1.

Results

The number of cfu/plate before washing from 11 males is presented to the left in Fig. 12.2; the number of cfu/plate after shower bath with a non-

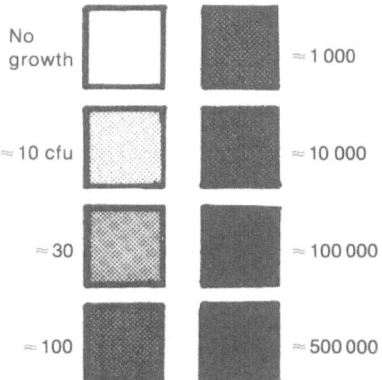

Fig. 12.1. Key of shadings representing numbers of colony-forming units (cfu) in Figs. 12.2 through 12.4.

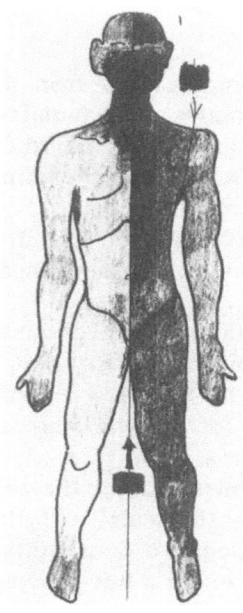

Before   After

Fig. 12.2. Density of cfu on a man after a shower bath using nonmedicated soap.

medicated soap is shown to the right. Corresponding results from females are shown in Fig. 12.3.

The amount of cfu/plate picked up from some of the skin areas before and after a shower bath with a nonmedicated soap is presented in Table 12.1.

The number of cfu/plate before and after showering with chlorhexidine soap is in Fig. 12.4. A similar reduction of cfu/plate was registered for females. Repeated sampling from 7 of the 11 males and 2 of the 5 females within 2 months showed similar numbers and kinds of bacteria from the indi-

Table 12.1. Amounts of cfu/plate measured from different skin areas before and after a shower-bath with a nonmedicated soap.

	Approximate number of cfu/plate		
Skin area	Before washing	After washing	After dis-infection
Forehead	0–10	≈ 10 000	0
Cheek	≈ 10 000	≈ 100 000	≈ 50
Chest	≈ 50	≈ 10 000	0–10
Armpit	≈ 500 000	≈ 10 000–100 000 (!)	≈ 50
Abdomen	≈ 10	1 000–10 000	0
Groin	≈ 50	1 000	0
Proximal inside of the thigh	≈ 500 000	1 000–10 000	50
Mid-thigh	0–10	≈ 100	0

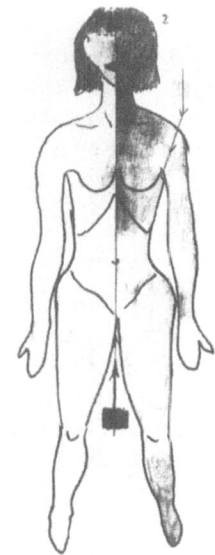

Before   After

Fig. 12.3. Density of cfu on a woman after a shower bath using nonmedicated soap.

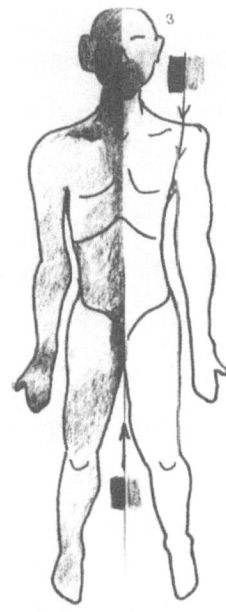

Before   After

Fig. 12.4. Density of cfu on a man after a shower bath using chlorhexidine soap.

vidual examined compared to the earlier findings during the period of a shower bath with a nonmedicated soap. Showering with chlorhexidine reduced the number of bacteria for a duration of 1 week. Coagulase-negative staphylococci were predominant. *Staphylococcus aureus* was a constant finding before disinfection in three of the males and one of the females.

Discussion

A shower bath with a nonmedicated soap will increase the contamination of aerobic bacteria by direct contact from most loci of the skin. Areas with high numbers of cfu (hair, armpit, and medial thigh) showed a decrease and are probably sources of contamination of other surfaces. Females do not have as high a number of aerobic bacteria as do males, but the number of cfu received on hygiene agar plates was higher after bath. Similar findings have been reported by Davis et al.[5] in a study of the effect of five consecutive daily baths. The loss of skin squamae increases conspicuously after washing. These results show the traditional bath or shower bath with a nonmedicated soap is a dubious preoperative routine for patients or staff members.

Shower baths with chlorhexidine soap show a conspicuous reduction of the transmission of and contamination with aerobic bacteria from both males and females.

References

1. Backelin B, Bergman BR, Brandberg Å (to be published) The panorama of infections in orthopaedic surgery.
2. Bergan T, Lystad A (1972) Evaluation of disinfectant inactivators. Acta Pathol Microbiol Scand [B] 80:507–510
3. Bröte L, Niléhn B (1976) Wound infections in general surgery with special reference to the occurrence of *Staphylococcus aureus*. Scand J Infect Dis 8:89–97
4. Burke JF (1963) Identification of the sources of staphylococci contaminating the surgical wound during operation. Ann Surg 158:898–904
5. Davies J, Babb JR, Ayliffe GAJ, Ellis SH (1977) The effect on the skin flora of bathing with antiseptic solutions. J Antimicrob Chemother 3:473–481
6. Ericson C, Lidgren L, Lindberg L (1973) Cloxacillin in the prophylaxis of postoperative infections of the hip. J Bone Joint Surg(Am) 55:808–813
7. Gierhake FW (1970) Postoperative Wundheilstörungen. Springer Verlag, Berlin, Heidelberg, and New York
8. Howe CD, Marston AT (1962) A study on sources of postoperative staphylococcal infection. Surg Gynecol Obstet 115:266–275
9. Jepsen OB (1972) Postoperative wound sepsis in general surgery, VII. Staphylococcal wound sepsis. Acta Chir Scand 138:343–348
10. Lindberg A (1977) *Staphylococcus aureus* i hudfloran. En metod för semikvantitativ bedömning - effekt av tvättning med 4 proc. klorhexidin. Lakartidningen 12:1185–1186
11. Lowbury EJL, Lilly HA (1973) Use of 4% chlorhexidine solution (Hibiscrub) and other methods of skin disinfection. Br Med J 3:510–515

12. Meers PD, Yeo GA (1978) Shedding of bacteria and skin squames after handwashing. J Hyg (Camb) 81:99–105
13. Thomsen FW, Olesen Larsen S, Jepsen OB (1970) Postoperative wound sepsis in general surgery. IV. Sources and routes of infection. Acta Chir Scand 136:251–260

Chapter 13
Postoperative Wound Infections in Vascular Surgery: Effect of Preoperative Whole Body Disinfection by Shower-bath with Chlorhexidine Soap

ÅKE BRANDBERG, JAN HOLM, JAN HAMMARSTEN
AND TORE SCHERSTEN

Bacterial flora on the patient's own skin is an important source of postoperative wound infections.[1,2] Washing with chlorhexidine solution reduces the skin flora. Data on this reduction by preoperative hand washing are published by Lowbury and Lilly[4] and by Smylie et al.[6]; reduction of anaerobic and aerobic skin flora after local disinfection of different skin areas, by Nielsen et al.;[5] and reduction of *S. aureus*, by Lindberg.[3]

This investigation studies the effect of preoperative total body disinfection by means of a shower-bath with chlorhexidine soap on the frequency of postoperative infections in vascular surgery.

Material and Methods

The study was prospective but not randomized. At Sahlgrens Hospital, Göteborg, Sweden, 341 patients from the vascular surgical units operated on with groin incision were studied. Data on sex, age, and number of patients with severe complications are in Table 13.1. All acute operations were omitted.

Two groups were studied: 171 patients were preoperatively prepared in the established standard routine at the clinic. This included shaving and preoperative local washing of the operation site with chlorhexidine soap (Hibiscrub), followed by local disinfection with chlorhexidine 0.5% in 70% alcohol. The other group, 170 patients, were preoperatively prepared by means of a shower-bath with chlorhexidine soap repeated 3–8 times prior to going through the same routine as the control group.

Except for the method with preoperative total body disinfection it was possible to keep most of the routines constant: the indications for operation were the same, as were the operating technique, equipment, and theater. The surgeons and the major part of the nursing staff were the same, pre- and postoperative care were performed in the same ward, antibiotic policy remained unchanged, and the same surgeon classified all infections. All

Table 13.1

	Unwashed (n = 171)	Washed (n = 170)
Males	128	126
Females	43	44
Age (average)	63	68
Diabetes	18	20
Threatening gangrene	60	58

wound reactions in the groin with pus formation were recorded, even those with pus in minimal quantities and spontaneous resolution. No prophylactic antibiotics were administered.

Results

Infection occurred in 36 of the 254 males and in 7 of the 87 females studied (Table 13.2, Fig. 13.1). In the unwashed group, 30 of the 171 patients (17.5%) subsequently became infected. The corresponding figure for the 170 washed patients was 13 (i.e., 8%) infected (Table 13.3).[1] The male/female infection ratio was 1.6:1 and 1.9:1 in the two groups, respectively.

The observed decrease in the wound infection rate occurred irrespective of threatening gangrene, diabetes or, as given in Fig. 13.2, age.

Discussion

The reduction in incidence of infections is significant (p < 0.05). Except for preoperative washings, no change in the routine handling of the patients in the wards or in the operating rooms had taken place during the investigation period. In both groups, males were more frequently infected than females. The number of males and females studied was the same in both groups. Postoperative infections are more common in the elderly. The average age in

Table 13.2

	Unwashed (n = 171)	Washed (n = 170)	Total
Males (254)	25	11	36
Females (87)	5	2	7
	30	13	43

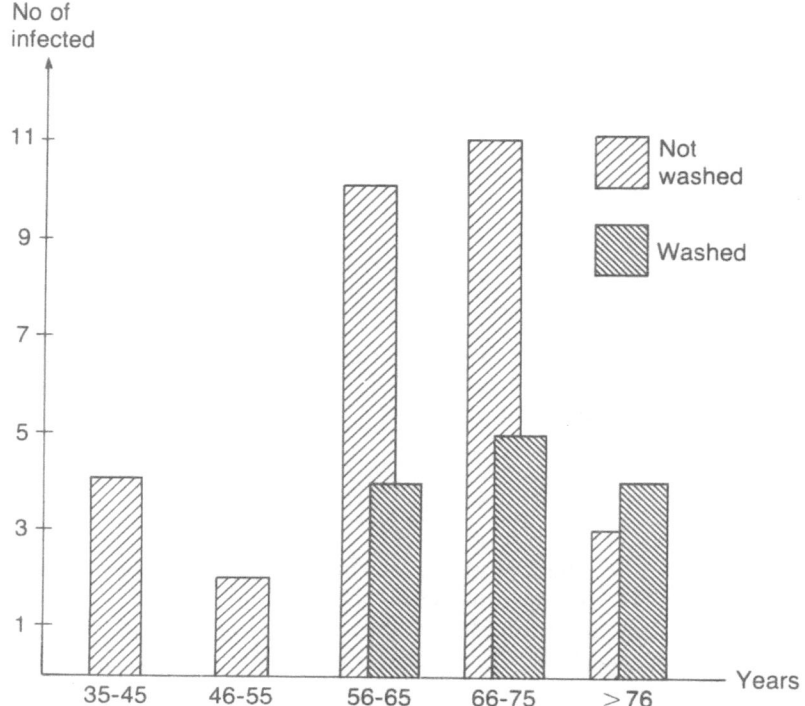

Fig. 13.1. Number of infections according to age group.

the washed group was higher and should consequently have increased in the number of the infected instead of contributing to the opposite, as shown. The study was prospective; the routines and the criteria were standardized and controlled. It was not possible to perform the study as a double-blind test because no repeated shower-bath was included as a preoperative measure in the standard routine of the clinic. However, the investigation was not randomized. Efforts were made to find other reasons for the decrease of infections. No important changes in the routine, indications for operation, nursing, equipment, rooms, surroundings, antibiotic policy, type of patients, ratio thrombectomia/grafts were found. At the start of the study, most of the grafts were holografts but we have successively changed over to monofilament sutures.

Table 13.3

	Unwashed (n = 171)	Washed (n = 170)
Males	19,5%	8,7%
Females	12%	4,5%
Total	17,5%	8%

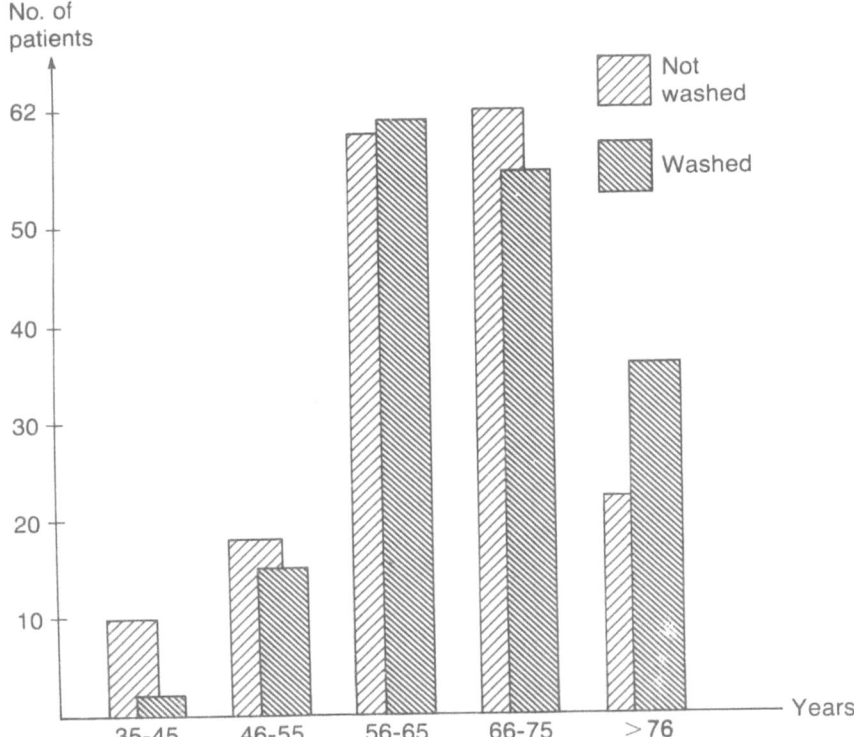

Fig. 13.2. Postoperative infection in washed and unwashed patients.

Summary

We studied 341 patients in the vascular surgical unit who had been operated on with groin incision. Preoperative skin disinfection was performed in 50% by means of repeated shower-baths with chlorhexidine soap. Incidence of infection was reduced from 17.5% in the nonprepared control group to 8% in the group treated with chlorhexidine.

References

1. Cruse P (1970) Surgical wound sepsis. Can Med Assoc J 102:251–258
2. Dineen P (1975) Influence of operating room conduct on wound infection. Surg Clin North Am 55:1283–1287
3. Lindberg A (1977) *Staphylococcus aureus* i hudfloran. En metod för semikvantitativ bedömning - effekt av tvättning med 4 proc. klorhexidin. Lakartidningen 12:1185–1186
4. Lowbury EJL, Lilly HA (1973) Use of 4% chlorhexidine solution (Hibiscrub) and other methods of skin disinfection. Br Med J 1:510–515

5. Nielsen ML, Raahave D, Stage JG, Justesen T (1975) Anaerobic and aerobic skin bacteria before and after skin disinfection with chlorhexidine: an experimental study in volunteers. J Clin Pathol 28:793–797
6. Smylie HG, Logie JRC, Smith G (1973) From Phisohex to Hibiscrub. Br Med J 4:586–589

Chapter 14
Newer Germicides: What They Offer

MARY BRUCH

Two developments have drastically changed FDA's approach to regulation of topical antimicrobial agents: the recognition of the systemic toxicity of small amounts of hexachlorophene when absorbed after topical application; and the subsequent review and recommendations concerning both safety and effectiveness of other antimicrobial ingredients as part of the FDA review of OTC drugs.

Nationally, attitudes toward topically applied antimicrobials really began to change with the introduction of hexachlorophene and other substituted bisphenols, substituted salicylanilides and carbanilides that displayed substantivity. The insolubility and substantivity of these ingredients allowed formulation in liquid and bar soaps. Whether the mechanism of persistent action is the same for all of these is unclear; these and new chemicals introduced into the United States, such as chlorhexidine, have produced new uses, approaches to effectiveness, and safety problems.

It may be useful to present a brief regulatory history of antimicrobial products, with particular reference to hand-washing agents. Topically applied antimicrobial products are used to reduce the microbial flora of the skin, prevent infection or treat infection by inhibiting, killing or removing pathogenic microorganisms on the skin. With hand-washing products the risk of infection may be reduced in the user and those individuals who are in contact with the user, such as in routine patient care and presurgery. These formulations are classified as drugs when formulations containing active antimicrobial ingredients are labeled for these uses.

The realization that topical antimicrobial applications were useful in the prevention of disease, of course, dates back to the work of Semmelweis, Koch, and Lister. The variety and uses of such products since that time are legion. Most of us are aware of the use of phenol and peroxide for such purposes. The advent of the modern synthetic antimicrobial agents for use in hand-washing and surgical-scrub products dates to the introduction of iodophores and hexachlorophene products post-World War II and in the early 1950's respectively. Prior to these products, the major one applied for surgical operating use was tincture of green soap and alcohol. During these early years, not much attention was paid to these products by FDA when the drug regulations were being legislated and revised. Events have changed the agency's attitute toward them.

The introduction of new antimicrobial-containing soaps and surgical scrubs promoted the development of multiple sterile basin, hand-washing tests to determine the effectiveness of removal and/or killing of bacterial cells on the skin. During the period from 1962–1972 the major test of effectiveness for these products was the conventional hand-washing test, including Price,[4] Cade,[1] and Quinn[5] procedures.

When the Over-The-Counter Drug Review was initiated by FDA in 1972 the NAS–NRC review of pre-1962 prescription drugs for effectiveness was completed and the implementation of the NRC recommendations was underway, the agency turned its energies to the review of the great profusion of products and labeling claims in the OTC drug market.

The accumulating evidence concerning the systemic toxicity of topically applied hexachlorophene was just becoming obvious. The uncertainties and problems faced by the agency led the directors of the OTC review to place the Antimicrobial Panel in a high priority position: the Antimicrobial I Panel was the second panel convened (June, 1972). The record of the panels' deliberations and decision are well known and their final report published in the *Federal Register*[2] and in the public and scientific press.

The coincidence of the recognition of hexachlorophene toxicity and the OTC review initiated many changes in the toxicity and effectiveness testing of topically applied antimicrobial products designed for daily use. At the same time as the OTC review was proceeding, within FDA, the necessity for guidelines for the testing of these same products, particularly surgical scrubs, became obvious. Development of the needed guidelines was started soon thereafter. The recommendations of the panel on hexachlorophene and subsequent FDA regulatory decisions initiated the submission of approximately 20 New Drug Applications (NDA's) for hexachlorophene products for hand-washing and surgical scrub indications. These two indications were the only permissable ones remaining and their sale was placed under prescription (Rx) status. Two lines of development—changes in internal testing guidelines at FDA and the development of definitions and testing procedures by the OTC Panel—ultimately resulted in the panel's report published Sept. 13, 1974. Many of these testing procedures have been incorporated into requirements for NDA's for antimicrobial agents not reviewed by the OTC Panel.

At the time the OTC review was initiated, only ingredients in products marketed for a material time and to a material extent in the United States were included in the review. Any other antimicrobial agent which a sponsor wishes to market for hand-washing, surgical scrub or first aid indications is regarded as a new drug and must now have an NDA.

Although there are a number of new products under investigation with an Investigational New Drug Exemption (IND), only one NDA for a new drug entity has been submitted and approved since the panel's review. This ingredient is chlorhexidine gluconate, approved as a surgical scrub, health-care personnel hand wash, and as a preoperative preparation in the form of a tincture.

An important area of change initiated by the panel was their attempt to separate the individual claims for the types of products being discussed. They defined several product areas: antimicrobial soap, health-care personnel hand wash, patient preoperative skin preparation, skin antiseptic, skin wound cleanser, skin wound protectant, and surgical hand scrub. The specific definitions are in Chapter 17. However, I include the three that this discussion will deal with in detail.

A *surgical hand scrub* is a nonirritating antimicrobial-containing preparation significantly reducing the number of microorganisms on intact skin. Surgical hand scrubs should be broad-spectrum, fast-acting, and persistent. A *skin antiseptic* is a non-irritating anitmicrobial-containing preparation preventing overt skin infection. A *patient preoperative skin preparation* is a fast-acting, broad-spectrum antimicrobial-containing preparation significantly reducing the number of microorganisms on intact skin. I comment on these three definitions, particularly as they relate to the testing of new compounds and the criteria which must be met in order that a compound can be approved and labeled.

I define a *health-care personnel hand-wash* as a nonirritating antimicrobial-containing preparation designed for frequent use: it reduces the number of transient microorganisms on intact skin to an initial baseline level after adequate washing, rinsing and drying; and is broad-spectrum, fast-acting, and if possible, persistent. The testing procedures have not been sharpened to the extent those for surgical scrubbing have been. The results indicate the necessity to run a vehicle control. Many products will show significant reduction but criteria are difficult to set for this product class. It is obvious that this is the place to use fast-acting antimicrobials. The question that remains is, how much reduction is necessary?

The skin antiseptic is presented as a definition . . . and a short one. Only one clinical trial to show prevention of infection has been done successfully to my knowledge (Chapter 15). It is encouraging to know that a prophylactic trial can be satisfactorily performed, since critics have protested that a definitive test could not be done.

Surgical Hand Scrub

As a cooperative effort between the Division of Anti-Infective Drug Products and the OTC Panel, the glove juice test, later published in the OTC Antimicrobial I Panel's report, was developed. Outside consultants contributed to its writing. In previous testing procedures the main focus has been on the number of bacteria removed after a subject has scrubbed with a formulation according to directions: the initial reduction of the count after scrubbing and prior to donning the gloves. This is basically only one aspect of the requirement for a surgical scrub. The second, and perhaps the most important aspect, is the microbial buildup with time, illustrated by the estimation of the reestablishment of the microbial count on the hands after partial removal by the initial scrubbing. This effect is encouraged by the occlusive, moist conditions and often elevated temperature within the glove. A brief summary of the test follows:

Procedures for the glove juice test
  1) 35 subjects—selected for count $-1.5-4 \times 10^6$.
  2) Determination of baseline after 2 weeks of pretest with no antimicrobial exposure. 3 determinations of baseline count using glove fluid procedures and a stripping solution. Same procedure as for test of products.
  3) The transient flora are removed with a 30-s wash with distilled water and Camay soap.
  4) The hands and 2/3 of forearm are scrubbed according to the directions on the label and the gloves are donned wet.
  5) The control hand on all subjects (right or left) are sampled with stripping solution in the glove at 1 min. The right hand of 5 subjects is sampled each hour up to 6 h.
  6) The test is repeated on day 2 and day 5 of the test.
  7) The statistical procedures for analysis of the baseline count, initial reduction and reestablishment of the count are specified.

Details of statistical procedures developed with the glove juice test permit the detection of the degree of effectiveness of a number of actions desirable in a surgical scrub. The entire testing procedure is repeated three times in 5 days. It is possible to determine the initial reduction in count on the hands; the kinetics of the microbial buildup, reestablishing the microbial flora on the skin over a 6-h period (at hourly intervals); the amount of substantive action demonstrated over the 6-h period in the glove; and whether there is a reduction in the initial flora between day 1 and day 2 and day 1 and day 5.

The initial reduction of the flora on these days is a measure of the resident hand flora, recorded as reduction from baseline at zero time (1 min). This test may appear deceptively simple: it should be emphasized that this is not a simple test, although once understood, it is relatively easy to execute. This test has been invaluable in providing a single test that can be performed and interpreted to show differences between various antimicrobial agents used as surgical scrubs. It represents a standard against which to measure the degree of effectiveness.

Many new antimicrobial agents have the characteristic of binding to the stratum corneum resulting in residual activity or persistence of the effect of the chemical. The effect varies with the chemical and its duration in time is dependent on concentration. Among the ingredients having substantive or persistent activity are hexachlorophene (HCP), triclocarban (TCC), triclosan, and chlorhexidine. Absorption of an antimicrobial chemical through the skin may be related to its substantivity: for example, HCP is absorbed at an established equilibrium rate from the material in the stratum corneum. Other substantive ingredients may bind to the protein so strongly that absorption does not take place, as is the case with chlorhexidine. Highly exaggerated exposure conditions in animals and humans have not resulted in detectable chlorhexidine absorption. Data are being accumulated on absorption through infant skin.

In the past, major emphasis has been on the amount of initial reduction after scrubbing and derives naturally from the reliance on phenol and alcohol and tincture of green soap when there was no possibility of substantive action in the glove. Not much thought was given to what was happening after the surgeon donned the glove. Emphasis was placed, and rightly so, on removing as many organisms as possible with the initial scrub. This of course is the historical precedent for the elaborate hand-scrubbing procedures still practised today in many hospitals. The goal was to remove or kill the highest number of organisms possible since none of the products then used were substantive.

With the advent of persistent, or substantive, antimicrobial agents such as hexachlorophene, the emphasis was changed and placed on the ability to reduce the total flora on the hands when used regularly. It cannot be overemphasized that this activity, rather than initial reduction, is the chief asset of hexachlorophene. The ideal surgical scrub would be one which gives a good initial reduction, is persistent in the glove by the standards of our testing requirements, keeps the count under baseline for 6 h, and reduces the total microbial flora of the hand when in use over several days.

When the glove juice test was published, there was difficulty in knowing where to set the criteria, since data for the various ingredients from tests performed according to glove juice protocol were not available. Since that time, a significant number of tests have been submitted as part of New Drug Applications to the Food and Drug Administration. The major portion has been to support the effectiveness of the various hexachlorophene formulations, for which the sponsors were required to submit NDA's. Glove juice tests with other ingredients have been submitted usually to provide comparative data in NDA's.

The data in Figs. 14.1–14.3 are an idealized representation of what is seen when the results of a test with an iodophor, a 3% hexachlorophene product, and a chlorhexidine formulation are examined. The initial reduction is usually two logs or better with the iodophor, but that this reduction lasts only about 2 h in the gloves before the reestablishment of the flora takes place at a rather precipitous rate. On the other hand (Fig. 14.2), a typical 3% hexachlorophene-containing product used as a surgical scrub would show only a 1-log reduction initially, but this reduction is held below baseline for the 6h period of the test in the glove. The liquid product has generally more easily met the criteria and statistical tests for an initial 1-log reduction that the sponge or sponge-brush product filled with the same liquid formulation. Exceptions have occurred that further emphasize the importance of formulation and release of the active from both formulation and materials.

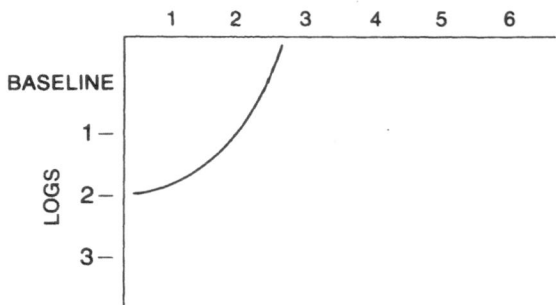

Fig. 14.1. Iodophor detergent scrubs. Initial reduction of bacterial flora was about 2-log$_{10}$ after iodophor detergent scrub, but the reduction lasted only 2 h and the bacterial flora were reestablished.

One needs to look carefully at the attributes and defects of the products available and choose the best on the basis of intended use and risk to the patient. In this light, it would be difficult to argue that iodophor products do not act rapidly against a variety of organisms with significant reduction, but in contrast, can be irritating and lack substantive action. HCP-containing ones reduce the resident flora of the user when regularly and repeatedly applied, are low in irritation potential, but lack rapid action or high initial reduction of flora. Chlorhexidine, approved in 1976, (Fig. 14.3) has good initial reduction, approximately 2-log$_{10}$, and substantivity in the glove and from day to day with repeated use. It has fairly good activity against gram-negative organisms.

Circumstances have made data available on chlorhexidine which do not exist for most other chemical antimicrobial ingredients. The criteria of effectiveness for solutions for disinfection of contact lenses requires the determination of D-values (the time, usually in minutes, required to kill 1-log$_{10}$ of cells under controlled conditions). This information is valuable in determining exposure times and the killing ability of antimicrobial agents, both physical and chemical (Table 14.1). As with most chemicals, the activity if not

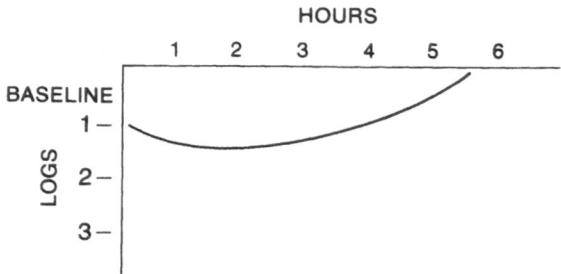

Fig. 14.2. Liquid detergent scrubs with HCP-3%. Initial reduction of bacterial flora was 1-log$_{10}$ after hexachlorophene treatment, and the reduction was held below the baseline for 6 h.

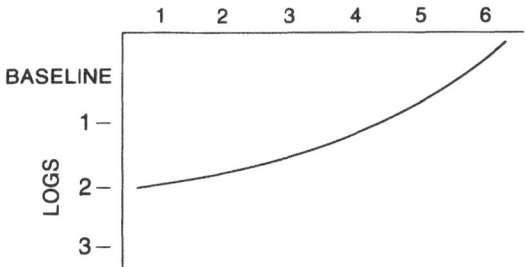

Fig. 14.3. Chlorhexidine detergent scrub. Initial reduction of bacterial flora was 2-$\log_{10}$ after chlorhexidine treatment and the reduction was held below the baseline for 6 h.

equivalent across the board but varies with the specific organism. Since the concentration of chlorhexidine in these formulations is .005% (50/10^6), it may be assumed that chlorhexidine is a relatively potent antimicrobial.

The prevailing concept among infection-control personnel that one product is bought for the hospital and used throughout may need revision. The costs or benefits need to be weighed against the deficits of each product for each specific use since in this case, as with most things, perfection does not exist. I plead for thoughtful selection of products for surgery, nurseries, intensive care facilities, public lavatories, and food-handling areas in institutions. The microbial flora of the skin and, most often, the hands is paramount but the risks for the patients and the hazards in the form of specific infecting organisms differ. For example, the organism presenting a risk to the patient in a burn ward is *Pseudomonas*, while staphylococci pose the most imminent risk for neonates in the nursery. One needs to weigh both sides; the effectiveness in various uses weighed against the risk of contact with organisms presenting a hazard.

The testing of antimicrobial formulations specifically for effectiveness as a preoperative preparation did not exist until the definition and testing

Table 14.1. D$_{19}$ Values for Chlorhexidine Gluconate Solution (1/20 000, or .005%, or 50/10^6)[a]

	Minutes
S. epidermidis	0.63
S. marcescens	5.15
P. aeruginosa	0.05
C. parapsilosis	30
A. fumigatus	340
Herpes simplex virus	8.8

[a]Data supplied by Burton Parsons Division of Alcon Laboratories. D-values equal the time (usually in minutes) required to kill of 1 log of cells under controlled conditions

requirements suggested by the Antimicrobial I Panel. Most often surgical-scrub products were marketed for this use which had been tested only with the routine hand-washing procedures then in use. Any suspected inadequacies in activity of the formulation were compensated for by elaborate, routinized steps prior to surgery. These ablutions and rituals may be more advantageous to the surgeon's psyche than to the patient.

In their attempt to carefully define the use and testing requirements to prove effectiveness of a preoperative preparation, the OTC Panel urged that testing be done on body areas commonly involved in surgery, such as the abdomen and groin. One may speculate whether surgical scrubbing prevents postsurgical infection, but there is no need to do this for preoperative preparation. The product most effective in reducing the flora of the skin in a specified time period is the ideal. The minimal criteria suggested by the panel for a preoperative preparation is a 3-$\log_{10}$ reduction in 30 min. Fortuitously one side of the body can be used for the test, and the other as a control.

Only one antimicrobial ingredient and two formulations of it have been tested with these procedures. These tests demonstrated that areas of the body such as the groin can easily be utilized and have sufficiently high counts to demonstrate a 3-log reduction with statistical assurance. However the abdomen varied in count and often did not show recovery of baseline counts much over 10^3 in the area of the template used for swabbing.

In one study, surgical preparation techniques involved covering the patient's abdomen with a sterile towel. The subjects revealed much higher base-line levels, around 10^5. The criteria selected should as proof of effectiveness for this indication, be considered minimal and formulations which will exceed this hopefully can be found.

Sponge and sponge-brush combinations impregnated with antimicrobial scrub products are undoubtedly convenient and at first seemed like a good idea. But as a consequence of NDA submissions it is now obvious that, what goes in doesn't always come out. Our estimate, for HCP at least, is that only about one-half the ingredient is available from the sponge. This may help explain the difficulty encountered by many sponsors who performed the glove juice test with 3% HCP products in sponges. It would not be an exaggeration to say that the most difficulty encountered in meeting the criteria for the test for surgical-scrub products have been with impregnated products. It has been especially difficult to demonstrate an initial 1-$\log_{10}$ reduction, particularly on day 1 and day 2 of the test.

Tests of other antimicrobials have also revealed difficulties in meeting the same effectiveness shown with the liquid product when it is used to impregnate sponges or sponge brushes. Significant variation in resting results have been noted. FDA has also wrestled with the problems encountered by so-called "sterile" sponges. First, there appears to be no overriding reason for sterility of these products since they are used outside the operating theater. In addition ethylene oxide is frequently used to achieve sterility and many of these formulations contain the chloride ion that results in residue formation.

The highest ethylene chlorhydrin residue level yet seen in a drug product by the Bureau of Drugs was with one of these sponges. The integrity of most of the packages is not compatible with a long shelf life in the maintenance of sterility. The kindest comment may be that the sterility claim has been a sales and advertising gimmick.

I believe it is possible to produce a convenient and effective impregnated sponge product but problems with complexation and adherence of the formulation to the sponge or plastic materials coupled with availability of an adequate concentration of active ingredient from various formulations must be solved first.

As more data from glove juice tests have been examined, differences in results of effectiveness with the same concentration of HCP have been observed. The specific formulation does make a difference as do the instructions for use. Revision of the directions for use and formulations have resulted in increased effectiveness when products were retested after failure to meet criteria.

Particularly with HCP products, FDA has taken issue with the labeling of concentrates for dilution on use. With these rather archaic products, the effective concentration has been 0.25% (formulated in soap). This concentration has only borderline effectiveness and, unfortunately, the dilution is usually made by untrained personnel and often by eye. This situation pertains even though any HCP product is now prescription and has demonstrated toxicity. This carryover concept from historical disinfectant use is difficult to justify. FDA has allowed continued marketing of a concentrate contingent upon revised instructions for dilution that will result in a concentration with leaway for misdilution. It is unclear what the expected effectiveness of this product might be since HCP does not offer the rapid activity required when used in lavatories or other service areas in hospitals.

FDA is aware of industrial interest in highly specific products, often for narrow use; for example, a preoperative or general prepping product for ophthalmic use only. Again, the weighing of effectiveness and safety changes to achieve removal of skin microflora around the eye area must be balanced against the high risk from gram-negative contamination of the area and the sensitivity of ocular tissue.

Another narrowly defined product of interest is one that can safely be used in body cavities and deep wounds. Speculation for an increased demand included a concern that, increasingly, antibiotics may be ineffective in prophylaxis of infection in surgery because of resistance to organisms causing postsurgical infection.

The questions of necessity, gimmickry, and competitive marketing should be posed. Is there a need? Do these more defined and refined uses offer any advantage for the patient?

There are increasing numbers of products designed to be used with an array of old and new medical devices of varying degrees of sophistication and penetration of body surfaces and cavities. These include chemical soaking solutions for disinfection or sterilization (between use) for instruments such as bronchoscopes and laproscopes, adjunctive solutions for disinfection of hard and soft contact lenses, solutions for intermittant catheterization, and an interesting selection of materials for superficial skin treatment prior to injection, catheterization or manipulation. What criteria of effectiveness should be applied and what are the interrelationships between indigenous flora and extraneous pathogens?

There is a general recognition that the development of new chemical entities for topical antimicrobial use is not a rapidly expanding area. The cost from development to marketing is great, and there are many pitfalls along the way. Consequently there are only one or two really new entities currently under investigation. There is apparent interest in compounds which release oxidizing agents and in those which are only slightly or moderately absorbed after topical application. One area in which there has not been much activity is novel formulation or means of delivery.

References

1. Cade AR (1951) A method for testing the degerming efficacy of hexachlorophene soaps. J Soc Cosmetic Chemists 2:281–290
2. OTC Antimicrobial Products and Drug and Cosmetic Products. Report of FDA OTC Antimicrobial I Panel. Sept. 13, 1974 Federal Register Vol. 39, No. 179, Part II
3. OTC Topical Antimicrobial Products, Tentative Final Monograph. Jan. 6, 1979 Federal Register, Vol. 43, No. 4
4. Price PB (1938) The bacteriology of normal skin: a new quantitative test applied to a study of the bacterial flora and the disinfectant action of mechanical cleansing. J Infect Dis 63:301–318
5. Quinn H, JG Voss, HS Whitehouse (1954) A method for the *in vivo* evaluation of skin sanitizing soaps. Appl Microbiol 2:202–204

Chapter 15
Antibacterial Soaps: Chlorhexidine and Skin Infections

DAVID TAPLIN

The 15 years that separate the publication of *Skin Bacteria and Their Role in Infection* and the emergence of this volume have been marked by significant contributions to our understanding of bacterial skin infections and their prevention. Most of this new knowledge was contributed by the authors whose writings appear elsewhere in this book.

Faced with devastating losses of manpower in Vietnam due to skin infections, the U.S. military sponsored much of the research conducted during those years. From 1965 until 1970, close to $1 million per year was expended by the armed forces on dermatologic research. It is appropriate, therefore, to look back at some of the rewards of those efforts, and to evaluate where we stand today.

History

In 1965, the "degerming" of skin of hospitalized patients, newborn infants, and the hands of hospital personnel was accomplished almost entirely by the use of Phisohex, a sudsing antibacterial detergent containing 3% hexachlorophene.

There is no doubt that this product was capable of greatly reducing the numbers of gram-positive bacteria on the skin, and that, when this flora contained potentially virulent strains of *Staphylococcus aureus*, the end result could be demonstrated as a lower incidence of clinical infections, at least in hospitals. As early as 1962, Miller et al.[7] suggested that this product was unsuitable for degerming the skin prior to surgery because of its narrow spectrum. They recommended instead, "agents with rapid bactericidal action affecting all pathogenic bacteria should be used for preparation of the skin for operation."

The experimental and epidemiologic evidence that agents active only against gram-positive organisms encouraged a shift in the cutaneous flora to gram-negative bacteria went largely unheeded in spite of the increasing incidence of nosocomial gram-negative infections. The demise of hex-

achlorophene came about in an unexpected manner, when evidence accumulated in the late 1960s that this agent could be absorbed through the
skin to produce significant and sometimes fatal central nervous system damage. A spirited discussion of this issue appeared in *Archives of Dermatology*.[2,6]

In the population at large, antibacterial soaps were already well established in 1965, their popularity due to the demonstrable deodorant effects. By 1968, 50% of all soap purchases were for antibacterial soaps.
Mackenzie[4] gives a useful listing of the ingredients used in the most popular
soaps at that time. All contained one or more halogenated carbanalides,
salicylanilides, with or without hexachlorophene. The mixtures were often
changed, and it is important to realize that data derived in the past using old
formulations may not be applicable to the same brand names on sale today.
In the late 1960s, two major companies became interested in promoting
claims for their products to prevent skin infections, thus bringing them into
the classification of drugs under the jurisdiction of the Food and Drug Administration.

Mackenzie initiated a controlled trial of a soap (Dial) containing 0.75%
hexachlorophene and 0.75% triclocarban at the U.S. Naval Academy, An-
napolis, in 1965.[5] During a 6-month period, he compared the incidence of
cutaneous infections among 602 men using the antibacterial soap against
those in 599 men given a control bar with no antibacterial agents. At the end
of 6 months there were 23 infections among the antibacterial soap users
versus 41 in the control group.

In 1966, Leonard[6] conducted a study at the United States Military
Academy West Point in which four companies comprising 474 men used an
antibacterial soap (Safeguard) for 2 months. This soap contained a 2% mixture of tribromsalan, triclocarban, and cloflucarban. During this period,
their infection rates were recorded against those among five companies of
609 men using a control bar of soap. At the end of 2 months, 17 infections
had occurred in the antibacterial soap group compared with 39 in the control
populations.

In the same year Duncan et al.[3] used the same soap and control bar to
study the effects of prophylactic use among 2500 workers on prison farms in
Texas. In two groups using the antibacterial soap under regulated conditions, a statistically significant decline in acquisition of new infection was
noted. In two other groups in which the daily use of the antibacterial soap
was not mandatory, no significant effects were seen.

These three studies indicated that Dial and Safeguard soap as then
formulated were useful in reducing the incidence of skin infections. A closer
evaluation of these studies shows that almost all of the conditions described,
and in one of the studies, cultured, the investigators were dealing with
primary staphylococcal infections, including furunculosis, staphylococcal
impetigo, infected lacerations, and felons.

In any event, these studies provided sufficient justification for the Armed
Forces Epidemiology Board to issue a recommendation in August 1967 encouraging the use of antibacterial soaps for the prevention of skin infections

in the U.S. military. This recommendation had little effect on preventing infections among U.S. forces in Vietnam.

In 1967, reports from field commanders in Vietnam indicated that skin infections were causing alarming numbers of lost man days. A team from the University of Miami surveyed troops in Vietnam, confirmed that fungal and bacterial infections were rampant among forward operational units, and suggested that *Streptococcus pyogenes*, rather than *Staphylococcus aureus*, might be the more prevalent offender.

This was confirmed by field investigations by Allen et al.[7] Their studies left no doubt that streptococcal pyoderma was the most common problem. They found that *Staphylococcus aureus* could be cultured as easily from normal skin of soldiers as from their purulent lesions, where they were considered secondary colonizers. Meanwhile, other authors were reporting streptococcal pyoderma as the most prevalent form of skin infections both in the United States and on a worldwide basis. As expected, the prevalence was higher among low income populations in the tropics.[11]

By 1973, the global epidemiology of common bacterial sores was well established. From Vietnam, Thailand, Uganda, Colombia, Venezuela, Israel, Brazil, Trinidad, and in the United States from Baltimore, Minneapolis, Miami, and Birmingham came reports of large numbers of cases of streptococcal infections. Major epidemics were reported, with disastrous results when the offending strains of the streptococci were nephritogenic.[9]

These studies were mostly conducted in populations at large, with the highest attack rates seen in the tropics, where biting insects, life-styles, and climate play important roles. Scabies also played a significant role as a prelude to infection.

Physicians serving affluent populations living in adequate housing were often unaware of the extent of the streptococcal problems in less fortunate local populations. An explanation for this phenomenon is offered at the end of this chapter.

The stage was thus set in 1973 for the first controlled trials of antibacterial soaps in a population at risk to streptococcal infection. Unknown to each group at the time, two investigations were in progress. In Trinidad, Sharett et al.[10] attempted to reduce the incidence of streptococcal infections by the introduction of a soap containing 0.75% hexachlorophene and 0.75% triclocarban (Dial). Two months of treatment with plain soap in 17 randomly selected families failed to reduce the incidence of infection. Neither did a further two months of treatment with the antibacterial soap. Admittedly, the use of the soaps was not controlled, but the results did not lend support to any recommendation to health authorities to encourage the use of antibacterial soaps.

In the same year, a more closely controlled study was conducted in a large boarding school in Arizona by investigators from the Department of Dermatology, University of Miami. This study is reported here in detail, and is followed by an account of a more recent evaluation of chlorhexidine gluconate for the prevention of common skin infections among Costa Rican schoolchildren.

The Arizona Study

From September 1973–October 1973, David Taplin, William Eaglstein, Richard Feinstein, Patricia Mertz, and Jane Walker conducted this study, in which all of 653 children aged 7–12 years attending a boarding school in Arizona were examined by two dermatologists. All skin defects were noted and all lesions considered by these physicians to be clinically infected were cultured on selective media for *S. aureus* and *S. pyogenes*. Clinical infection was judged to be any defect in the skin exhibiting pain, pus, and erythema. With one exception, all infections seen in 2 months were clinically streptococcal pyoderma. We saw no staphylococcal impetigo and only one furuncle. Some 80% of all lesions cultured during the baseline examination yielded Group A *S. pyogenes*, and 60% yielded *S. aureus*. Following the baseline examinations and microbiology, seven dormitories were issued an antibacterial soap (Safeguard) containing 1.5% 2:1 triclocarban:cloflucarban. Children in the remaining seven dormitories were assigned identical bars containing no antibacterial agents. For 2 months the children showered once per day under supervision of our monitors, and the clinical and microbiological studies were repeated after 1 and 2 months.

Table 15.1 shows that in the entire school population, no beneficial effect of the antibacterial soap in the prevention of infections could be detected. The higher prevalence at the end of two months in both groups reflects the usual summer increase which had been noted consistently in previous years. There is a strong possibility that the prevalence of infection ended up lower than normally seen in this population as a result of the regimented daily hygiene. Dormitory supervisors reported that this summer represented the lowest incidence of sores in recent memory. As in any study of this size, a percentage of those enrolled failed to complete the study or missed treatments for reasons beyond the control of the research team.

We, therefore, examined the results among children known to have completed 2 full months of showers as recorded on the daily check lists. The record of these completers (Table 15.2), demonstrates no reduced prevalence which could be ascribed to the antibacterial soap. In addition to the microbiology of the lesions, we made contact cultures of normal skin of the wrist and mid-back of each child. Table 15.3 shows that there was initially a lower number of children from whom *S. aureus* could be recovered in the group assigned to the antibacterial soap.

Table 15.1. Point prevalence of clinical infections in boarding school children—Arizona 1973[a]

	Baseline	One month	Two months
322 children assigned to antibacterial soap	4.9	6.3	9.0
311 children assigned to plain soap	4.2	6.2	8.1

[a] Results of 2-month trial based on entire school enrollment.

Table 15.2. Point prevalence of clinical infections—Arizona 1973[a]

	Baseline	One month	Two months
206 on antibacterial soap	5.3	6.7	9.7
195 on plain soap	4.1	6.6	8.7

[a]Results of trial among children completing 2 months of daily supervised showerings.

There was a similar baseline difference in the two groups in the recovery of *S. pyogenes* from normal skin, again showing a lower prevalence in those who were to become antibacterial soap users. At the end of 2 months, the recovery of *S aureus* remained lower in the antibacterial soap group; this result was as we might have expected. To our surprise, the difference in prevalence of *S. pyogenes* in the groups was reversed after 2 months, with twice as many children yielding streptococci in the antibacterial group compared with those on placebo bars, despite the fact that they entered the trial with half the prevalence. This gave rise to the interesting possibility that the use of commercially available antibacterial soaps on a regular basis might encourage the survival of *S. pyogenes* on normal skin.

Several clues supported this theory. First, there are many examples of interactions between skin bacteria, including inhibition of streptococci by *S. aureus*.[8] Our experiences in Vietnam and among U.S. Army Rangers in the United States did not suggest that antibacterial soaps offered effective prophylaxis against streptococcal pyoderma. Some 70% of all men, with or without infections, claimed to be habitual users of such soaps; we could never elicit any differences in prevalence between those who did and did not regularly use them. Our next surprise came when we tested the Arizona strains of staphylococcus and streptococcus against dilutions of the two study soaps in distilles water. We detected activity against *S. aureus* with as little as $1/10^6$ of the antibacterial soap, whereas the placebo bar was almost devoid of activity. Against *S. pyogenes*, there was activity in the plain bar soap diluted $1/10^6$, but no increase in activity demonstrated by the antibacterial soap. We presume that the antistreptococcal activity of the plain soap is due to fatty acids, to which streptococci are known to be highly sensitive. On the basis of these in vitro tests, we would not have expected the active bar to be superior to the plain soap, and such was the case.

A further clue is suggested by Aly et al. (Chapter 4), in which the natural defenses of the skin surface against colonization by streptococci appear to be related, at least in part, to substances probably produced by the action of normal skin bacteria. Putting this all together, the laboratory data suggests that TCC:TFC soap would have no more chance of preventing streptococcal infections than plain soap, and that if the normal flora of the skin plays any role against streptococci, the reduction of this flora might well encourage the acquisition of the streptococci. This possibility would gain more credence if we accept that Group A streptococci colonize the skin prior to the emergence of clinical infection.

Table 15.3. Recovery of *S. aureus* and *S. pyogenes* from normal skin—Arizona 1973[a]

	Baseline		One month		Two months	
	Staphylococcus	*Streptococcus*	*Staphylococcus*	*Streptococcus*	*Staphylococcus*	*Streptococcus*
Antibacterial soap	5.7	9.5	13.3	17.0	7.6	26.6
Plain soap	10.0	20.0	17.1	21.0	12.4	10.5

[a]Recovery of group A streptococcus from normal skin was three times higher than baseline in the antibacterial soap users, but only half as likely in the plain soap users.

In our studies, we could just as easily recover streptococci by contact sampling of the clidren's bedsheets as we could from normal skin. Moreover, when we sampled the normal skin of 50 children who had been cultured 2 days before, we identified about the same percentage (25%–30%) of "carriers." However, only two cases were "repeaters," suggesting transient contamination rather than colonization of the skin. It is virtually impossible to colonize normal intact skin with Group A streptococci, even with occlusive techniques. The only correlations we could find between infection and recovery from normal skin related to the presence of an active lesion on the child sampled, or the presence of several active cases in the dormitory in which the child lived.

The Arizona study was conducted in a closed population, with ample opportunity for close contact. As will be seen in the chapter on epidemic furunculosis, I believe intimate body contact to be an important aspect in the transmission of both staphylococci and streptococci. In the population at large, it is rare to find Group A streptococci on the normal skin of schoolchildren without clinical infection.

All of the studies so far described used groups of subjects in which treatments were divided by dormitories, families, barracks or other convenient units. In some respects they were comparisons of these units rather than a measure of individual responses. Indeed, in several of the studies, trends in infection rates in relation to the use of soaps differed markedly between these administrative units.

To my knowledge, until 1976 there had been no studies performed in a normal population at large in which treatments were randomized throughout the populations at risk. In that year, an opportunity came to us to put this format to test. A year previously, we and our collaborators from the University of Costa Rica had identified a small, isolated town in southern Costa Rica as an endemic area for skin infections. At the time of this survey, 40% of all children in the local schools had laboratory-confirmed streptococcal pyodermas. The agent chosen for study was chlorhexidine gluconate.

The Costa Rican Study

David Taplin, Bryan Frederick, Gwen Scott, Jan Mason, Susan Dunitz, and Sandra Breiterman, of the University of Miami and Roger Bolaños, Ines Santisteban, and Bernardo Caliva, University of Costa Rica conducted a study in which all children enrolled in two schools were examined, and every defect on the skin was recorded, whether or not infected. For the next 6 weeks, they were sprayed from neck to feet, excluding genitalia and buttocks, with either 2% chlorhexidine gluconate in distilled water containing 4% isopropanol or distilled water containing 4% isopropanol. The distribution was randomized throughout the population of 304 children and the procedure was done on Monday through Friday.

Twice per week we reexamined all children and repeated the microbio-

Table 15.4. Baseline profile of the populations—Costa Rica 1976[a]

	Chlorhexidine treated	Placebo treated
Total subjects	151	153
Sex (male/female)	68/83	68/85
Age (years)	8.8	8.9
Height (in.)	50.7	51.2
Weight (pounds)	60.1	61.4
% with lesions	62.2	55.6
% with infected lesions	10.6	10.4

[a]There were no significant differences in any of the above parameters between the two study groups at the start of the trial.

logy, culturing all clinical infections on selective media, using paired swabs, one of which was dipped in a chlorhexidine neutralizer before plating. No other antibacterial soaps or shampoos were permitted during the study and neither subjects nor investigators were aware of the identity of the treatments or randomization.

Results

Analysis of the baseline data (Table 15.4) showed that both groups were appropriately matched in demographic characteristics, number of lesions, and percent of lesions infected at the time of entry to the study. Table 15.5 demonstrated the acquisition of new lesions and infected lesions during the 6-week trial and the total number of separate infection episodes. Note that a "lesion" denotes any minor skin trauma, including insect bites, abrasions, scratches, minor lacerations, and burns.

Table 15.5. Result of treatments—Costa Rica 1976[a]

	Chlorhexidine, n = 152	Control, n = 153
Subjects with one or more new lesions	150	153
Subjects with one or more newly infected lesions	15*	38*
Number of new lesions during trial	1 401	1 526
Number of newly infected lesions during trial	16[b]	91[b]
Total number of separate infection episodes including repeaters	15[b]	65[b]

[a]During the 6-week trial, there was a highly significant difference in favor of chlorhexidine for number of subjects with newly infected lesions, total number of infected lesions, and number of separate infection episodes.
[b]$p < 0.001$

It was clear from these results that the chlorhexidine treatments significantly reduced the chances of infection occurring as a result of minor skin trauma (p > 0.001). This was true whether analyzed for the number of subjects infected one or more times during the 6-week study, for the total number of infected lesions, or for the number of separate infection episodes, which included children reinfected after previous sores were healed.

Randomization was constructed before the study for the total population at risk, and we were fortunate in achieving almost perfect matching of the two populations. Moreover, the two groups were equally at risk to infection throughout the study as measured by the acquisition of new lesions (minor trauma). The microbiology was equally satisfying. At the baseline 87%–88% of infected lesions yielded *S. pyogenes* in both groups, and the same recovery (87%–88%) was obtained in both groups for *S. aureus*.

During the study period, 80% of the clinical infections in the placebo group yielded *S. aureus*, and 80% grew *S. pyogenes*. In the chlorhexidine group, the yield during the trial was 58% for both organisms, but with only 16 children needing culture in this group, caution in interpretation is in order. One child developed transient irritation in the sprayed area after 2 weeks of treatment, and two of the six assistants who sprayed the children developed similar irritation on the dorsum of the hands. The design of the trial simulated a maximum exposure test under conditions of high temperature and humidity.

This completed the first trial of a topical antibacterial prophylactic agent in a population at large suffering from the usual form of skin infections, under a properly randomized double-blind protocol in which the investigational team applied the treatments, and is the first product with a valid claim as an "antiseptic" as proposed by the FDA. Previous to this study, objections were raised by some manufacturers and investigators suggesting that such studies were logistically impossible to conduct, and therefore constituted an unreasonable requirement by the FDA.

Such a study does require enormous manpower by a trained team, selection of suitable populations at risk, and a high degree of cooperation from parents, school authorities, and local health agencies, and is very expensive. However, through this work the possibility of successful prevention of common skin infections by topical means was shown to be feasible.

In developing nations in the tropics, a therapeutic approach will never solve the problem and is far too expensive for serious consideration. A cheap means of prevention in the form of a soap or lotion offers the best approach. Twenty years of use as a personal hand-care product in temperate climates suggest that chlorhexidine is only slightly irritating and has almost no potential for allergenicity. Products for use in the tropics may require particular care in formulation and further testing. Our experience indicates that benefit-to-risk ratio is high.

Our work in Costa Rica greatly added to our understanding of common infections. Since we saw over 300 children every day for 6 weeks, and minutely examined them twice per week, we witnessed the emergence of infections. Not a single infection arose de novo. All 107 new sores began with trauma to the skin, with excoriated insect bits as the most significant initiating factor. Minor abrasions were the next most likely lesion to be infected, and none of the thorn or kitten scratches became purulent. Scabies was not present in this population. The point prevalence in the two study schools during the 6 weeks of the study never rose above 5%. Families were given bland soaps and the children were sent to school scrupulously clean. In two schools within 5 km, the point prevalence remained at 25%, about what we expected for the time of year. This confirms the preventive value of good daily hygiene that we noted in Colombia.[11] A further contribution to the reduced incidence of infection may have been the removal of potential reservoirs by preventing a large number of infections in the chlorhexidine-treated group. Wound-feeding eye gnats of the genus *Hippelates* were abundant. Half of them yielded *S. aureus* 3 days after capture, and 30% contained *S. pyogenes* in the gut.

Epidemiology

If one single fact has emerged from the last 15 years of our efforts, it is that children overwhelmingly represent the population at greatest risk, with those under 9 years of age having the highest incidence. The number of children worldwide with skin infections must be several millions and constitutes a significant load on medical services.

Many investigators conducting community surveys, as opposed to reports of clinical encounters, have shown that streptococcal pyoderma vastly outweighs primary staphylococcal disease in prevalence and incidence. I add the results of our own studies as an example in Table 15.6. In point prevalence surveys over 6 000 children, we found only six furuncles and three cases of staphylococcal impetigo. A rough figure from our surveys suggests that streptococcal infections outweigh staphylococcal infections by 500:1.

Why, then, do some dermatologists, and not a few pediatricians, believe that "most" skin infections are caused by *Staphylococcus aureus*? The answers are not as obscure as might appear. To begin with, many of the Group A streptococci are not recovered on culture due to faulty techniques. Staphylococci are easily cultured on most media, thus giving the clinician a false impression of the cause of the infection seen. The streptococci hide in the depths of lesions. Crusts must be lifted, swabs used aggressively and plated on selective media without delay, such as crystal violet blood agar. My team has tripled the recovery of streptococci from lesions compared with our efforts in 1965, simply by improved techniques.

Streptococcal infections are found in greatest numbers among low income, overcrowded populations. In many such communities they are accepted as one more aspect of their impoverished life-style, rarely to be taken to the doctor. Dermatologists are not sought as the first choice, because parents perceive them as specialists for skin problems such as rashes, warts, acne, and psoriasis. Their infections are just ordinary sores which "every

Table 15.6. Culture results of infected lesions using selective enriched media—University of Miami Teams

Location	Year	Climate[a]	Number cultured[b]	% Recovery, S. aureus	% Recovery, S. pyogenes
Medellin, Colombia, S.A.	1973	temperate	50	62	84
Miami	1968–1970	subtropical	252	59	95
Uganda	1971	subtropical	94	57	76
Arizona	1973	arid, hot	40	60	80
Parrita, Costa Rica	1976	tropical	115	68	86
Apartado	1973	tropical	91	90	90
Panama	1974–1978	tropical	185	88	84
Total			827	61	85

[a]Denotes climate at study site.
[b]Clinically infected lesions only.

child gets" in the summer. Seasonal variations in prevalence are more pronounced in areas with large differences in winter and summer climates.

Bullous impetigo, on the other hand, appears mostly on the face, has no particular respect for social status, can occur in epidemic clusters and because of its visibility and comparative rarity, is more often brought to or referred to dermatologists. Furuncles are painful, may require surgical drainage, and are short-lived compared with streptococcal sores. Patients seek treatment, or the boil spontaneously ruptures; the chances of finding them in a point-prevalence survey are slim. In the two studies in military academies mentioned before, only 184 staphylococcal infections were found in 280 000 man days at risk, so the chances of finding infected cases on any one survey day were very small.

The problems of nosocomial infections represent yet another type of ecosystem. Little wonder that we each have different impressions of the problems. The development and evaluation of products designed to influence the bacterial flora of the skin, whether to make us more socially acceptable, to prevent infection, or to prevent us from infecting others, should take these factors into account. A product for the prevention of boils, while of great value to the patient with recurrent furunculosis, would have a limited market potential if that were its only indication. A safe, nonirritating product in a bar soap which could prevent streptococcal pyoderma would have an immediate and worldwide application.

References

1. Allen AM, Taplin D, Twigg L (1971) Cutaneous streptococcal infections in Vietnam. Arch Dermatol 104:271–280
2. Catalano PM (1975) Hexachlorophene—not a cry of "Wolf". Arch Dermatol 111:250–251
3. Duncan WC, Dodge BG, Knox JM (1969) Prevention of superficial pyogenic skin infections. Arch Dermatol 99:465–468
4. Leonard RR (1967) Prevention of superficial cutaneous infections. Arch Dermatol 95:520–523
5. MacKenzie AR (1970) Effectiveness of antibacterial soaps in a healthy population. JAMA 211:973–976
6. McCarl WG (1976) In support of hexachlorophene. Arch Dermatol 112:1031–1033
7. Miller JM, Jackson DA, Collier CS (1962) The microbicidal property of pHisohex. Milit Med 127:576–578
8. Noble WC, Somerville DA (1974) Microbiology of human skin. W.B. Saunders Company, Ltd, London, Philadelphia, Toronto, pp 66–72
9. Poon-King T, Potter EV, Svartman M, Achong J, Mohammed I, Cox R, Earle DP (1973) Epidemic acute nephritis with reappearance of M-type 55 streptococci in Trinidad. Lancet 1:475–479
10. Sharett RA, Finkelea JF, Potter, EV, Poon-King T, Earle DP (1974) The control of streptococcal skin infections in south Trinidad. Am J Epidemiol 99:408–413
11. Taplin D, Lansdell L, Allen AM, Rodriguez R, Cortes A (1973) Prevalence of streptococcal pyoderma in relation to climate and hygiene. Lancet 1:501–503

Chapter 16
Antimicrobial Soaps: Benefits Versus Risks

Francis N. Marzulli and Mary Bruch

Although soaps have been used for centuries, whole body washing with soap to cleanse and deodorize is a development of the past 300 years, and the use of germicides in soaps dates back two or three decades.

Fifty years ago, one measure of the civilization level achieved by a nation was said to be its per capita consumption of soap. A 1979 measure of the civilization level achieved by a nation might be the production of toxic and carcinogenic chemicals in that nation. This is a sobering reflection; however, it does not imply a direct relation between these measures. Rather it is a reminder that pervasive manmade chemicals have a potential for harm that must be monitored.

The benefits of using soaps are accompanied in varying degrees by disadvantages. The defatting action of soaps may cause chapping, an overabundance of alkali may cause itching and discomfort, certain fragrance ingredients of soaps may cause allergic reactions, and toxic substantive antimicrobials may cause systemic effects.

At the height of the hexachlorophene toxicity tragedies[15] Dr. Paul Stolley, an epidemiologist member of FDA's Antimicrobial I OTC Review Panel, asked: "How can one justify using a toxic chemical on the entire body for a lifetime?" Nevertheless, the incorporation of safe and effective antimicrobial ingredients in soaps can be a boon to those few individuals whose skin surface film provides a breeding ground for offensive, odor-forming microorganisms, despite reasonable hygiene and personal cleanliness.

Bathing is largely a personal choice; what is satisfactory and necessary for some may be distasteful or unhealthy for another. Odors that are inoffensive and even pleasant to some, such as pheromones, may be abhorrent to others. Having briefly alluded to some of the benefits of effective antimicrobial soaps, one must balance the good they do against the harm they may cause if they are toxic.

History of Developments in Antimicrobial Soaps

Antimicrobial soaps began with the introduction of Lifebuoy soap, a carbolic-smelling body cleanser containing cresylic acid, an antiseptic whose main purpose was to mask odor rather than inhibit microorgansims. The next development was the use of hexachlorophene (HCP) as a true antimicrobial ingredient in soaps whose main purpose was to reduce the number of odor-forming microorganisms on skin.[20] As it is substantive for skin—that is, not washed off the skin during bathing—repeated use of HCP is capable of reducing the numbers of resident gram-positive microorganisms on skin. Bar soap containing HCP was removed from the market following recognition of its toxic potential by FDA's Antimicrobial I Panel.[5]

The introduction of tetrachlorosalicylanilide (TCSA) as an antimicrobial ingredient in soaps in England resulted in the first recognized problem of photosensitization of significant numbers of consumers and a completely new entity, the persistent light reaction. Afflicted patients exhibited a light-activated skin reaction that persisted for long periods after exposure to the chemical had ceased, possibly because the antimicrobial ingredient remained firmly bound to components of the living epidermis or dermis.[21]

Bithionol, a bisphenolic structure that resembles hexachlorophene except that it has a sulfur rather than a methylene bridge, was introduced as an antimicrobial ingredient of soaps in the 1950s. It was removed from the market when it became known that, like TCSA, it caused persistent light reactions.[4]

Tetramethyl thiuram disulfide (thiram) had a short-lived use in marketed antimicrobial soaps. It was withdrawn from the U.S. market because of odor and contact allergy problems[16] although it is still used in other countries.

Tribromosalicylanilide (tribromsalan, or TBS) was the next major antimicrobial ingredient used in soaps. It was removed from the market when FDA's Antimicrobial I Panel became convinced that there were safer drugs to control odor. The benefit-risk ratio did not support its continued use.[5] This material was thought to cause photoallergy in certain users long after cessation of exposure to the product. It cross-reacted with other salicylanilides and there was inadequate data demonstrating its safety.[5]

Dibromosalicylanilide (DBS) and tetrachlorosalicylanilide (TCSA) were much more potent photosensitizers and cross-sensitizers than TBS and were considered unsafe for drug or cosmetic use.[5] At present, trichlorocarbanilide and triclosan (Irgasan DP-300) are the major germicidal ingredients of currently marketed antimicrobial soaps.[5]

Other candidate antimicrobial ingredients have serious limitations that may preclude their use. Cationic chemicals such as chlorhexidine are not compatible with soaps, which are anionic. Simple phenolics such as p-chloroxylenol are not substantive for skin; only bisphenolics have this property. Zinc pyridinethione may cause toxicity problems when used in bar soap. A chlorofluorocarbon, such as cloflucarban, is not used because the cost of establishing its safety as an ingredient alone or in combination may not warrant additional study at this time. Dantoin is a relatively new antimicrobial whose effectiveness for this purpose is not yet known.

Discussion

To be suitable for over-the-counter (OTC) consumer use, an antimicrobial bar soap should be safe and effective. Ideally, it should be substantive for skin and it should both reduce and inhibit growth of odor-forming gram-positive organisms. Further, chronic use should not result in local or systemic effects in significant numbers of consumers. This includes irritation, sensitization and photosensitization of skin, as well as retention in or damage to internal organs.

In 1973, the OTC Antimicrobial I Drug Review Panel established certain criteria for evaluating OTC drug products containing halogenated salicylanilides and other antimicrobial ingredients for topical use.[3] This panel defined an antimicrobial soap as "a soap containing an active ingredient with in vitro and in vivo activity against skin microorganisms." It concluded that there is no special value in eliminating resident gram positive organisms to the extent that harmful gram-negatives may become established. A bar soap should be aimed at odor control in contrast to a surgical scrub, which is intended to reduce the capacity of contaminated skin of an examining physician, operating surgeon, pediatric nurse, or other health care personnel to produce infection.

The OTC Antimicrobial I Panel classified antimicrobial ingredients in three categories.[5] These are:

Category I—Safe and effective for OTC use
Category II—Not safe and effective for OTC use
Category III—Available data insufficient to make safety judgment (The legality of this category is now under court resolution.)

Table 16.1 summarizes the findings of the panel with regard to four candidate antimicrobial soap ingredients reviewed for safety and efficacy at that time and found inadequate.

Recently three types of ingredients have emerged for use in antimicrobial soaps based on their meeting most of the chemical and physical requirements needed for effectiveness. These include the salicylanilides and the closely related carbanilides (Table 16.2). The salicylanilides have two benzene rings linked by CONH. The carbanilides have two benzene rings linked by NHCONH. Important members of the salicylanilide group are the brominated compound, tribromsalan, with three bromine atoms on the benzene rings, and fluorosalan, with two bromine atoms and one fluorine atom. The carbanilide series includes trichlorocarbanilide (triclocarban) with three chlorine atoms and cloflucarban, or Irgasan CF_3, with two chlorine atoms and one fluorine atom. Triclosan, or DP-300, is also currently used. It is a diphenyl ether containing two benzene rings linked with oxygen and bearing three chlorine atoms and one hydroxyl group on the benzene ring (Table 16.2).

Table 16.1. Category II ingredients (Not safe and effective for OTC use) as determined by Antimicrobial I Panel

Substance	Effectiveness	Safety	Toxicity	Benefit
Hexachlorophene	Reduces gram +	No	Systemic (CNS)	Poor
Tribromsalan	Reduces gram + Reduces odor	No	Photodermatitis (persistent)	Poor
Fluorosalan	Inadequate data	Not demonst.	Resembles TBS	Inadequate data
Phenol	Ineffective at use concentration	No	Kidney and liver on chronic use	Poor

Table 16.2. Important recent germicidal ingredients of antimicrobial soaps

Generic name	Formula	Abbreviation	FDA category
1. 3,4',5-Tribromosalicylanilide (Tribromsalan)		TBS	2
2. 3,5-Dibromo-3'-trifluoro-methylsalicylanilide (Fluorosalan)		—	3
3. 3,4,4'-Trichlorocarbanilide (Triclocarban)		TCC	3
4. 4,4'-Dichloro-3-(trifluoro-methyl) carbanilide (Cloflucarban)		TFC	3
5. 5-Chloro-2-(2,4-dichloro-phenoxy)phenol (Triclosan) (Irgasan DP-300)		—	3

The benefits and risks of using antimicrobials are summarized below:

Benefits
  1) Reduces body odor
  2) May reduce incidence of pyogenic infections

Risks
  1) Local or systemic toxicity
  2) Change in long-time beneficial resident flora
  3) Overgrowth of gram-negative bacteria

Odor reduction, the principal benefit of antimicrobial soaps, has been studied to some extent. It is not entirely clear which microorgranisms are responsible for odor and the amount of reduction in microorganism population necessary for a deodorant effect. Currently one or more gram-positive species—such as *Staphylococcus epidermidis*, propionibacteria, or aerobic diphtheroids—are believed involved.[5] Some have suggested that a 70% reduction in microbial flora might provide a reasonable deodorant effect. Bar soaps that achieve a 90% or greater reduction of gram-positive organisms may be too active, i.e., they allow overgrowth of gram-negative organisms. This effect was not observed, however, when explored by the Center for Disease Control (CDC) in monitored hospitals reporting a comparison of nursery infections related to infant bathing procedures with HCP and with soap and water.[2]

In 1953, Shelley et al.[18] were the first to show that axillary apocrine sweat is sterile and odorless when secreted. The characteristic axillary odor appeared related to a degradation of apocrine sweat involving gram-positive organisms. The odor was abolished when the axilla was treated with HCP. Shehaden and Kligman[17] agreed with this assessment. They reported an overgrowth of gram-negative organisms when gram-positives were eliminated with neomycin and aluminum chlorhydroxide. No odor emanated from the gram-negative-colonized armpit. More recently, breakdown of secreted androgenic steroids has been suggested as responsible for some axillary odor production.[13]

HCP was patented in 1941.[7] Early studies showed it was toxic by single oral or i.v. dose. In 1952, tests conducted by South Shore Analytical and Research Laboratory[8] for the manufacturers showed toxic potential by skin application to animals (Table 16.3).

These early warnings of a toxic percutaneous potential were virtually ignored. Eventually it became difficult for those responsible for the manufacture of HCP to accept animal toxicity data in the face of increasing apparent safe use by humans. Thus occasional but recurring animal and human toxicity problems with HCP were viewed only as a minor annoyance until infant deaths caused by a product inadvertently contaminated with HCP dramatically focused attention on HCP as a causative agent. A growing number of separate concerns about the safety of HCP led to the formation of

Table 16.3. Cutaneous toxicity of HCP in animals[a]

Animal	Concentration, %	Application	Effects
2 Rabbits	15	6 ml	Systemic toxicity 1/2
2 Rabbits	15	9 ml	Lethal
6 Guinea pigs	15	9 ml	Lethal
2 Rabbits	3	6 ml for 3 & 17 days	1/2 dead at 3 days
2 Rabbits	3	3 ml for 13 days	Toxic

[a]Partial data for South Shore Analytical & Research Laboratories 1952, from Gump (1969).

FDA's OTC Antimicrobial 1 Panel in 1972 and the banning of HCP in soap.[5] The panel continued to meet with substitutions of original members; their more recent deliberations have been reported as Antimicrobial 2 Panel.[6]

The continuing search for additional information has led to HCP's being the most studied antimicrobial ingredient for toxic effects. There is now an extensive literature documenting systemic effects from topical application with special reference to the CNS as a target site. Premature babies appeared to be especially vulnerable because of the inadequacy of the skin barrier in early life. The variability in permeability of human skin is another factor not receiving adequate attention. Some individuals have exceedingly permeable skins and would be especially vulnerable to topically applied toxic substances. As regional variations in skin permeability are considerable, this may be an important consideration.

Now that HCP is known to traverse the placental barrier, the possibility has been considered that HCP resembles thalidomide in being teratogenic in man at lower doses than in other mammals.[12] Swedish epidemiologists report an association between HCP exposure and malformation in children.[9]

As the use of HCP in multiple products (over 200 at one time) may have contributed to the restriction in its use, it appears wise that the manufacturers of triclocarban (TCC), a currently marketed antimicrobial ingredient, have limited its use to one product—soap.

One of the risks that we have not dealt with adequately is the risk to consumers from marketing products without prior adequate safety testing. The classical case is HCP. At present, antimicrobial soaps containing TCC alone or in combination with triclosan (Irgasan DP-300) are marketed. These germicides are listed under Category III, safety not yet demonstrated, because reviewers questioned the validity of one of the basic long-term toxicity studies.[6] Under such circumstances, one must ask what risks these products impose on consumers, especially when one considers that it takes four or more years to complete new work.

Procter and Gamble markets Safeguard soap as a drug. It contains 1.5% TCC. They also market other antimicrobial soaps with deodorant (cosmetic) claims. Zest and Coast contain unspecified amounts of TCC.

Armour-Dial markets Dial, which contains 1.5% TCC, as a drug. Colgate markets Irish Spring as a cosmetic: it contains 0.75% TCC and 0.25% triclosan. Their Palmolive Gold contains unspecified amounts of TCC alone. Lever Brothers markets Lifebuoy and Phase III, containing 0.25% triclosan, as cosmetics.

TCC has been studied by predictive and clinical test procedures for its local toxic effects by members of the International Contact Dermatitis Research Group (ICDRG) and was reported to be without significant effects when used in bar soap at use concentrations. Parameters examined in these studies included skin irritation, skin sensitization, phototoxicity, and photoallergy.[11]

Recent work on TCC performed for Procter and Gamble was filed with the FDA hearing clerk.[10] These studies appear to demonstrate the safety of 1.5% TCC in a bar soap in long-term use. They include oral and bathing studies on infant monkeys; metabolism studies on rats and monkeys; absorption, distribution, and excretion studies on humans, rats, and monkeys; teratology and reproduction studies in rats which show the reproductive system as an important target; and blood levels in humans and rats that appear to demonstrate a significant safety factor. A bar soap tested in 1975 for efficacy containing 1% TCC and 0.5% cloflucarban showed significant reduction in total aerobic flora of the back, chest, forearm, calf, and foot. The axilla, which is the most important site for odor control, was not found significantly affected, however,[21] which poses a question about the efficacy of this combination under continued-use conditions. The lack of appreciable substantivity of TCC for skin,[1] which diminishes the amount available for skin penetration, may enhance its safety but reduce its efficacy.

Triclosan is a relatively stable, broad-spectrum antimicrobial agent for use in germicidal soaps. Although many studies have been conducted with triclosan, reviewing authorities have not found them acceptable, nor has its use been restricted to soap, like TCC; hence the available data should be considered exploratory.[5,6] The acute oral toxicity in various species suggests it is only slightly toxic by this route; however, this may be due to poor gastrointestinal tract absorption as indicated by a comparison of its acute IV and oral toxicities.[3]

Metabolic conversion, mainly to the glucuronide, but also to the sulfate, occurs in the liver, which is the main target organ. There is a possibility that triclosan may possess a nephrotoxic potential. In dogs, which may be a susceptible species, liver damage occurs at 67/106 total triclosan in blood, with a no-effect level of 36/106. Human plasma levels of up to 2.5/106 total triclosan were achieved in a study of five subjects who bathed with soap containing 1% triclosan for 45 days.[2]

Triclosan does not appear to be allergenic, photoallergenic, phototoxic, or cross-reacting.[3] It was not mutagenic in short-term tests; however, reproduction, teratology, and carcinogenic studies have not been thoroughly evaluated. Although said to have a broad-gram-negative and gram-positive antimicrobial action, *P. aeruginosa*, viruses, and fungi are resistant.[13] The margin of safety of triclosan and its antimicrobial effectiveness under conditions of use are not yet entirely clear (compare the margin of safety with the OTC Antimicrobial Panel estimates[5,6]). The expanding uses of triclosan under these circumstances warrants continued monitoring and review by the FDA.

Acknowledgments

The assistance of the OTC Division, Bureau of Drugs, in providing reference assistance is acknowledged.

References

1. Black JG, Howes D, Rutherford T (1975) Skin deposition and penetration of trichlorocarbanilide. Toxicology 3:253–264
2. Center for Disease Control (1974) National Nosocomial Infections Study, Epidemiology Notes and Reports. Fourth Quarter 1973. U.S. Dept. of HEW, Washington, D.C.
3. Ciba-Geigy Corp., Ardsley, NY (1978) Irgasan DP-300 (Triclosan). Reply to tentative final monograph on OTC topical antimicrobials. Submission to FDA Hearing Clerk under 75 N-0183, SUP 019
4. Federal Register (Feb. 14, 1968) Cosmetics containing bithionol. Vol. 33, No. 31, p. 2934, par. 3.60
5. Federal Register (Sept. 13, 1974) O-T-C topical antimicrobial products and drug and cosmetic products. Vol. 39, No. 179, p. 33102, Part II, Dept. of HEW, Washington, D.C.
6. Federal Register (Jan. 6, 1978) OTC topical antimicrobial products. Vol. 43, No. 4, p. 1210, Part II, Dept. of HEW, Washington, D.C.
7. Gump WS (July 29, 1941) Dihydroxyhexachlorodiphenylmethane and method of producing same. U.S. Patent 2,250,480
8. Gump WS (1969) Toxicological properties of hexachlorophene. J Soc Cosmet Chem 20:173–184
9. Halling H (1977) Suspected link between exposure to hexachlorophene and birth of malformed infants. Lakartidningen 74:542–546
10. Herrmann KW, Procter & Gamble Co. (March, 1978) Toxicity data supporting the safety of unrestricted marketing of 1.5% TCC in bar soaps. Filed with FDA Hearing Clerk under 75 N-0183, CP002 OTC Antimicrobial Products.
11. International Contact Dermatitis Research Group (ICDRG) (1978) Maibach H et al. Triclocarban: Evaluation of contact dermatitis potential in man. Contact Dermatitis 4:283–288
12. Kalter H (Jan. 25–30, 1965) A collection of lectures and demonstrations from the Second Workshop on Teratology. Supplement to Teratology Workshop Annual, Berkeley, Calif.
13. Labows J, Preti G, Hoelzle E, Leyden J, Kligman A Steroid analysis of human apocrine secretion (to be published from University of Pennsylvania, Dept. of Obstetrics & Gynecology and Dept. of Dermatology.)
14. Marzulli F, Maibach H (1973) Antimicrobials: Experimental contact sensitization in man. J Soc Cosmet Chem 24:399–421
15. Marzulli F, Maibach H (1975) Relevance of animal models: the hexachlorophene story in animal models. In: Maibach H (ed) Dermatology. Churchill Livingstone, Edinburgh, London, New York, pp 156–167
16. Merck Index 8th edn (1968) Merck & Co., Rahway, New Jersey, p 1047

17. Shehaden N, Kligman A (1963) The bacteria responsible for axillary odor. J Infect Dis 41:3.
18. Shelley WB, Hurley H, Nichols A (1951) Axillary odor. Arch Dermatol Syphilol 68:430
19. Toxicology Data. Dept. of Industrial Medicine, Ciba-Geigy Corp., Ardsley, New York
20. Traub EF, Newhall CA, Fuller JR (1944) The value of a new compound used in soap to reduce the bacterial flora of the human skin. Surg Gynecol Obstet 79:205
21. Voss JG (1975) Effects of an antibacterial soap on the ecology of aerobic bacterial flora of human skin. Appl Microbiol 30:551–556
22. Wilkinson DN (1962) Further experiences with halogenated salicylanilides. Br J Dermatol 74:295

Chapter 17
Antimicrobials: Regulatory Aspects
Heinz J. Eiermann

In a discussion of the regulatory aspects of antimicrobials, a distinction must be made between their use as active ingredients of drugs, as inactive ingredients of drugs, and as cosmetic ingredients. It must be understood that, from a regulatory viewpoint, the term "use" generally means the use claimed in labeling. An unintentional effect does not necessarily change the legal status of an ingredient or the product in which the ingredient is used.

The Food, Drug, and Cosmetic Act defines the term "drug" as "articles intended for use in the diagnosis, cure, mitigation, treatment or prevention of disease ... and articles intended to affect the structure or any function of the [human] body." Cosmetics are intended to cleanse, beautify, or promote attractiveness without affecting the body's structure or functions.[12]

If an antimicrobial is intended to kill microorganisms or inhibit their growth on the skin, and this effect is stated in labeling, it is legally a drug because it has a therapeutic effect. If the same antimicrobial is claimed to prevent body odor, it is not a drug because prevention of body odor is considered a cosmetic effect. It is also not a drug if the antimicrobial is intended to prevent the growth of microorganisms in the product, not on the skin; antimicrobials that function as preservatives are viewed as "inactive" ingredients.

Typical examples of products which may be either drugs or cosmetics, or both, are deodorants and antiperspirants. A deodorant spray or soap claiming antibacterial activity is a drug; it may even be a drug requiring a new drug application if its active ingredient or its use is new, or if it is not generally recognized as safe and effective. A product of the same composition whose antibacterial effect is not mentioned in labeling is a cosmetic. The deodorant action does not affect a body function; it is a cosmetic claim. A product claiming reduction in wetness and prevention of axillary odor is a drug as well as a cosmetic. The antiperspirant action is considered a drug claim because it affects the perspiration function of the human body.

This discussion addresses the regulatory aspects of antimicrobials used as active ingredients of topical over-the-counter (OTC) drugs, as antibacterial agents of deodorants, and as preservatives of drugs and cosmetics.

Antimicrobial Drug Ingredients

Bithionol

Bithionol [2,2'-thiobis (4,6-dichlorophenol)] was the first antimicrobial ingredient to attract the attention of the Food and Drug Administration (FDA). It was used in several topical drug products under approved new drug applications (NDA). In October 1967, the FDA revoked these approvals. Clinical experience and photopatch testing had shown that bithionol was a strong photosensitizer. In some instances, the photosensitization persisted for prolonged periods. There was evidence to indicate that it may produce cross-sensitization with halogenated salicylanilides and hexachlorophene.[13] In 1968, bithionol was also prohibited as an ingredient of cosmetic products.[6,14]

Hexachlorophene

Hexachlorophene [HCP; 2,2'-methylenebis (3,4,6-trichlorophenol)] became a regulatory issue when Kimbrough and Gaines[27] and others detected in 1971 that its oral administration to rats and other animals produced lesions in the brain, brain stem, and spinal cord, and Curley et al.[11] and others determined that it readily penetrated human and animal skin.

The toxicological data on HCP were published at about the time the FDA initiated its OTC drug review program. One of the appointed panels, the Antimicrobial I Panel, was requested to evaluate the safety and effectiveness of topical antimicrobials.[5,15,16]

To curtail the widespread and uncontrolled use of HCP and to reduce the risk of harm to consumers while the Antimicrobial I Panel was reviewing the relevant toxicological data, the agency proposed regulations restricting the use of HCP in drugs and cosmetics. In OTC drugs, the level of use of HCP was to be limited to a concentration not exceeding 0.75%. In cosmetics, it was to be used only where absolutely needed as a preservative and at a level not greater than 0.1%.[19]

When new evidence was received about the neurotoxicity of HCP—including the news from France about the poisoning of more than 40 infants from use of HCP-contaminated baby powder—and when the panel had come to the conclusion that this substance could not be used safely in OTC drugs, the FDA prohibited almost all uses of HCP. In OTC drugs and cosmetics, it is now permitted only where absolutely necessary as a preservative and only at concentrations not exceeding 0.1%. In prescription drugs, its use is limited to bacteriostatic skin cleansing and to control gram-positive infections where other infection-control procedures have been unsuccessful. When HCP is used, FDA regulations require extensive label warnings.[3,15]

Halogenated Salicylanilides

Halogenated salicylanilides, namely, tribromsalan (TBS; 3,4',5-tribromosalicylanilide), dibromsalan (DBS; 4',5-dibromosalicylanilide), me-

tabromsalan   (MBS;   3,5-dibromosalicylanilide),   and   3,3',4,5'-tet-
rachlorosalicylanilide (TCSA), as a group, became the subject of the third
regulatory action taken by the FDA against antimicrobials for topical human
use. TBS was identified by the OTC Antimicrobial I Panel as a potent photosensitizer with the ability to produce cross-sensitization with other
halogenated salicylanilides and cause skin reactions in sunlight even after
exposure to it was discontinued. DBS, MBS and TCSA were even stronger
photosensitizers and cross-photosensitizers than TBS.[20]

In October 1975, the FDA prohibited further use of halogenated
salicylanilides in OTC drug products without an approved new drug application. All previously approved NDA's, mostly for antimicrobial soaps, had
either been withdrawn or amended to delete the halogenated salicylanilides
from the product formulation. The regulations prohibit their use in cosmetic
products.[4,7,22]

The Antimicrobial I Panel completed its work in 1974. Its findings and
regulatory recommendations were published in the *Federal Register* as a
"Proposal to Establish a Monograph for OTC Topical Antimicrobial Products."[21] Numerous comments were received in response to the proposed
monograph. Several recommendations were adopted by the FDA and incorporated in the tentative final regulation published in January 1978.[24]

The panel's and agency's findings are summarized in Tables 17.1–17.7.
Each table represents a separate product category. The seven categories are
defined as follows:

Antimicrobial soap. A soap containing an active ingredient with both in
vitro and in vivo activity against skin microorganisms.

Health-care personnel hand wash. A nonirritating antimicrobial-containing preparation designed for frequent use; it reduces the number of transient
microorganisms on intact skin to an initial baseline level after adequate
washing, rinsing, and drying; and it is broad-spectrum, fast acting, and, if
possible, persistent.

Patient preoperative skin preparation. A fast-acting broad-spectrum antimicrobial-containing preparation that significantly reduces the number of
microorganisms on intact skin.

Skin antiseptic. A nonirritating, antimicrobial-containing preparation that
prevents overt skin infection.

Skin wound cleanser. A nonirritating, liquid preparation (or product to be
used with water) that assists in the removal of foreign material from small superficial wounds, does not delay wound healing, and that may contain an antimicrobial ingredient.

Skin wound protectant. A nonirritating antimicrobial-containing preparation applied to small cleansed wounds; it provides a protective physical barrier and a chemical (antimicrobial) barrier that neither delays healing nor
favors the growth of microorganisms.

Surgical hand scrub. A nonirritating antimicrobial-containing preparation
that significantly reduces the number of microorganisms on intact skin. A
surgical hand scrub should be broad-spectrum, fast acting, and persistent.

The reviewed ingredients are classified into three categories. The three categories express the respective conditions of safety and effectiveness the products containing a classified ingredient would pose in the judgement of the panel and the agency. They are:

Category I. Conditions under which antimicrobial products are generally recognized as safe and effective and are not misbranded.

Category II. Conditions under which antimicrobial products are not generally recognized as safe and effective or are misbranded.

Category III. Conditions for which available data are insufficient to permit final classification.

The following general comments may be made about the proposed classification of the reviewed antimicrobials:

None of the ingredients reviewed for use in antimicrobial soaps are generally recognized as safe and effective (Table 17.1). Six antimicrobials, i.e., cloflucarban, triclorcarban, triclosan, para-chloro-meta-xylenol (PCMX), 1.5% phenol solution, and povidone-iodine complex, are classified in Category III and require additional toxicological and efficacy testing to permit final classification. Quaternary ammonium compounds, hexylresorcinol, iodine tincture, and the detergent-based iodophers are placed in Category II for reasons of chemical incompatibility with soap.

Of the ingredients reviewed for use in health-care personnel hand wash products, none are classified in Category I (Table 17.2). Six ingredients—i.e., three quaternary ammonium compounds, 1.5% phenol solution,

Table 17.1. Classification of Antimicrobial Ingredients for Use in Antimicrobial Soaps

Active ingredient	Category	Remarks
Incompatible ingredients	II	Quaternary ammonium compounds, most iodophors, iodine tincture, hexylresorcinol and triple dye.
Cloflucarban	III	Maximum use level 1.5%. Additional safety data required.
Triclocarban	III	Maximum use level 1.5%. Additional safety data required.
Triclosan	III	Maximum use level 1.0%. Additional safety and efficacy data required.
Para-chloro-meta-xylenol	III	Additional safety and efficacy data required.
Phenol (1.5% max.)	III	Aqueous/alcoholic solution. Efficacy data required.
Povidone-iodine complex	III	Additional safety and efficacy data required.

Table 17.2. Classification of Antimicrobial Ingredients for Use in Health-care Personnel Hand Wash Products

Active Ingredient	Category	Remarks
Triclosan	II	
Tincture of iodine	II	
Iodophors[a]	II	
Cloflucarban	II	Category III when used in bar soap.
Triclocarban	II	Category III when used in bar soap.
Para-chloro-meta-xylenol	II	Additional safety efficacy data required.
Hexylresorcinol	III	Efficacy data required.
Phenol (1.5% max.)	III	Aqueous/alcoholic solution. Additional safety and efficacy data required.
Quaternary ammonium compounds[b]	III	Additional safety and efficacy data required.

[a] Iodine complexed with phosphate ester of alkylaryloxy polyethylene glycol, poloxamer-iodine complex, povidone-iodine complex, nonyl phenoxypoly (ethyleneoxy) ethanol-iodine, undecoylium chloride-iodine complex.
[b] Benzalkonium chloride, benzethonium chloride, methylbenzethonium chloride.

hexylresorcinol, and PCMX—require further safety and efficacy studies for final classification. Nine ingredients—i.e., five iodophors, iodine tincture, triclosan, triclocarban, and cloflucarban—are in Category II because they are not recognized as safe and effective. However, triclocarban and cloflucarban, when used as antimicrobials of bar soap-type products, are classified in Category III.

Iodine tincture is generally recognized as safe and effective as an active ingredient of patient preoperative skin preparations (Table 17.3). Cloflucarban, triclocarban, and triclosan are not so recognized and are placed in Category II. All other ingredients reviewed for use in patient preoperative skin preparations require additional testing to determine their final classification.

With the exception of cloflucarban and triclocarban, all listed ingredients reviewed for use in skin antiseptics are classified in Category III because of incomplete safety or effectiveness data (Table 17.4). Some require only minor testing. Iodine tincture requires further study of its delaying effect on wound healing and to determine minimal effective dose. Cloflucarban and triclocarban are not generally recognized as safe and effective for use in skin antiseptics.

Generally recognized as safe and effective for use in skin-wound cleansers are the three quaternary ammonium compounds at a use concentration not greater than 1/750 in water, hexylresorcinol at a maximum dose of 1/1000, and poloxamer 188, a polyoxyethylene polyoxypropylene block polymer known by the trade name Pluronic F-68 (BASF-Wyandotte), when used as a 20%–40% active aqueous solution (Table 17.5). Cloflucarban and triclocarban are listed in Category II except when used in bar soap products, where

Table 17.3. Classification of Antimicrobial Ingredients for Use in Patient Preoperative Skin Preparations

Active Ingredient	Category	Remarks
Tincture of iodine	I	1.8–2.2% iodine and 2.1–2.6% sodium iodide in water or in hydroalcoholic vehicle.
Cloflucarban	II	
Triclosan	II	
Triclocarban	II	
Para-chloro-meta-xylenol	III	Additional safety and efficacy data required.
Hexylresorcinol	III	Efficacy data required.
Phenol (1.5% max.)	III	Aqueous/alcoholic solution. Additional safety and efficacy data required.
Iodophors[a]	III	Additional safety and stability data required.
Quaternary ammonium compounds[b]	III	Additional safety and efficacy data required.

[a] Iodine complexed with phosphate ester of alkylaryloxy polyethylene glycol, poloxamer-iodine complex, povidone-iodine complex, nonyl phenoxypoly (ethyleneoxy) ethanol-iodine, undecoylium chloride-iodine complex.
[b] Benzalkonium chloride, benzethonium chloride, methylbenzethonium chloride.

Table 17.4. Classification of Antimicrobial Ingredients for Use in Skin Antiseptics

Active Ingredient	Category	Remarks
Cloflucarban	II	
Triclocarban	II	
Tincture of iodine	III	Wound healing delay to be determined.
Para-chloro-meta-xylenol	III	Additional safety and efficacy data required.
Hexylresorcinol	III	Efficacy data required.
Triclosan	III	Maximum use level 1.0%. Additional safety and efficacy data required.
Phenol (1.5% max.)	III	Aqueous/alcoholic solution. Efficacy data required.
Iodophors[a]	III	Additional safety and efficacy data required.
Quaternary ammonium compounds[b]	III	Additional safety and efficacy data required.
Triple dye[c]	III	Additional safety and efficacy data required.

[a] Iodine complexed with phosphate ester of alkylarloxy polyethylene glycol, poloxamer-iodine complex, povidone-iodine complex, nonyl phenoxypoly (ethyleneoxy) ethanol-iodine, undecoylium chloride-iodine complex.
[b] Benzalkonium chloride, benzethonium chloride, methylbenzethonium chloride.
[c] Crystal violet, brilliant green, proflavine hemisulfate in water.

Table 17.5. Classification of Antimicrobial Ingredients for Use in Skin-wound Cleansers

Active Ingredient	Category	Remarks
Quaternary ammonium compounds[a]	I	0.133% aqueous solution.
Hexylresorcinol	I	0.1% max.
Poloxamer 188[b]	I	20–40% aqueous solution.
Cloflucarban	II	In category III when used in soap.
Triclocarban	II	In category III when used in soap.
Tincture of iodine	III	Wound healing delay to be determined.
Triclosan	III	Maximum use level 1.0%. Additional safety and efficacy data required.
Para-chloro-meta-xylenol	III	Additional safety and efficacy data required.
Phenol (1.5% max)	III	Aqueous/alcoholic solution. Efficacy data required.
Iodophors[c]	III	Additional safety and stability data required.

[a] Benzalkonium chloride, benzethonium chloride, methylbenzethonium chloride.
[b] Pluronic F-68 by BASF-Wyandotte.
[c] Iodine complexed with phosphate ester of alkylaryloxy polyethylene glycol, poloxamer-iodine complex, povidone-iodine complex, nonyl phenoxypoly (ethyleneoxy) ethanol-iodine, undecoylium chloride-iodine complex.

they are classified in Category III. All other antimicrobials reviewed for use in skin-wound cleansers require further toxicity and efficacy testing to permit final classification.

Cloflucarban and triclocarban are not generally recognized as safe and effective for use in skin-wound protectants (Table 17.6). All other ingredients are listed in Category III for lack of adequate safety or effectiveness data. None of the antimicrobials is listed in Category I for use in skin-wound protectants.

The classification of ingredients for use in surgical hand-scrub products is almost identical to the classification of ingredients for use in skin-wound protectants (Table 17.7). The only exception is triclosan, which is not generally recognized as safe and effective for this use.

The next step in the rule-making process, normally, would be the publication of a final order to make the proposed requirements effective. Those interested in the reclassification of ingredients from Category III to Category I would then have two years to complete the required testing and prove safety and effectiveness. But before publication of a final regulation, the agency is required to act on the written objections and requests for oral hearing received in response to the tentative final order. Several such requests were made, and other comments requested that the FDA reopen the administrative record so that further scientific data could be submitted. The comments argued that more than three years had passed since publication of the

Table 17.6. Classification of Antimicrobial Ingredients for Use in Skin-Wound Protectants

Active Ingredient	Category	Remarks
Cloflucarban	II	
Triclocarban	II	
Tincture of iodine	III	Wound healing delay to be determined.
Para-chloro-meta-xylenol	III	Additional safety and efficacy data required.
Hexylresorcinol	III	Efficacy data required.
Triclosan	III	Maximum use level 1.0%. Additional safety and efficacy data required.
Phenol (1.5% max)	III	Aqueous/alcoholic solution. Efficacy data required.
Iodophors[a]	III	Additional safety and stability data required.
Quaternary ammonium compounds[b]	III	Additional safety and efficacy data required.

[a] Iodine complexed with phosphate ester of alkylaryloxy polyethylene glycol, poloxamer-iodine complex, povidone-iodine complex, nonyl phenoxypoly (ethyleneoxy) ethanol-iodine, undecoylium chloride-iodine complex.
[b] Benzalkonium chloride, benzethonium chloride, methylbenzethonium chloride.

proposed monograph and several additional studies had been conducted on the safety and effectiveness of the reviewed antimicrobials pertinent to their regulation.

In 1979 the agency reopened the administrative record for a period of 4 months.[26] After review of the newly submitted data, the agency will publish a second tentative final order.

Table 17.7. Classification of Antimicrobial Ingredients for Use in Surgical Handscrub Products

Active Ingredient	Category	Remarks
Triclosan	II	
Cloflucarban	II	
Triclocarban	II	
Tincture of iodine	II	
Para-chloro-meta-xylenol	III	Additional safety and efficacy data required.
Hexylresorcinol	III	Efficacy data required.
Phenol (.5% max)	III	Aqueous/alcoholic solution. Efficacy data required.
Iodophors[a]	III	Additional safety and stability data required.
Quaternary ammonium compounds[b]	III	Additional safety and efficacy data required.

[a] Iodine complexed with phosphate ester of alkylaryloxy polethylene glycol, poloxamer-iodine complex, povidone-iodine complex, nonyl phenoxypoly (ethyleneoxy) ethanol-iodine, undecoylium chloride-iodine complex.
[b] Benzalkonium chloride, benzethonium chloride, methylbenzethonium chloride.

The regulatory aspects of the antimicrobial ingredients reviewed by the Antimicrobial II or other OTC panels are not being discussed. Most of these ingredients are outside the scope of this discussion, and the review of the ingredients of immediate interest has not yet been completed.

Antimicrobial Cosmetic Ingredients

Many antimicrobial drug ingredients are used as cosmetic ingredients. Most deodorant bar soaps and underarm deodorants contain antimicrobials. They may also be present in foot powders, mouthwashes, breath freshners, and feminine deodorants.

The regulation of antimicrobials as drugs will continue to influence their use in cosmetics. The actions taken against hexachlorophene and halogenated salicylanilides are good examples of the impact of drug review panel recommendations on the regulatory status of antimicrobials when used as cosmetic ingredients. However, not every reclassification from Category III to Category II will cause an ingredient to be prohibited in cosmetics.

The Food, Drug, and Cosmetic Act gives the FDA much less authority to regulate antimicrobials as cosmetic ingredients than when used as ingredients of drug products. The agency cannot order a cosmetic manufacturer to conduct toxicological testing when the safety of a product or ingredient has come into question, and it cannot remove a product from the market, or prohibit further use of an ingredient, unless it has toxicological data or other factual information demonstrating that the product or ingredient is hazardous to consumers under the conditions of use. Reclassification of an ingredient to Category II for failure of the drug industry to conduct the required testing to demonstrate safety does not mean that the respective ingredient is now unsafe. As a cosmetic ingredient, its regulatory status remains essentially unchanged.

If, for example, no testing were carried out to permit the agency to conclude that triclosan or triclocarban is safe for its intended use and the agency placed it into Category II, it could still be used as a cosmetic ingredient until such time that toxicological test data or consumer experience demonstrate that it is hazardous. But the cosmetic manufacturer using such an ingredient may then be required to warn consumers in labeling that "[t]he safety of this product has not been determined."[8] This label warning is required when the safety of a product, including the safety of each ingredient, has not been adequately substantiated. A product not bearing this label warning may be deemed misbranded, i.e., its labeling may be considered false or misleading.

The agency's authority to regulate cosmetic label claims is as limited as its authority to regulate safety. A case in point is the labeling of cosmetics containing antimicrobial ingredients and claiming deodorancy. The term "deodorant" is not defined by regulation. Its dictionary definition is somewhat ambiguous, and a statistically meaningful consumer perception study that could serve as a basis for regulating this term has not been carried out. And even if the agency were successful in regulating the term "deodorant," it may still be faced with the prospect of conducting its own performance studies to determine a product's deodorant efficacy if it wanted to take action against a misbranded product. Persistent resources constraints make it unlikely that such an enforcement program could be undertaken in the forseeable future.

Antimicrobial Preservatives

The regulatory requirements for antimicrobials used as preservative ingredients differ substantially from those applicable to antimicrobials used as drugs. The monograph on OTC topical antimicrobial drug products makes some mention of the use of antimicrobials as preservatives. In the January 1978 tentative final order, the agency proposes the following definition of the term antimicrobial preservative ingredient:

> *Antimicrobial preservative (inactive) ingredient.* A compound or substance that kills microorganisms or prevents or inhibits their growth and reproduction and is included in a product formulation only at a concentration sufficient to prevent spoilage or prevent growth of inadvertently added microorganisms, but does not contribute to the claimed effects of the product in which it is included.

Only a small number of the antimicrobials reviewed by the Antimicrobial I Panel appear to be used as preservatives, and their frequency of use is quite low. According to a 1977 tabulation of the frequency of use of preservatives in cosmetic formulations voluntarily disclosed to the FDA, less than 4% of the approximately 16 000 preservative uses involved antimicrobials reviewed by the Antimicrobial I Panel, and many of these registrations are likely to represent deodorant uses.[28] A review of some of the registered formulations in which these antimicrobials are present showed that the majority are associated with deodorant products.

The ten most frequently used preservatives, in descending order of frequency of use, are: methyl paraben, propyl paraben, imidazolidinyl urea, formaldehyde, quaternium 15, butyl paraben, sorbic acid, 2-bromo-2-nitro-1, 3-propanediol, sodium dehydroacetate and sodium benzoate. The tabulation also lists 8 uses of hexachlorophene and 68 uses of organic mercurial compounds as preservatives.

Mercurial ingredients may only be used as a preservative of eye-area cosmetics for which no other effective and safe nonmercurial preservative is available, and only at a level not exceeding $65/10^6$ of mercury, calculated as the metal.[9,19] All other uses of mercury-containing compounds are prohibited by regulation because of their potential to cause skin irritation and sensitization and their ability to penetrate the skin and produce various systemic toxic effects.

For determining the effectiveness of antimicrobial agents as preservatives, the tentative final order proposes two testing methods, namely: (1) a test similar to the antimicrobial preservative effectiveness test described in the United States Pharmacopeia;[32] and (2) a test which is almost identical to the preservation testing method suggested by the Cosmetic, Toiletry and Fragrance Association (CTFA) Inc.[10]

USP XIX procedures were modified to include rechallenge after two weeks and addition to the test samples of an organic load consisting of heat-killed yeast cells and inactivated horse or bovine serum. The modified CTFA procedures identify the specific microorganisms to be used in the test, namely, those listed as challenge microorganisms of the USP XIX test. The CTFA method already requires rechallenging of the test samples.

The tentative final order proposes that a product be considered effectively preserved if the yeast concentrations do not increase and the concentrations of viable bacteria are reduced by the 14th day after the last challenge to not more than 1% of the initial concentrations. Additionally, during the subsequent 2 weeks of storage, the concentration of each test microorganism must not exceed the density determined on the 14th day.

The tentative final order characterizes the term "effectively preserved" as preservation "sufficient to prevent spoilage or prevent growth of inadvertently added microorganisms."[24] This statement must be interpreted to mean that preservation should be adequate to protect a product during manufacture as well as consumer use. The Antimicrobial I Panel and the USP XIX make specific reference to protection from the growth of microorganisms introduced "as a result of customer use" or "subsequent to the manufacturing process."[21,32]

While there is general agreement on the level of preservation a product should enjoy, uncertainty has been voiced about the reliability of the proposed testing methods and standards in predicting such protection. Of particular concern is preservation of products that are used daily and may last for several weeks or even months.

The proposed procedures require only one rechallenge, and the same microorganisms are used for the two challenges. Mixed-culture challenges or mixed-sequential challenges are not stipulated in these tests. Wilson and Ahearn[30] and Ahearn et al.,[1,2] however, have shown that repeated challenge of mascaras with *Staphylococcus epidermidis* increases the incidence of microbial contamination of marginally preserved products and that such mascaras, when first challenged with *S. epidermidis* become more susceptible to *Pseudomonas* spp. contamination.

Neither the proposal of the Antimicrobial I Panel nor the agency's tentative final order provide factual data supporting an assumption that a product meeting the proposed testing criteria is effectively preserved as defined by regulation. Apparently, no comparative studies have ever been conducted (or at least none have been reported in the scientific literature) comparing the results of microbiological challenge tests with those obtained from testing of the same products under conditions of consumer use.

In the aforementioned studies, Wilson and Ahearn demonstrated that topically used products do not easily become contaminated with pathogenic microorganisms during normal use. Among the many study groups involved in the testing of commercial mascara products to determine preservative efficacy under use conditions, only one mascara of one commercial product showed contamination with *Klebsiella pneumoniae* and one sample of another product yielded *P. aeruginosa* after more than six months of testing.[29] However, when the pathogenic microorganisms are given the opportunity to invade the cornea, the adverse effects can be devastating. Over 7 years, Wilson[31] and Wilson and Ahearn[30] identified 12 clinical cases of bacterial corneal ulcer associated with *P. aeruginosa* contaminated mascaras and three cases of fungal corneal ulcer caused by the use of eye cosmetics containing *Fusarium* spp. In all instance, partial or complete blindness occurred in the injured eye.

The high risk and severity of injury associated with the use near the eye of cosmetics that have become contaminated with pathogenic microorganisms prompted the FDA to publish in 1977 in the Federal Register a notice of intent to propose regulations requiring adequate preservation of these cosmetics under use conditions.[23] To establish a scientific basis for the regulatory requirements, interested persons were invited to submit information concerning microbiological testing methods and standards of performance which assured that the cosmetics would not become contaminated with microorganisms during manufacturing or consumer use. Unfortunately, no useful factual comments were received in response to this notice.

The agency is now in the process of obtaining the necessary data through a cooperative study with the University of California.[25] The project encompasses comparative testing in the laboratory and under simulated-use conditions of experimental products containing preservative systems of varying efficacy. It is anticipated that some experimental formulations will become contaminated under both testing conditions while others will not show microbial growth. The performance standard for predictive laboratory testing will then be represented by the antimicrobial activity level of the experimental product that was found to be adequately preserved to withstand microbial challenges during simulated use testing.

The study is expected to be completed in 2 years. Regulations establishing microbiological testing procedures and preservation standards may then be proposed. These preservation requirements are intended to become part of all inclusive regulations delineating good manufacturing practice for cosmetics.

References

1. Ahearn DG, Wilson LA, Julian AJ, Reinhardt, DJ Ajello G (1974) Microbial growth in eye cosmetics: contamination during use. Developments in Industrial Microbiology 15:211–215.
2. Ahearn DG, Sanghivi J, Haller GJ Wilson LA (1978) Mascara contamination: in use and laboratory studies. J Soc Cosmet Chem 29:127–131.

3. Code of Federal Regulations. Title 21, Part 250, sec. 250.250. Hexachlorophene, as a component of drug and cosmetic products.
4. Code of Federal Regulations. Title 21, Part 310, sec. 310.508. Use of certain halogenated salicylanilides as an inactive ingredient in drug products.
5. Code of Federal Regulations. Title 21, Part 330, sec. 330.10. Procedure for classifying OTC drugs as generally recognized as safe and effective and not misbranded, and for establishing monographs.
6. Code of Federal Regulations. Title 21, Part 700, sec. 700.11. Cosmetics containing bithionol.
7. Code of Federal Regulations. Title 21, Part 700, sec. 700.15. Use of certain halogenated salicylanilides as ingredients in cosmetic products.
8. Code of Federal Regulations. Title 21, Part 700, sec. 740.10. Labeling of cosmetic products for which adequate substantiation of safety has not been obtained.
9. Code of Federal Regulations. Title 21, Part 700, sec. 700.13. Use of mercury compounds in cosmetics including use as skin-bleaching agent in cosmetic preparation also regarded as drugs.
10. Cosmetic, Toiletry and Fragrances Association, Inc. (1970) A guideline for the determination of adequacy of preservation of cosmetics and toiletry formulations. TGA Cosmetic Journal 2:20–23
11. Curley A, Hawk RE, Kimbrough RD, Nathenson G, Finberg L (1971) Dermal absorption of hexachlorophene in infants. Lancet II:296–297
12. Federal Food, Drug and Cosmetic Act, As Amended (1938). Section 201 (g) (i) (U.S. Code, title 21, section 321 (g) (i)). Approved June 25, 1938.
13. Food and Drug Administration (1967) Drugs for Human Use Containing Bithionol. Notice of withdrawal of approval of new drug applications. Fed Reg 32:15046. October 31, 1967
14. Food and Drug Administration (1968). Cosmetics containing bithionol. Fed Reg 33:2934–2935. February 14, 1968.
15. Food and Drug Administration (1972a) Over-the-Counter Drugs. Proposal establishing rule making procedures for classification. Fed Reg 37:85–89. January 5, 1972
16. Food and Drug Administration (1972b) procedure for classification of over-the-counter drugs. Fed Reg 37:9464–9475. May 11, 1972
17. Food and Drug Administration (1972c) Antibacterial ingredients in drug and cosmetic products for repeated daily human use. Fed Reg 37:219–220. January 7, 1972
18. Food and Drug Administration (1972d) Hexachlorophene as a component in drug and cosmetic products for human use. Fed Reg 37:20160–20164. September 27, 1972
19. Food and Drug Administration (1973) Use of mercury in cosmetics including use as skin-bleaching agent in cosmetic preparations also regarded as drugs. Fed Reg 38:853–854. January 8, 1973
20. Food and Drug Administration (1974a) Certain halogenated salicylanilides as active or inactive ingredients in drug and cosmetic products. Notice of proposed rule-making. Fed Reg 39:33102–33103. September 13, 1974
21. Food and Drug Administration (1974b) Over-the-Counter Drugs. Proposal to establish a monograph for OTC topical antimicrobial products. Fed Reg 39:33103–33141. September 13, 1974
22. Food and Drug Administration (1975) Certain halogenated salicylanilides as active or inactive ingredients in drug and cosmetic products. Fed Reg 40:50527–50531. October 30, 1975
23. Food and Drug Administration (1977) Preservation of cosmetics coming in contact with the eye. Intent to propose regulations and request for information. Fed Reg 42:54837–54838. October 11, 1977
24. Food and Drug Administration (1978a) Over-the-counter drugs generally recog-

nized as safe, effective and not misbranded. OTC Topical Antimicrobial Products. Fed Reg 43:1210–1249. January 6, 1978

25. Food and Drug Administration (1978b) Contract for development and validation of a preservative efficacy testing method for cosmetic products. Contract No. 223-78-2015. September, 30, 1978

26. Food and Drug Administration (1979) Topical antimicrobial products for over-the-counter human use. Reopening of the administrative record. Fed Reg 48:13041–13042. March 9, 1979

27. Kimbrough RD, Gaines TB (1971) Hexachlorophene effects on the rat brain. Arch Environ Health 23:114–118

28. Richardson EL (1977) Preservatives: frequency of use in cosmetic formulas as disclosed to FDA. Cosmetics and Toiletries 92:85–86

29. Wilson LA, Julian AJ, Ahearn DG (1975) The survival and growth of microorganisms in mascara during use. Am J Ophthalmol 79:596–601

30. Wilson LA, Ahearn DG (1977) Pseudomonas-induced corneal ulcers associated with contaminated eye mascaras. Am J Ophthalmol 84:112–119

31. Wilson LA (1979) Quarterly Summary Report, March 1 to May 31, 1979. FDA Contract No. 223-2016 MOD 13. Submitted May 10, 1979.

32. United States Pharmacopeia XIX (1975) Antimicrobial preservatives-effectiveness. United States Pharmacopeial Convention, Inc. Mack Publishing Company, Easton, pp 587–588

Chapter 18
Antimicrobial Efficacy in the Presence of Organic Matter

John A. Ulrich

Disinfectants, germicides, and other microbial inhibitors may have their antimicrobial activities greatly altered by the presence of organic compounds. A spectrum of partial antagonism to complete inactivation of the disinfectants as well as fortification of the microbial inhibitory action have been amply recorded. These interactions may be produced by direct interaction between the organic substance and the antimicrobial compound. Indirect interaction may be brought about by changes in the reaction mixture (such as variation of pH, viscosity, surface tension activity, competition for binding sites, and other physical, electrical or biological effects).

This paper will comprise mainly the direct interactions between the newer skin germicides and the types of organic compounds found on normal skin and in wounds.

The types of direct interaction between microbial inhibitors and other organic compounds are summarized in Table 18.1. These are usually first-order reactions. A disinfectant may react and become irreversibly bound to an organic compound resulting in complete inactivation because inhibitory substances usually are active only in the "free" form.

In other instances, the disinfectant may react reversibly with the organic compound so that an equilibrium between free and bound is established. The disinfectant-organic compound complex will serve as a reservoir to maintain the equilibrium level of the free form. If the percentage of the free form is high enough to be inhibitory or in the germicidal range, a favorable

Table 18.1. Type Reactions of Germicides with Organic Substrates

"A" $\underset{\text{(A)}}{\overset{\text{B}}{\rightleftharpoons}}$ "AB"	Overall reaction
Free Bound	
0% $\longrightarrow$ 100%	Completely bound (inactivation)
100% $\longleftarrow$ 0%	No binding
1.0% $\rightleftharpoons$ 99%	Reversible equilibrium

situation results. This concentration will be maintained for a longer period of time since only the free form is inactivated or eliminated and the rate of inactivation is concentration dependent. The higher the concentration, the more rapid the inactivation or elimination.

Reactions With Complex Naturally Occurring Organic Mixtures

Whole Blood

Disinfectants are commonly applied in the treatment of traumatic or exudative wounds and the presence of blood, plasma, and serum may greatly affect the activity of the antimicrobials.

Lowbury and Lilly[7] studied the antagonistic effect of whole blood on a variety of skin disinfectants. Experiments were conducted by washing the hands with inhibitory agents and subsequently applying 2 ml of whole blood to the hands and immediately gloving. Quantitation of the normal skin flora was determined 1 h after gloving. Controls on the same subjects were set up by moistening the hands with 2 ml of sterile water, gloving, and repeating the quatitation in 1 h. The results are gathered in Table 18.2; blood differentially affects the disinfectants. Iodophores exhibited the greatest loss of activity in the presence of whole blood. Chlorhexidine was effected the least, with hexachlorophene in an intermediate position.

The effect of whole blood or bacterial broth in a suspension of organism exposed to chlorhexidine was investigated by Calman and Murray.[3] Water was used as the control material. The results in Table 18.3 show that gram-negative bacteria represented by *Escherichia coli* and *Pseudomonas pyocyaneus* and gram-positive organisms such as *Staphylococcus aureus* and β-hemolytic streptococci required higher concentrations of chlorhexidine to kill standard populations in the presence of broth. Even higher concentrations of the germicide are required in the presence of 25% whole blood. *Pseudomonas pyocyaneus* in 25% blood required a 40-fold increase in concentration to kill the same size population suspended in water. *Escherichia coli* needed only a fourfold increase to accomplish the same purpose. The gram-positive cocci exhibited intermediate increases of 24- and 32-fold for staphylococcus and streptococcus respectively.

Table 18.2. Mean Per Cent Reduction of Human Skin Populations by Various Germicides in Presence of Blood (Lowbury and Lilly 1974)

Menstrum	Per cent reduction[a]			
	Hibiscrub Chlorhexidine	Disfex hexachlorophene	Disodene iodophore	Control soap
Water (2 ml)	93.3	62.8	92.8	30.9
Blood (2 ml)	90.4	50.8	68.1	46.2

[a]One hour after application

Table 18.3. Dilutions of Chlorhexidine to Inhibit Organisms in 2.5 min in Various Menstra (Calman and Murray 1956)

	Inhibitory dilution (X1000)			
Menstrum	*E. coli*	*P. pyocyaneus*	*S. aureus*	*Beta strep.*
Water	20	80	60	320
Broth	30	20	30	40
Blood 25%	5	2	2.5	10

Serum

The antagonistic effect of human serum against four different types of antimicrobial preparations was studied by Ayliffe et al.[2] Hycolin (a phenol), Hibitane (a chlorhexidine), Roccal (a quaternary ammonium compound), and Wescodyne (an iodophore) were included. The results of the activities of the agents are shown in Table 18.4. With *Staphylococcus aureus* as the test organism, Wescodyne 1-100 and Hibitane at concentration of 0.02% did not kill all organisms in 10 min. Hibitane at a concentration of 0.2% and Roccal were not affected by the presence of serum. Hycoline 1-100 required 2.5 min in the presence of serum to destroy the population. When *Pseudomonas aeruginosa* was employed as the test organism, Hibitane 0.02% and Wescodyne 1-100 were ineffective after a 10 min exposure. Hycolin 1-100 required 5 min to kill the population, and Hibitane at 0.2 was not affected by the presence of the serum.

Hennessey[6] looked at the effect of 10% horse serum in the test medium with *Pseudomonas aeruginosa* as the test organism. It is evident from the slopes of curves in Fig. 18.1 that the presence of the serum does not affect the rate of kill by chlorhexidine but that a higher concentration is required to affect kill.

Pus

Pus has an antagonistic effect against the activity of Hibiclens and Betadine scrub preparations. Sheikh[10] tested pus naturally infected with *Proteus*

Table 18.4. Effect of Serum on the Activity of Antimicrobial Agents (Ayliffe et al. 1966)

	Min. to destroy population			
	S. aureus		*P. aeruginosa*	
Agents and conc.	No serum	20% serum	No serum	20% serum
Hycolin 1-100 (phenol)	1	215	1	5
Hibitane 0.02%	5	>10	5	>10
Hibitane 0.2% (chlorhexidine)	1	1	1	1
Roccal 0.1% (quaternary)	1	1	1	10
Wescodyne 1-100 (iodophore)	1	>10	5	>10

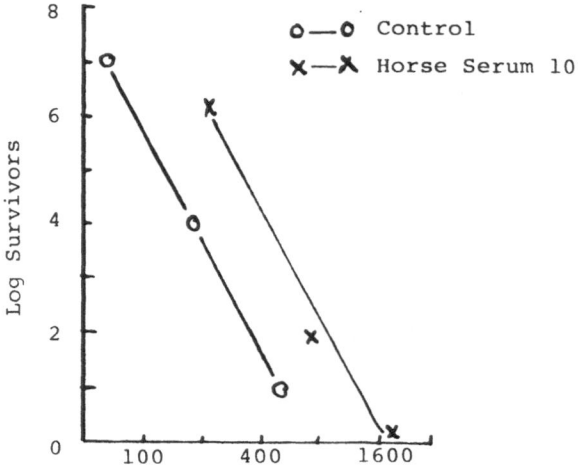

Fig. 18.1. Effect of horse serum 10% on chlorhexidine against *Pseudomonas aeruginosa* (Hennessy 1973).

morganii. As a control, he inoculated the growth medium with the same population of the isolated organism as found in the pus. Both disinfectants exhibited loss of activity when the pus was present at a 10% concentration in the growth medium (Table 18.5). Hibiclens was active at high dilution both in the absence or presence of pus but lost 81.3% of its effectiveness when 10% pus was present. Betadine was not as effective on a dilution basis as Hibiclens against *Proteus morganii* but lost 50% of its activity in the presence of pus.

Human Sebum and Free Fatty Acids

Parran and Brinkman[8] extracted sebum from skin with lipid solvents and suspended it in medium containing antimicrobial compounds. The test organism was the common skin isolate, *Staphylococcus epidermidis.* The antimicrobials were added to filter paper disks, which were then placed on inoculated agar Petri plates with and without sebum. The endpoint was the lowest concentration of the antimicrobial showing a zone of inhibition of bacterial growth around the disk. The ratios listed in Table 18.6 are the amount of germicide required to inhibit the organism in the plate without

Table 18.5. Effect of Pus on the Activity of Antimicrobial Agents Against *Proteus Morganii* (Sheikh 1978)

Agent	Highest inhibitory dilution	
	Without pus	With pus
Hibiclens	1:400	1:75
Betadine	1:8	1:4

Table 18.6. Effect of Human Sebum on the Activity of Antimicrobial Agents Against *Staphylococcus epidermidis* (Parran and Brinkman 1965)

Germicide	Ratio of inhibitory concentrations without sebum: with sebum
Salicylic acid	1:2
Hexachlorophene	1:1,024
Bithionol	1:256
Benzalkonium Cl⁻	1:8

sebum (listed as 1) to the amount required to inhibit in the presence of sebum. Note that all the germicides were antagonized in the presence of sebum. Hexachlorphene required a 1 024-fold increase over the control plates when sebum was a constituent of the medium. Salicylic acid was least antagonized.

Parran and Brinkman also determined the effect of free fatty acids in a simulated sebum on a series of germicides. The results are gathered in Table 18.7. The presence of the free fatty acids produced a potentiating effect for hexachlorophene, benzalkonium chloride, cetylpyridinium chloride, and bithionol but phenol and sodium undecylinate were antagonized.

Triglycerides and Natural Oils

Saggers and Stewart[9] examined the effect of triglycerides and natural oils on the activity of chlorhexidine and an iodophore against *Staphylococcus aureus* and *Pseudomonas aeruginosa*. They determined the time in minutes required to kill all the organisms present (Table 18.8). When the iodophore was tested, as the length of the carbon chain of the fatty acids increased, the antagonistic effect also increased. This stepwise effect was not as obvious with chlorhexidine. Both coconut oil and olive oil more effectively antagonized chlorhexidine and the iodophore than the pure triglycerides. The activity of the iodophores was reduced to a greater extent than that of the chlorhexidine for both organisms.

Table 18.7. Effect of Free Fatty Acids in Simulated Sebum on Antimicrobial Agents Against *Staphylococcus epidermidis* (Parran and Brinkman 1965)

Germicide	Ratio required to inhibit— no lipid in medium No F.F.A.'S	lipid in medium With F.F.A.'S
Hexachlorphene	1:512	1:256
Benazlkonium Cl⁻	1:128	1:8
Cetypridinium Cl⁻	1:64	1:8
Phenol	1:1	1:2
Na-undecylinate	1:1	1:8
Bithionol	1:256	1:128

Table 18.8. Effect of Fats and Oils on the Activity of Antimicrobial Agents (Sagger and Steward 1964)

	Min. to destroy population			
	S. aureus		*P. aeruginosa*	
Agent	$PVP - I_2$	Chlorhexidine	$PVP - I_2$	Chlorhexidine
Triacetin	2	<0.5	<0.5	<0.5
Tributyrin	3	0.5	1.0	0.5
Triolein	5	0.5	2.0	0.5
Coconut oil	>10	2	3.0	1.0
Olive oil	>10	2	2.0	1.0

Surface Active Agents

Compounds having surface-tension reducing properties may profoundly affect the activity of antimicrobial substances. The activity may be increased or decreased depending on the combinations tested. The organisms used in the tests also affects the results. Gershenfeld and Perlstein[5] examined the interaction of a 1:500 dilution of aerosol OT at pH 7.0 with several antimicrobial compounds. The test organism was *Staphylococcus aureus* (Table 18.9). The inhibitory activity of phenol and zonite were increased in the presence of aerosol OT but hexylresorcinol lost 32.5% of its antimicrobial activity.

Davis et al.[4] looked at a variety of substances, including surface-tension depressants, for their activity on chlorhexidine. Lubrol W at a concentration of 1% was decidedly antagonistic to chlorhexidine but Tween 80 appeared to have no effect (Table 18.10). This latter observation has not been confirmed by more recent reports. The same status applies to the data on egg lecithin (0.1%), which has been shown to have high antagonism to many antimicrobial substances. Other materials that reduced the effectiveness of chlorhexidine included whole milk (10%), rabbit serum (10%), Suramin (1%), and especially egg yolk (1 yolk/100 ml), which exhibited the greatest repressive effect.

A study designed to test the antagonistic effect of combinations of surface-tension-suppressing compounds with egg lecithin or azolectin was reported anonymously (Table 18.11).[1] Lubrol W at a 1% concentration was combined singly with the above substances, and each combination was effective in reducing the inhibitory effect of chlorhexidine against gram-positive and

Table 18.9. Effect of Aerosol OT[a] on the Activity of Antimicrobial Agents Against *Staphylococcus aureus* (Gershenfeld and Perlstein 1941)

	Dilution to sterilize in 10 min	
Agent	Without aerosol OT	With aerosol OT
Phenol	80	200
Hexylresorcinol	4,000	2,500
Zonite	3,000	4,000

[a]Dilution of aerosol OT 1-5000 pH 7.0.

Table 18.10. Effect of Organic Materials on the M.I.C. of Hibitane Against *Staphylococcus aureus* (Davis et al. 1954)

Substance and conc.	Minimal inhibitory concentration
Control	500
Yeast nucleic acid 1%	500
Suramin 1%	200
Milk (whole) 10%	100
Rabbit serum 10%	200
Lubrol W 1%	200
Tween 80 1%	500
Egg lecithin 0.1%	500
Egg yolk (1 yolk/100 ml)	4

[a]Reciprocal of dilution ÷ 1 000

gram-negative bacteria. *Staphylococcus aureus* was least affected by the combinations, while *Bacillus subtilis* exhibited the greatest range of variation. Among the gram-negative organisms, *Escherichia coli* was afforded the poorest protection by the antagonists; *Pseudomonas aeruginosa*, the greatest.

Carbohydrates

Carbohydrates have been most commonly found to effect germicides indirectly by the formation of fermentation acids, with a resulting fall in pH. Bacterial inhibitors most active at alkaline pH's are adversely affected, while those favored by a lowering of the pH are fortified.

Hennessey[6] reported what appears to be a direct effect of sucrose on chlorhexidine with *Streptococcus mutans* as the test organism. In the presence of 0.02% chlorhexidine and 5% sucrose, a population of greater than 1×10^8 of the organism is completely destroyed in slightly more than 60 min while the control lacking only sucrose had less than one log drop in the same period of time (Fig. 18.2). In the work of Davis et al.[4] chlorhexidine loses activity as the pH is lowered (Table 18.12). *S. mutans* can ferment sucrose, with a resulting decrease in pH, but the sugar has an antagonizing effect on the chlorhexidine.

Table 18.11. Effect of Organic Antagonists on Chlorhexidine (Anon 1979)

Organism	Inhibitory concentration, ng/ml		
	Control	Control + 1[a]	Control + 2[b]
S. aureus	1.95	500	125
S. faecalis	3.9	500	500
B. subtilis	3.9	1,000	1,000
E. coli	3.9	500	500
P. aeruginosa	31.25	1,000	1,000
P. vulgaris	78	1,000	1,000

[a]3% egg lecithin + 1% Lubrol W
[b]0.3% Azolectin + 1% Lubrol W

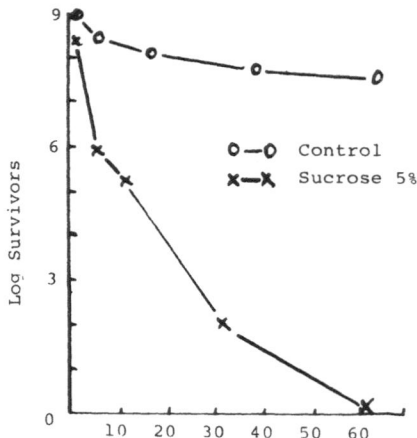

Fig. 18.2. Effect of sucrose on the activity of chlorhexidine against *Streptococcus mutans* (Hennessy 1973).

Summary

Each germicide has a set of optimal conditions under which it best exhibits its ability to inhibit or destroy a microbial population. Rarely are all these conditions present in an "in use" situation where tissue, body fluids, exudative materials, normal flushing action of saliva or mucus are present.

All of the major classification of substances (proteins, lipids, and carbohydrates) alone or in combination may have a profound effect on the efficiency of the antimicrobial compound. Proteins generally and in particular tend to decrease the ability of germicides to inhibit or kill bacteria. Lipids have a broad spectrum of activity from complete antagonism to fortification of the inhibitory activity of the germicide. Carbohydrates have been less well studied in their direct activity on the antimicrobials although direct inhibition has been demonstrated.

Each germicide should be tested under "in use" conditions or in a system that simulates as closely as possible the parameters known to exist on or in tissue.

Table 18.12. Effect of pH on the Activity of Hibitane Against *Staphylococcus aureus* (Davis et al. 1954)

pH	Survivor × 10^3	
	5 Min	10 Min
5.25	1,000	1,000
6.22	2,300	50
7.08	90	5
7.97	17	0.9

References

1. Anon. (1979) Hibitane (chlorhexidine): A review of toxicology, biochemistry and bacteriology. ICI Americas Report
2. Ayliffe GAJ, Collins BJ, Lowbury EJL (1966) Cleaning and disinfection of hospital floors. Br Med J 2:442–451
3. Calman RM, Murray J (1956) Antiseptics in Midwifery. Br Med J 2:200–204
4. Davis GE, Francis J, Martin AR, Rose FL, Swan G (1954) 1:6-Di4' Chlorophenyldiguanidohexane (Hibitane). Laboratory investigation of a new antibacterial agent of high potency. Br J Pharmacol 9:192–196
5. Gershenfeld L, Perlstein D (1941) The effect of aerosol OT and hydrogen ion concentration on the bacteriocidal efficiency of antiseptics. Am J Pharm 113:237–255
6. Hennessey TS (1973) Some antibacterial properties of chlorhexidine. J Peridont Res 12: Supplement, 61–67
7. Lowbury EJL, Lilly HA (1974) The effect of blood on the disinfection of surgeons' hands. Br J Surg 61:19–21
8. Parran JJ, Brinkman RE (1965) The effect of human skin surface lipids upon the activity of antimicrobial agents. J Invest Dermatol 45:89–92
9. Saggers BA, Stewart GT (1964) Polyvinyl-pyrrolidone-iodine: an assessment of antibacterial activity. J Hyg (Camb) 62:509–518
10. Sheikh AW (1978) Antibacterial activity of Hibiclens vs. Betadine in the presence of infected pus. ICI Americas Report (personal communication)

Chapter 19
Topical Antimicrobials: Perspectives and Issues

E. J. L. Lowbury

Topical antimicrobials relate to two distinct situations: (1) the application of antimicrobial agents to unbroken skin, where they should not be appreciably absorbed or exposed to the inactivating effects of blood and exudate; and (2) the application of such agents to burns and open wounds, traumatic or surgical, from which absorption and inactivation is to be expected—also where, by contrast with the situation on unbroken skin, systemically administered chemotherapy can be very effective. The application of antimicrobial agents to mucous membranes and their injection or instillation into hollow viscera are additional situations which fall between those represented by unbroken skin and open wounds.

In Birmingham we have made a number of studies on disinfection of the skin and on topical chemoprophylaxis of burns. As far back as 1951,[8,9] controlled prophylactic trials in our burns unit had shown a significantly lower incidence of infection by *Streptococcus pyogenes, Pseudomonas aeruginosa,* and other bacteria, with better skin-grafting results and shorter healing times, when burns were given topical applications of appropriate antimicrobials than when given no such protection. Improved prophylaxis, introduced later by workers in several centers using silver compounds, mafenide, and chlorhexidine, was shown in our trials to cause bacteriological and clinical benefits, including a reduction in mortality.[14] Though these methods have greatly improved our control of infection, none is entirely satisfactory, and all are subject to difficulties arising from bacterial resistance and from side effects in the patient; fortunately they can be effectively supplemented by other methods.

Disinfection of the Skin

Principles and Objectives

Forty years ago, Price[23] classified skin microflora into two groups: (1) resident organisms that grow on the skin; and (2) transient organisms that are picked up by the skin but do not grow on it. The resident flora are mainly organisms harmless to healthy people and healthy tissues, e.g., micrococci and coryneform bacilli. But in operations on patients or tissues with deficient antimicrobial resistance such commensal organisms can behave like opportunist pathogens. Occasionally *Staphylococcus aureus* is present as a resident, and in damaged or waterlogged skin gram-negative bacilli may become resident. Any organism that can survive after being deposited on the skin may be regarded as a member of the transient flora; some (notably *Streptococcus pyogenes*) are quickly reduced to small numbers either by the action of unsaturated fatty acids or through the lethal effects of drying (e.g., gram-negative bacilli). See Chapter 4.

Microbes on the skin may cause *self-infection*, due to contamination of an operation wound with skin bacteria at the operation site, or *cross-infection* by various channels (e.g., a nurse may transfer pathogens that her hands have picked up from one patient to another patient—these organisms are transients, for even if they belong to species capable of colonizing the healthy skin, they could not do so in the time that it takes to move from one patient to the next). A surgeon whose glove develops a pinhole leak (which happens in 20%–30% of the rubber gloves used during operations) may squeeze resident or transient bacteria through the hole into the operation wound; as transients are more readily removed by washing than are the residents, the latter are the ones most likely to be transmitted in this way during an operation.

A disperser of *S. aureus* carries these organisms as part of his resident flora and disperses them into the air, where they become a potential source of airborne infection. Clearly there is a need to remove, or at least greatly reduce, the transient flora on nurses' hands, and the resident as well as the transient flora on the hands of surgeons, the skin of operation sites, and the bodies of other dispersers working in hospitals.

The numbers of bacteria on the skin may be reduced (a) by removing them physically—mainly on desquamating epithelium—with a detergent and water (*skin cleansing*), and (b) by killing them on the skin through the application of bactericidal agents or antiseptics (*skin disinfection*). Transients can be removed more readily than residents by disinfection or by cleansing. Indeed, cleansing with soap and water is so inadequate against resident flora that the practice of combined cleansing and disinfection does not seem to have any advantage over simple disinfection, unless the hands are physically dirty; it has the disadvantage of making it necessary to dilute and wash away the antiseptic with fresh running water poured on to the hands from a tap to achieve detergent cleansing.

Removal of Transient Organisms

It is commonly assumed that a suspension of bacteria spread on the skin provides a satisfactory model of transient bacteria from which the effects of cleansing or disinfection can be assessed; this method has been used by a number of workers.[2,4,7,19,21,24,25] A recent study in our unit, however, has shown that the effectiveness of cleansing is much smaller against bacteria which have been rubbed on to the skin than it is against those which have been laid on the skin and allowed to dry.[10] In these experiments the suspension of bacteria was applied on the four fingertips of each hand after thoroughly cleansing them) and either allowed to dry or rubbed against the same finger of the other hand; the corresponding fingertips of the two hands were treated as a unit. Viable counts were made from washings obtained by rubbing the fingertips for 3 min against glass beads in tubes of Ringer's solution with neutraliser. A mean percentage reduction of 70% was obtained by a 30-s wash with soap and water against bacteria rubbed into the skin, compared with a reduction of 98% against bacteria deposited and allowed to dry on the skin.

By contrast with these effects of detergent cleansing, rubbing an antiseptic (70% ethanol) on hands with fingertips deliberately contaminated caused as great a reduction in the yield of skin bacteria when the bacteria had been rubbed onto the skin as when they had been spread and allowed to dry; in both cases the effect of disinfection (about 99.7% reduction) was much greater than the effect shown by detergent cleansing (Table 19.1).

Paradoxically, the use of a detergent antiseptic preparation and water from a tap gave better removal of bacteria that had been rubbed onto the skin than of those that had been allowed to dry on the surface. This was probably due to the longer time of exposure to undiluted antiseptic of the bacteria that were rubbed on to the skin, compared with the time of exposure of those lying on the surface. In agreement with this hypothesis, the use of an antiseptic detergent preparation without addition of water from a tap gave similar results, whether the bacteria were dried on or rubbed onto the skin.

Removal of Resident Organisms: Disinfection of Surgeon's Hands

In our comparisons of various antiseptics and antiseptic detergents we used a hand-wash viable-count sampling technique, the hands being rinsed in a standard way with Ringer's solution containing 1% Tween 80, 1% Lubrol W, and 0.5% lecithin to prevent inhibition by antiseptic carried over—tests

Table 19.1. Effects of Soap and Water and of Disinfection on Transient Flora Applied with and without Friction. Mean Log Reduction Factors (Log Pretreatment Count Minus Log Posttreatment Count) (Lilly and Lowbury 1978)

Treatment (method of cleansing or disinfection)	Bacterial suspension		Significance of difference
	Dried on	Rubbed on	
Soap and water	1.76	0.53	$t = 9.439$ $p < 0.001$
70% ethyl alcohol	2.62	2.60	$t = 0.188$ $p > 0.1$ (NS)
Chlorhexidine detergent with water	1.35	2.07	$t = 5.557$ $p < 0.001$
Chlorhexidine detergent without added water	1.96	1.81	$t = 1.174$ $p > 0.1$ (NS)

for carry-over were also made.[15,20] Groups of volunteers who took part in the study used each agent on a different day. At least 7 days were allowed to elapse between the end of one experiment and the beginning of the next, so that the natural skin flora could return to its equilibrium density. In studies on antiseptic detergent preparations after the pretreatment sample, the hands were disinfected and cleansed by a thorough 2-min wash (without a brush) covering the whole area, with care to avoid leaving small areas untreated, and using the agent in the manner recommended by the manufacturer with addition of running warm water from a tap. The hands were rinsed and dried before the immediate post-treatment sample was taken, and the effects of repeated disinfection were tested by similar samplings after six successive hand preparations, three on one day and three on the next.

An alternative method of disinfection by 70% or 95% alcohol, with or without addition of 0.5% chlorhexidine, involved rubbing 10 ml two lots of 5 ml) of the solution (which also contained 1% glycerol to prevent excessive drying of the skin) into the hands until they were dry.[1,18] A Latin-square statistical design was adopted, each agent or method being used by each volunteer and all the agents or methods being used on each experimental day.

Several antiseptic detergent preparations caused a progressive reduction of 99% or more in the yield of natural skin flora after six successive treatments: these included detergent preparations of 4% chlorhexidine (Hibiclens), 10% povidone iodine (Betadine, Disadine), triclosan (Zalclense) and 2% or 3% hexachlorophene (pHisohex, Ster-Zac). After a single application the greatest mean percentage reduction was obtained with the chlorhexidine detergent preparation. Even larger reductions were obtained after a single application by the use of 0.5% chlorhexidine in concentrated (70% or 95%) ethanol, and by ethanol alone, if 10 ml of the fluid was rubbed vigorously into the skin of both hands and forearms until they were dry.[18] This method is suitable for physically clean hands, but when hands are soiled (e.g., with blood or pus) they should be washed with a detergent or antiseptic detergent preparation and water.

Progressive reduction in the yield of the natural skin flora on repeated disinfection leads to the establishment of an equilibrium between removal of bacteria from the surface and the emergence of bacteria from the deeper layers. This was convincingly shown by an experiment in which different antiseptic preparations were compared with each other.[11] The reduction in numbers of bacteria on repeated disinfection, beyond which there was no further reduction on further disinfection, varied with the antiseptic; when a weak preparation (e.g., chlorocresol detergent solution) was used, the density of skin bacteria shown at equilibrium was much greater and the reduction in numbers from the initial level was much smaller than it was with highly effective preparations, such as alcoholic chlorhexidine (Hibiclens).

When a nonantiseptic detergent wash was used on the hands after treatment with an antiseptic had brought the yield of bacteria on sampling down to equilibrium level, there was a sharp *increase* in the numbers of bacteria yielded on sampling (Figs. 19.1–19.3). This illustrates how far we are from

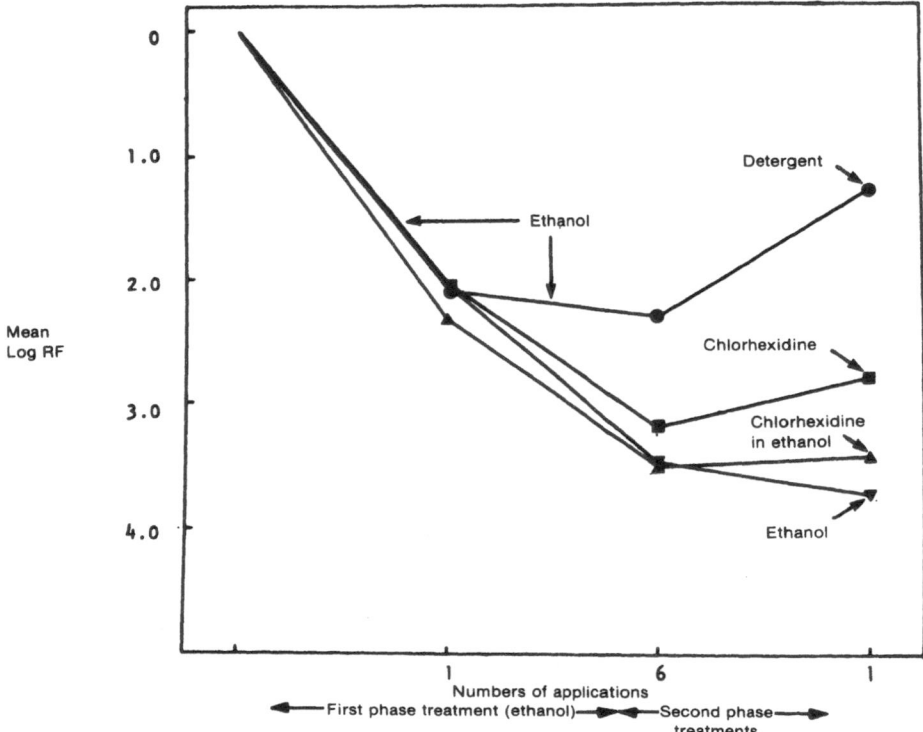

Fig. 19.1. Mean log reduction factors of bacteria in samplings from hands after one and six treatments with 95% ethanol, and after an immediate further treatment with (1) 95% ethanol (2) 0.5% chlorhexidine in 95% ethanol, (3) chlorhexidine detergent, and (4) a nonantiseptic detergent solution (the base of chlorhexidine detergent); four volunteers were used in each group.

sterilizing the skin, even on repeated use of the most effective agents. These and other experiments showed that while it was possible to bring about a further reduction in yield of skin bacteria by disinfecting with alcoholic chlorhexidine after reaching the low equilibrium level with chlorhexidine or another antiseptic detergent (two-phase disinfection[15,22]), it was not possible to cause a further reduction by using chlorhexidine on hands that had reached the low equilibrium attainable with alcohol or alcoholic chlorhexidine. A smaller number of treatments was required to reach equilibrium with the alcoholic preparations than with antiseptic detergent preparations.

The Operation Site

Tests for the yield of bacteria on sampling the surface of the skin are appropriate in assessing the disinfection of surgeons' or nurses' hands. As regards disinfection of the operation site, it is be more relevant to find how big a reduction in the actual skin flora can be obtained by such measures. But

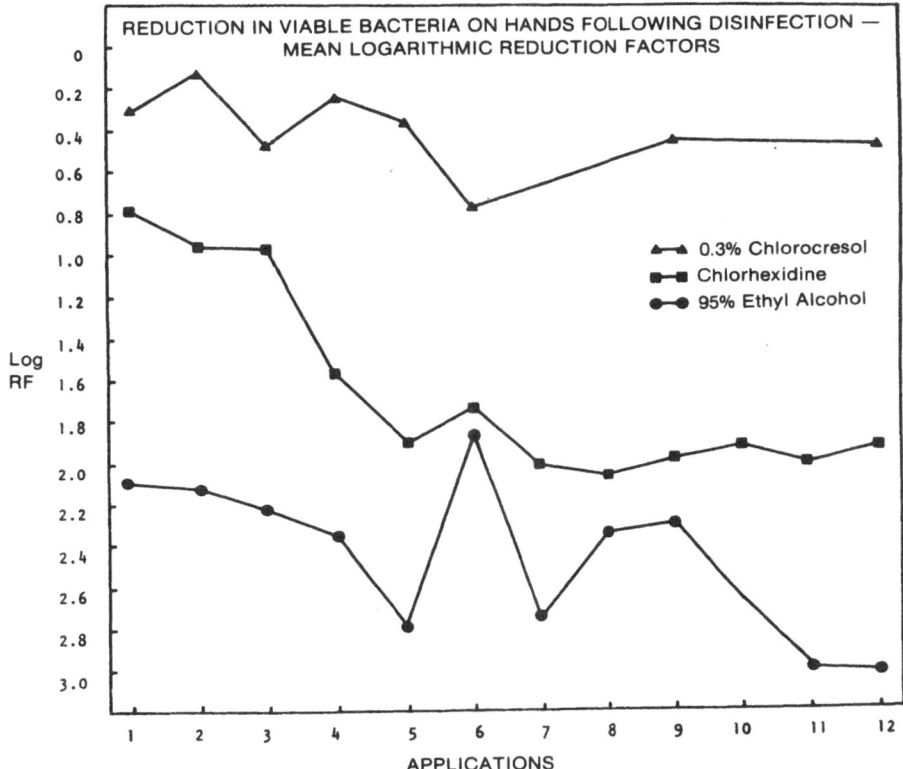

Fig. 19.2. Mean percentage reduction in yield of skin bacteria on repeated treatment of hands with soap, chlorhexidine detergent, and chlorocresol (0.3%) detergent solution and water, and with 95% ethanol.

direct measurement by bacterial counts on skin biopsy is ethically unacceptable in human subjects, as a pretreatment biopsy would be required, and, in any case, a biopsy could only be performed on a small and not necessarily representative area of skin. However, there is probably some association between the reduction, after disinfection, in the numbers of bacteria in the skin and the reduction in the yield of bacteria on sampling, and tests by surface sampling methods are generally accepted as giving useful information on the relative effectiveness of different methods of operation-site disinfection. We have therefore used a technique of assessment for disinfection of the operation site similar to that which we used in assessing disinfection of the hands.[19,22]

Previously, I showed that a 2-min application on gauze swabs of 0.5% chlorhexidine or of 1% iodine in 70% ethanol caused about 80% reduction in the yield of skin bacteria. This was significantly greater than the reduction obtained with any of the other agents tested, but much smaller than the effects of the same 0.5% alcoholic chlorhexidine when this was used for disin-

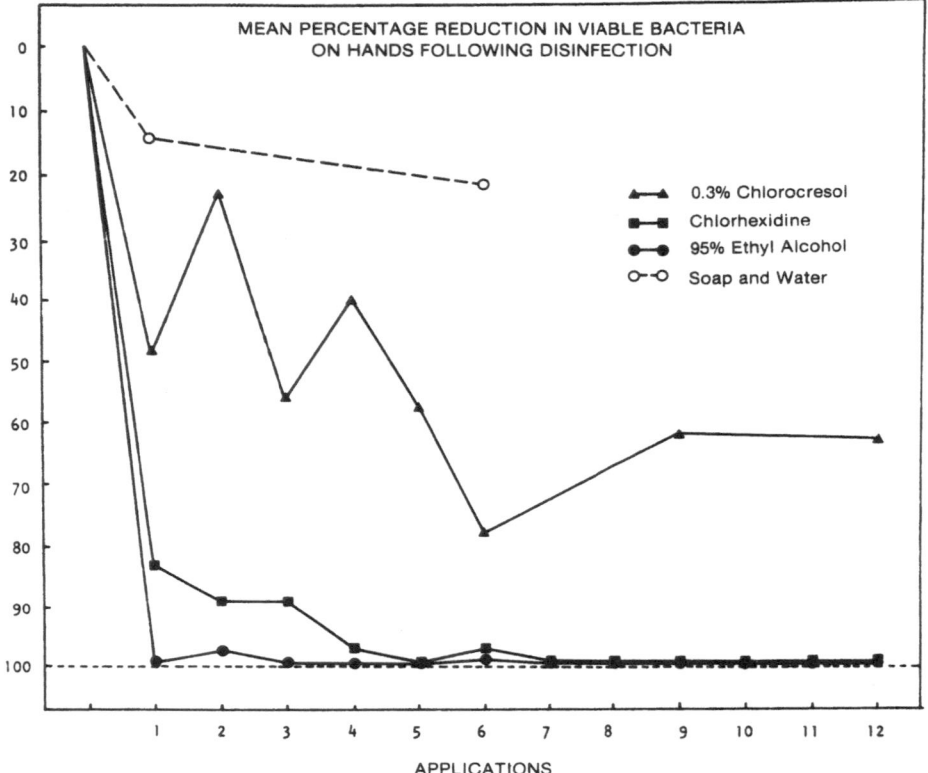

Fig. 19.3. Log reduction factors of yield of skin bacteria after 1-12 treatments with chlorhexidine and chlorocresol (0.3%) detergent solution and water, and with 95% ethanol.

fection of the hands by the method I have described (i.e., rubbing the solution into the skin by hand-to-hand friction). This difference is clearly due to differences in the mode of application; the greater the friction, the more effective the disinfection[17] (see Table 19.2). In earlier experiments, rubbing the antiseptic on with gauze was shown to cause better disinfection than spraying the antiseptic on to the skin.[22]

Table 19.2. Comparison of Gauze with Gloved Hand as Applicator for Disinfection of Skin (Lowbury and Lilly 1975)

Antiseptic	Mean percentage reduction in yield of skin bacteria on sampling after disinfection of hand			
	With gloved hand	With gauze swabs	t	p
0.5% chlorhexidine in 70% ethanol	98.8 ± 0.84	90.7 ± 2.12	3.58	< 0.01
0.5% chlorhexidine in 95% ethanol	99.9 ± 0.024	93.5 ± 2.15	2.98	< 0.02
0.5% chlorhexidine in distilled water	83.5 ± 5.82	53.7 ± 5.23	3.80	< 0.01

In these experiments, for convenience the hand was used as the test site for disinfection. Other sites, however, can behave differently. For example, Ayliffe[6] has studied disinfection of the abdominal skin and found reductions of 99% or more with aqueous as well as with alcoholic solutions of chlorhexidine or povidone iodine, and no significant difference between 70% alcohol and 70% alcohol containing a nonvolatile antiseptic 0.5% chlorhexidine.

Residual Action

The continuing effects of residues of antiseptic left on the skin after disinfection are commonly cited as an argument in favor of non-volatile antiseptics like chlorhexidine or hexachlorophene. The effect can easily be demonstrated by putting a suspension of bacteria on the skin after disinfection or (in control subject) after washing with soap and water. Bacterial counts from surface samples obtained an hour later showed few bacteria remaining on the skin disinfected with detergent or alcoholic solutions of nonvolatile antiseptic.[15,18,21]

Cole and Bernard[3] have shown some inactivation of the residual action of hexachlorophene when blood seeps into a surgeon's glove through pinhole punctures. We confirmed such an effect with hexachlorophene and with povidone iodine. but there appeared to be no inactivation by blood of the residual effects of chlorhexidine.[16]

During the course of a 3-h mock "operation" following skin disinfection there was a further fall in the yield of bacteria from the gloved hand, compared with numbers of bacteria yielded immediately after disinfection.[18] This applied not only to nonvolatile antiseptics that left a residue on the skin, but also to ethanol, which leaves no residue. The paradox of residual action by a volatile alcohol is apparently due to the death, during 3 h wearing of gloves, of many bacteria which can be resuscitated if they are transferred to culture medium immediately after exposure to alcohol.[12]

Disinfection in the Bath

Volunteers at our Hospital Infection Research Laboratory compared the use of several antiseptic detergents and, as a control, ordinary soap in four successive daily baths. Contact plate counts showed a considerable reduction in immersed areas when a 4% chlorhexidine detergent solution was used; the effects were, predictably, less good than those obtained when the same preparations were used for hand cleansing (Fig. 19.4).[5]

Spores

Of the skin disinfection agents mentioned, only one (iodine) has been shown to have some value in killing bacterial spores on the skin, and only when it is allowed to remain on the skin for 15–30 min. To avoid irritation of the skin, which occurs when alcoholic or Lugol's iodine is applied for such a period, an aqueous povidone iodine preparation should be used as a compress.[18,22]

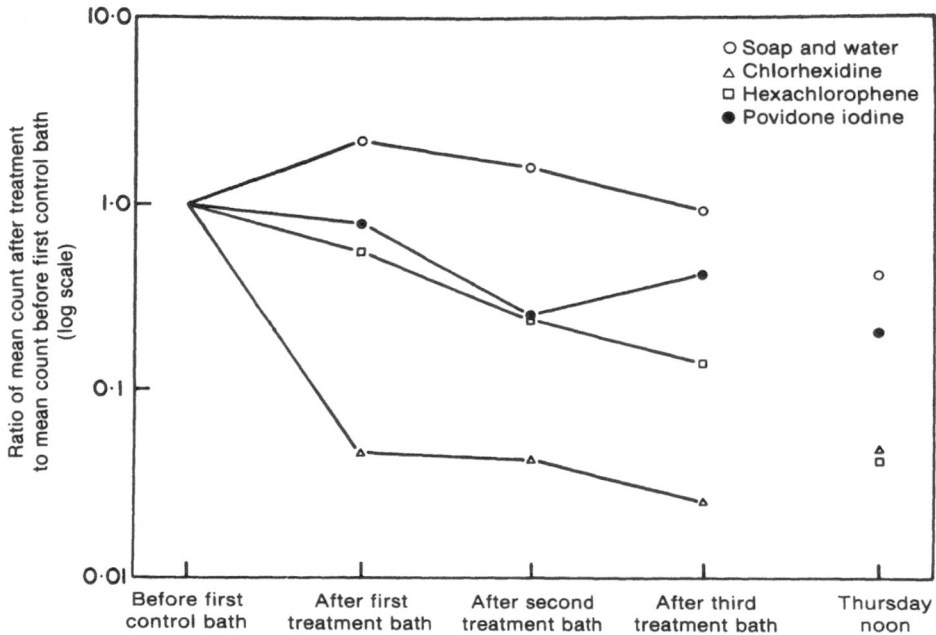

Fig. 19.4. Effect of soap and antiseptic preparations on bacterial counts from skin of inner side of thigh.

Risks of Inadequate Hand Cleansing

By the use of alcohol containing a dye that stained the skin, my nursing colleague, Miss Lynda Taylor, has shown that nurses commonly fail to cover the whole area of skin (particularly the tips of their fingers) when they disinfect their hands by a "free style" of hand washing. This error can be avoided by a systematic coverage of all areas during the wash.

Comments

The clinical significance of skin disinfection was illustrated in the 1840s (before the era of bacteriology) by Semmelweis, when he introduced the practice of hand washing with chlorinated lime for students before going to work in maternity wards after they had been working in the anatomy dissecting rooms. Following this innovation, there was a dramatic fall in mortality from puerperal sepsis. Many years ago we showed, at Birmingham Accident Hospital, a near-significant reduction in major sepsis after introducing skin disinfection methods for the surgeon's hands and for the operation site that had previously been found, on bacteriological criteria, to be more effective than the discarded methods.[13] Unfortunately this was a sequential study, not a controlled trial, so it is possible that other factors might have contributed to the lowering of the sepsis rate.

Because of the ethical objection to a controlled trial, the results of bacteriological assessment must be accepted as providing the most practical guidance in our choice of skin disinfecting methods in surgery. Moreover, since bacteria can reach wounds from many sources through many channels, even a controlled trial might well fail to provide conclusive evidence of the clinical value of skin disinfection, or of the relative value of different forms of skin disinfection, when its role is to block one of a number of collateral channels of access. However, convincing data has come from the adoption or the abandonment of methods of disinfecting the skin of neonates, which has led, respectively, to a reduction and a rise in staphylococcal sepsis.

Skin disinfection must be regarded as a rational component of asepsis and hygiene in hospitals, and it should be used rationally. In the case of nurses' hands, the removal of recently acquired transient organisms is the main requirement; for this, washing with soap and water has considerable value, but disinfection with alcohol or alcoholic chlorhexidine is much more effective.[1] The choice of method can reasonably be related to the degree of hazard, disinfection being reserved for high-infection-risk patients. Surgeons' hands may contaminate wounds with resident organisms that escape through holes in rubber gloves, and for the removal of these some form of disinfection—by alcoholic or detergent solutions required. Though most residents are what we used to dismiss as commensals, such organisms can be opportunist pathogens in immunodeficient patients or tissues, and *S. aureus* can sometimes behave as a resident, so it is rational to use effective antisepsis for the surgeons' preoperative hand preparation. Disinfection of operation and venepuncture sites is also rational. Whole body disinfection in the bath is a rational, though imperfect, procedure for members of hospital staff who are dispersers of *S. aureus*; it is more important that such persons should be kept out of aseptic areas. The indications for disinfection at sites of subcutaneous injection are more marginal, but this, like many procedures, may be regarded as an insurance policy against a sporadic, and probably remote, hazard. Some would call it a ritual, especially when the quick but usual "dab and jab" technique is used. But if one abandons it, one must, at least, make sure that the skin is physically clean before injecting.

References

1. Ayliffe GAJ, Babb JR, Bridges K, Lilly HA, Lowbury EJL, Varney J, Wilkins MD (1975) Comparison of two methods for assessing the removal of total organisms and pathogens from the skin. J Hyg Camb 75:259–274
2. Ayliffe GAJ, Babb JR, Quoraishi AH (1978) A test for "hygienic" hand disinfection. J Clin Pathol 31:923–928
3. Cole WR, Bernard HR (1964) Inadequacies of present methods of surgical skin preparations. Arch Surg 89:215–222
4. Colebrook L, Maxled WR (1933) Antisepses in midwifery. Br J Obstet Gynaecol 40:966–990
5. Davies J, Babb JR, Ayliffe GAJ, Ellis SH (1977) The effect on the skin flora of bathing with antiseptic solutions. J Antimicrob Chemother 3:473–481
6. Davies J, Babb JR, Ayliffe GAJ, Wilkins MD (1978) Disinfection of the skin of the abdomen. Br J Surg 65:855–858

7. Gardner AD (1948) Rapid disinfection of clean unwashed skin. Lancet II:760–763
8. Jackson DM, Lowbury EJL, Topley E (1951) *Pseudomonas pyocyanea* in burns. Its role as a pathogen, and the value of local polymyxin therapy. Lancet II:137–147
9. Jackson DM, Lowbury EJL, Topley E (1951) Chemotherapy of *Streptococcus pyogenes* infection of burns. Lancet II:705–711
10. Lilly HA, Lowbury EJL (1978) Transient skin flora. J Clin Pathol 31:919–922
11. Lilly HA, Lowbury EJL, Wilkins MD (1979) Limits to progressive reduction of resident skin bacteria by disinfection. J Clin Pathol 32:382–385
12. Lilly HA, Lowbury EJL, Wilkins MD, Zaggy A (1979) Delayed antimicrobial effects of skin disinfection by alcohol. J Hyg 82:497–500
13. Lowbury EJL (1965) Removal of bacteria from the operation site. In: Maibach HI, Hildick-Smith G (eds), McGraw-Hill, New York pp 263–275
14. Lowbury EJL (1979) Wits versus genes: the continuing battle against infection. J Trauma 19:33–45
15. Lowbury EJL, Lilly HA (1973) Use of 4 percent chlorhexidine detergent solution (Hibiscrub) and other methods of skin disinfection. Br Med J I:510–515
16. Lowbury EJL, Lilly HA (1974) The effect of blood on disinfection of surgeons' hands. Br J Surg 61:19–21
17. Lowbury EJL, Lilly HA (1975) Gloved hand as applicator of antiseptic to operation sites. Lancet II:153–156
18. Lowbury EJL, Lilly HA, Ayliffe GAJ (1974) Preoperative disinfection of surgeons' hands: use of alcoholic solutions and effect of gloves on skin flora. Br Med J IV:369–372
19. Lowbury EJL, Lilly HA, Bull JP (1960) Disinfection of the skin of operation sites. Br Med J II:1039–1044
20. Lowbury EJL, Lilly HA, Bull JP (1963) Disinfection of hands: removal of resident bacteria. Br Med J I:1251–1256
21. Lowbury EJL, Lilly HA, Bull JP (1964) Disinfection of hands: removal of transient organisms. Br Med J II:230–233
22. Lowbury EJL, Lilly HA, Bull JP (1964) Methods for disinfection sites. Br Med J II:531–536
23. Price PB (1938) The bacteriology of normal skin; a new quantitative test applied to a study of the bacterial flora and the disinfectant action of mechanical cleansing. J Infect Dis 63:301–318
24. Rotter M, Mittermayer H, Kunde M (1974) Untersuchungen zum modell der Künstlich Kontaminierten Hand-Vorschlag für eine Prüfmethode Zentralbl Bakteriol (Orig B) 159:560–581
25. Story P (1952) Testing of skin disinfectants. Br Med J II:1128–1130
26. Taylor LJ (1978) An evaluation of handwashing techniques-1 Nurs Times 74:54–55 An evaluation of handwashing techniques-2 Nurs Times 74:108–110

Chapter 20
Staphylococcus Aureus Adherence to Nasal Epithelial Cells: Studies of Some Parameters

RAZA ALY, H.R. SHINEFIELD, AND HOWARD I. MAIBACH

In many infectious processes, the initial event is the binding of organisms to host epithelial cells. The capacity to colonize the mucosa is proportional to the ability of bacteria to adhere.[7,12] In the past, the emphasis has been on understanding the factors involved in the pathogenesis of a disease, i.e., after the organism has invaded the host and clinical symptoms have become apparent. While the pathogenic aspects are eminently worthy of scientific inquiry, the mechanisms involved in the initial binding of bacteria to the epithelia were not seriously investigated. The lack of knowledge in this field has been partly due to the lack of suitable models for studying the natural ecology of several pathogenic bacteria. Some experimental animals are susceptible to infection by human pathogens and, as a result, most of the studies in this field have relied upon systemic infections caused by intravenous or interperitoneal inoculation of the pathogens in question. Such studies have not taken into consideration the natural routes by which pathogenic organisms usually initiate an infectious cycle. Consequently, this aspect of biology has not been well-studied. The current in vitro model utilizing isolated epithelial cells and bacteria represents procedures that most closely resemble the human system, thus enabling us to study the bacterial and host factors independently.

For bacteria to colonize the mucosal surfaces, they must first become firmly attached and then proliferate under existing conditions. Consequently, it is logical to assume that their colonization may be influenced by parameters that affect not only their growth, but also their adherence. Attempts to prevent colonization require an understanding of bacterial factors, their interaction with each other (bacterial interference), and the involved host factors. If the mechanism of bacterial adherence to mucosal surfaces is understood, it should be possible to prevent colonization before the bacteria have had the chance to damage the host.

Bacteria vary in their ability to attach to epithelial cells. *Streptococcus salivarius* and *Streptococcus sanguis* found abundantly on the oral epithelial surfaces demonstrate greater affinity for these cells.[7] In contrast, *Streptococcus mutans*, which are found in small numbers on oral epithelial sur-

faces, exhibit feeble adherence. *S. mutans* prefer teeth surfaces.[7] Studies
with a variety of animal cells have established that fimbriate gonococci are
significantly more adhesive than nonfimbriate gonococci.[13] Strains of *Neisseria gonorrhoeae* that adhere strongly to mucosal surfaces are more virulent than strains that adhere less well.[11]

Microbial flora in the intestinal canal selectively colonize the mucosal
epithelia.   Lactobacillus harbor on the keratizing stratified squamous
epithelial cells of the nonsecreting portion of the stomach in normal mice,[12]
while the columnar epithelium of the secreting portion of the stomach are
colonized by yeast of the genus *Torulopsis*. The latter appear to adapt and
grow well in the stomach mucin.

Bacterial Adherence to Adult Nasal Epithelial Cells

Recently, we have developed methods by which bacterial adherence to
human nasal mucosal cell was demonstrated.[3] We selected the nasal
epithelial cells because for many bacteria, particularly *S. aureus*, the nose is
a site for multiplication and dissemination. Vulvar skin is another reservoir
for *S. aureus*.[1]

Nasal epithelial cells were collected by gently scraping epithelia from the
anterior nares with a wooden sterile applicator. The epithelial cells were
washed in phosphate buffer on a Millipore filter apparatus. The epithelial
cell (1×10^3 cells/ml) and *S. aureus* (18 h culture 1×10^8 bacteria/ml) were
incubated together at 35°C for 90 min. After incubation, the cell mixture
was washed free of unattached bacteria. Direct smears were prepared from
epithelial cell suspension and stained for 15 s with gram crystal violet. The
number of bacteria adhering to epithelial cells were examined under the light
microscope. Fig. 20.1 demonstrates adherence of *S. aureus* under electron
microscope.

We demonstrated the selective ability of bacteria to adhere to nasal
mucosal cells. Significant adherence occurred with *Pseudomonas aeruginosa, Staphylococcus epidermidis, S. aureus, Streptococcus pyogenes,* and
diphtheroids but less with *Streptococcus mutans,* and *Klebsiella pneumoniae* (Table 20.1). Staphylococci, which constitute the major flora of the anterior nares, possess a distinct advantage over *S. mitis* and *S. mutans*. They
are the predominant flora of the buccal mucosa and teeth. The persistent
nasal carrier status of *S. aureus* was investigated. The adherence of *S.
aureus* was significantly greater ($p < 0.005$) for the carriers of *S. aureus* than
that for the noncarriers, i.e., 132 ± 82 for carriers and 67 ± 70 for noncarriers. This suggested that the greater affinity of bacteria to mucosal cells of
staphylococcal carriers might be a property of the mucosal cells or host environment rather than of the bacteria, since *S. aureus* was a common denominator in both carriers and noncarriers. By utilizing this model, we may gain
further insight in determining why some people become staphylococcal
carriers and others do not.

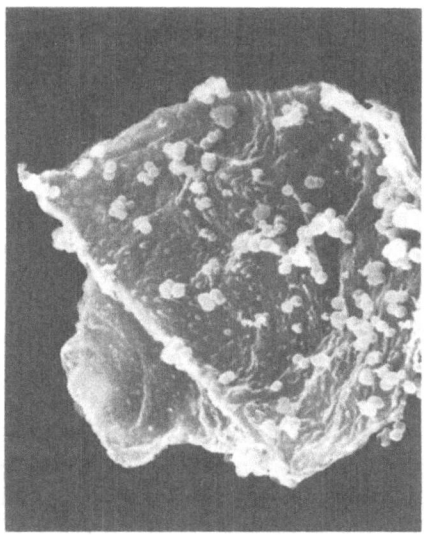 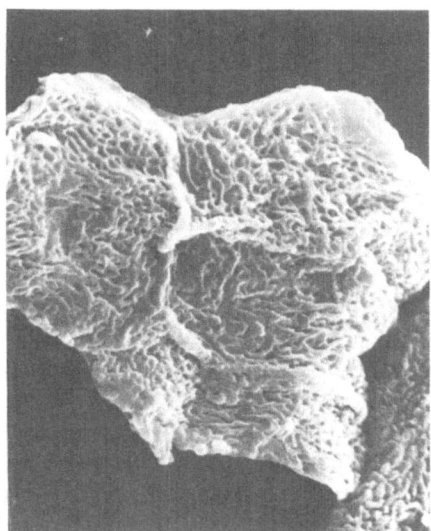

*Fig. 20.1.***a** *S. aureus* adherence to nasal epithelial cells (× 4700); **b** control, epithelial cells were not mixed with *S. aureus*. Scanning electron microscopic studies (× 5900).

Bacterial Adherence in Dermatitic Skin

S. aureus colonization is extremely high in patients with atopic dermatitis.[2] The high prevalence of *S. aureus* is noted not only in eczematous skin but also in the noninvolved skin of the adjacent area and the anterior nares. Although not as high as in atopics, high *S. aureus* counts are also noted in

Table 20.1. Selective Adherence of Bacteria to Nasal Epithelial Cells (Adapted from Aly et al. 1977)

Bacteria	Average bacterial count/cell	Average background count/cell[b]
S. aureus (20)[a]	53 ± 24	5 ± 2
S. epidermidis (5)	58 ± 16	3 ± 2
S. pyogenes (5)	120 ± 85	3 ± 2
S. mitis (5)	10 ± 2	3 ± 2
S. mutans (5)	3 ± 1	3 ± 2
S. salivarius (5)	19 ± 9	3 ± 2
Diphtheroids (6)	47 ± 41	2 ± 1
K. pneumoniae (6)	4 ± 3	2 ± 1
P. aeruginosa (6)	153 ± 92	2 ± 1

[a]The number in parenthesis is the number of subjects used.
[b]The mucosal cells were treated with phosphate buffer instead of bacteria.

Table 20.2. Adherence of *S. Aureus* to Epithelial Cells in Normal, Psoriatic, and Atopic Populations

Epithelial cells[a]	Average *S. aureus* counts /cell			
	Normal	Psoriatic	Normal	Atopic
Nasal	54	60	48	71
	P > 0.48		P < 0.05	
Skin (forearm)	27	33	38	51
	P > 0.16		P < 0.05	

[a]Fifty epithelial cells were counts for adherence in each group. Epithelial cells not treated with *S. aureus* (buffer saline) had less then 1.5 bacterial counts/cell.

psoriatic plaques. *S. aureus* rarely becomes part of the resident flora on normal skin.

We examined epithelial cells of patients with eczema or psoriasis for their binding capacity to *S. aureus*. The adherence of *S. aureus* to the epithelial cells of atopic subjects was significantly greater when compared with the normal subjects (Table 20.2). However, the difference in adherence between psoriatic epithelial cells and normal epithelial cells was not significant. The greater affinity of *S. aureus* to epithelial cells obtained from eczema patients correlates with the high density of *S. aureus* on the skin of these patients. Further studies are required in this area.

Bacterial Adherence in Infants

We know that bacterial colonization of the human oral cavity is at a low level for the first 24 h after birth and increases rapidly after this period.[8,14] The reason for this paucity of bacteria during early neonatal life is not known.

Ofek[9] demonstrated that the binding capacity of buccal epithelial cells to group A streptococci was minimal on days 1 and 2 and reached the adult level on day 3. This reduced binding capacity of epithelial cells has been attributed to immature receptor sites or other host factors.

Similarly, adherence of *S. aureus* to nasal mucosal cells is markedly low during the first 4 days of life (P < 0.001), reaching adult levels on the 5th day. The percent adherence of *S. aureus* to infant nasal epithelial cells was compared with that to adult nasal epithelial cells (Table 20.3). Adult nasal epithelial samples were obtained simultaneously for comparison. A reduced binding of S. aureus to nasal epithelial cells during the first 4 days of neonatal life and a rapid increase towards adult level on day 5 was demonstrated.

It is too early to speculate on the mechanism of reduced adherence of neonatal epithelial cells to *S. aureus*. The epithelial receptor sites for staphylococcal binding have not been clearly defined and our knowledge of physiological conditions that interfere in the interaction of bacteria and epithelial cells is lacking. It has been suggested in other situations that the reduced binding of buccal mucosal cells to groups A and B streptococci is due to the scant capacity of neonate epithelial cells to bind lipoteichoic acid.

Table 20.3. Binding of *S. Aureus* by Nasal Epithelial Cells from Newborns at Various Ages[a]

Age of infants (hours)	Percent adherence[b]
1 (24 h ± 3)	22 P < 0.001
2 (48 h ± 4)	25 P < 0.001
3 (74 h ± 4)	38 P < 0.001
4 (96 h ± 3)	35 P < 0.001
5 (120 h ± 4)	98 P > 0.873

[a] Fifty epithelial cells were counted for each age group.
[b] Adherence of *S. aureus* to adult nasal mucosal cell level was considered as 100%.

Streptococci binding appears to be due to lipoteichoic acid found on the surface of the organisms.[5]

A better understanding of the mechanisms involved in the adherence of *S. aureus* to infant nasal epithelial cells should provide useful information regarding colonization of the mucous membranes in newborns.

Role of Teichoic Acid in Staphylococcal Adherence

Teichoic acids (TA) are major cell-wall components of staphylococci. This compound is made up of a phosphate and ribitol backbone with D-alanine and N. acetyl D-glucosamine side chains. The molecule itself is composed of eight repeating units of ribitol phosphate (Fig. 20.2).

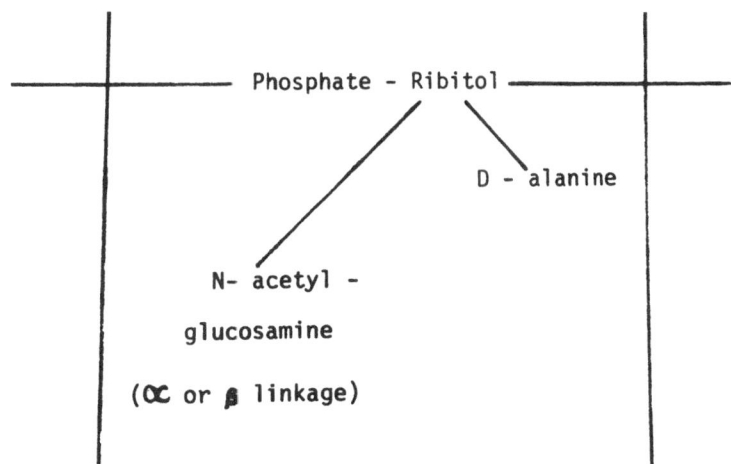

Fig. 20.2. Teichoic acid consists of polymers of ribitol connected by phosphate diester linkage, with side chain of D-alanine and varying proportions of N-acetylglucosamine. The polymer is composed of 8 repeating units of ribitol phosphate.

Table 20.4 Bacterial Adherence to Nasal Mucosal Cells[a]

Epithelial cells mixed with:	Bacterial Counts		Inhibition of adherence %
	Test	Control[b]	
Cell walls from *Staphylococcus aureus* and *S. aureus*	33	65	49 p < 0.01
Teichoic acid and *S. aureus*	13	45	71 p < 0.001
Lipoteichoic acid (from streptococci) and *S. aureus*	18	45	60 p < 0.001
Teichoic acid and *Streptococcus pyogenes*	61	74	17 p > 0.36

[a] Fifty epithelial cells for each treatment were counted.
[b] Epithelial cells were mixed with test organisms only.

We investigated the role of TA in the binding capacity of nasal epithelial cells to *S. aureus*. TA was extracted according to the methods of Baddily et al.[4] Nasal epithelial cells washed with phosphate buffer saline were preincubated for 30 min at 35°C with 1 mg/ml of TA fraction or lipoteichoic acid (LTA) of group A streptococci. The control epithelial cells were treated with phosphate buffer only. The treated epithelial cells were mixed with 10^8 colony-forming units/ml of bacteria as described previously.[2] Smears were prepared from bacterial epithelial cell preparations stained with crystal violet and examined under the light microscope for adherence. The adherence of *S. aureus* to epithelial cells pretreated with whole cell wall of *S. aureus* was also investigated.

Epithelial cells treated with TA and LTA demonstrated reduced binding to *S. aureus*; 71% and 60% reduction respectively (Table 20.4). Epithelial cells that had been treated with whole cell wall demonstrated 49% reduction of *S. aureus* binding when compared with the controls. Minimal reduction in the binding ability of group A streptococci was noted when the epithelial cells were treated with TA (17% reduction). This suggests that LTA combines with the epithelial cell receptor sites, which are common to both *S. aureus* and group A streptococci, thus reducing their binding to these cells; while TA is specifically bound to the receptor sites for staphylococci and not for streptococci. As a result, TA treatment of epithelial cells did not significantly reduce streptococcal binding.

We have provided evidence that LTA or TA binds to nasal epithelial cells and inhibits *S. aureus* adherence to these cells. The data suggests that TA on the *S. aureus* surface mediates the binding of these organisms to epithelial cells.

Electron Microscopic Studies

Electron microscopic studies with streptococci and various gram-positive studies have demonstrated the presence of polysaccharide fibers on the surface of bacteria. These have been shown to be involved in the binding of

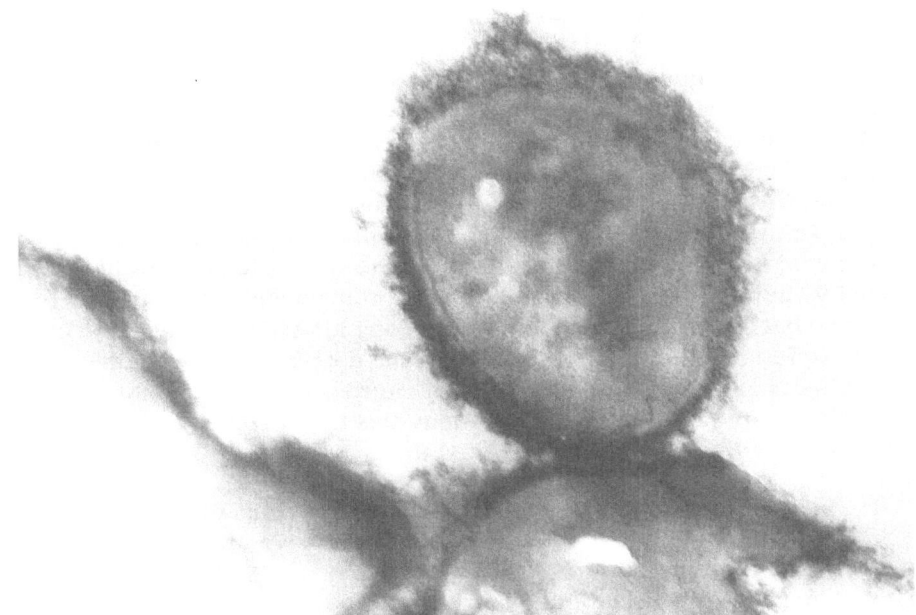

Fig. 20.3. Electron micrograph of ruthenium red stained preparations demonstrating *S. aureus* adherence to nasal epithelial cells by polysaccharide fibers ($\times$ 113,333).

bacteria to epithelial cells.[7] We have investigated the presence of similar polysaccharide-type material on *S. aureus*. This was done by staining the cells with ruthenium red, which reacts strongly with acid polysaccharides and other polyanionic of high-charge density. The adherence of *S. aureus* to human nasal epithelial cells is mediated by these polysaccharide fibers (Fig. 20.3). Some gram-negative bacteria have been observed to possess pili that impart adhesive qualities.[10] In view of their obvious ecological significance, the surface components of bacteria involved in adherent interactions with host surfaces deserve attention.

Mechanisms of Adherence

One of the most appealing theories about the mechanisms of bacterial adherence to mucosal surfaces was that of Costerton et al.[6] The epithelial cells of animals and bacterial surfaces possess polysaccharide fibers with exquisite chemical specificity. The polysaccharide fibers of bacteria, which for the most part are negatively charged, can form a polar bond with host-cell polysaccharide by way of divalent positive ion, such as magnesium, in the environment. Lectin, a protein with a specific attraction for the bacterial host-cell polysaccharides, can also form a bridge between them. This interaction is specific; bacteria whose fibers can bind neither directly to those of the host cells nor to suitable divalent ions present in the system simply to do not adhere.

Bacterial infection may begin with a specific adhesion of bacteria to a particular mucosal surface. The closely packed polysaccharide fibers from the host-parasite relationship provide an environment within which the bacterial toxins and enzyme diffuse and digest the contents of the animal cells—while the bacteria are still being attached to mucosal surfaces.

The persistent carriage of *S. aureus* in some individuals may be explained as follows; before aging epithelial cells of the skin desquamate, they are invaded by adhering bacteria that digest pits in them and slowly invade them. The persistence of *S. aureus* on the human skin could be due to fact that these bacteria are hard to eliminate from the deep pits. The attached bacteria in these pits may also derive their nutrients from the host. Experimentation is required to confirm or deny this hypothesis.

In summary, we have demonstrated that:

1) Bacteria have a selective ability to attach to nasal epithelial cells;

2) Nasal epithelial cells of *S. aureus* carriers have greater affinity for *S. aureus* than do noncarriers;

3) The ability of nasal epithelial cells to bind *S. aureus* is markedly low during the first 4 days of neonatal life;

4) Teichoic acid binds to nasal epithelial cells and thus inhibits the adherence of *S. aureus* to these cells; and

5) The adherence of *S. aureus* to human nasal epithelial cells is mediated by polysaccharide fibers.

An understanding of the molecular basis of host-parasite adherent interaction will lead to the discovery of agents that control microbial infections by interfering with bacterial adherence. Very little is known about the chemical basis of the mechanisms involved. The identification of such surface components should enable us to take measures at the initial event of the infectious process before damage to the host occurs.

References

1. Aly R, Britz M, Maibach HI (1979). Quantitative microbiology of human vulva. Br J Dermatol. 101:445–448
2. Aly R, Maibach, HI, Shinefield HR (1977) Microbial flora of atopic dermatitis. Arch Dermatol 113:780–782
3. Aly R, Shinefield HR, Strauss WG, Maibach HI (1977) Bacterial adherence to nasal mucosal cells. Infect Immun 17:546–549
4. Baddiley J, Buchanan JG, Rajbhandary UL, Sanderson AR (1962) Teichoic acid from the walls of *S. aureus* H structure of the N-acetylglucosaminylribitol residues. Biochem J 82:439–448
5. Beachey EH (1975) Binding of group A streptococci human oral mucosal cells by lipoteichoic acid. Trans Assoc Am Physicians 88:285–292
6. Costerton JW, Geesey GG, Cheng KJ (1978) How bacteria stick. Sci Am 238:86–95

7. Gibbons RJ, Van Houte J (1975) Bacterial adherence in oral microbial ecology. Ann Rev Microbiol 29:19–44

8. McCarthy ML, Snyder ML, Parker RB (1965) The indigenous oral flora of man. To newborn to the one year old infant. Arch Oral Biol 10:67–70

9. Ofek I, Beachey EH, Eyal F, Morrison JC (1977) Postnatal development of binding of streptococci and lipoteichoic acid by oral cells of humans. J Infect Dis 135:267–274

10. Okuda K, Takazoe I (1974) Haemagglitinating activity of *Bacteroides melaninogenicus*. Arch Oral Biol 19:415–416

11. Punsalang AP, Sawyer WD (1973) Role of Pili in the virulence of *Neisseria gonorrhoeae*. Infect Immun 8:255–263

12. Savage DC (1972) Survival on mucosal epithelia, mucosal epithelial penetration and growth in tissue of pathogenic bacteria. In: Smith H, Pearch JH (eds) Microbial pathogenicity in man and animal. Camb. Univ. Press, London, pp. 25–57

13. Swanson J (1973) Studies on gonococcus infection. IV. Pili: Their role in attachment of gonococci to tissue culture cells. J Exp Med 137:571–589

14. Torrey JC, Reese MK (1945) Initial aerobic flora of newborn infants: selective tolerance of the upper respiratory tract for bacteria. Am J Dis Child 69:208–214

INFECTIONS
AND EPIDEMIOLOGY

Chapter 21
Epidemiology of Skin Infections: Strategies Behind Recent Advances

ALFRED M. ALLEN AND DAVID TAPLIN

Epidemiologic information on skin infections has increased greatly in both quantity and quality during the past decade. Prior to that time, quality suffered from poor population sampling procedures and quantity was limited to the results of a few small surveys. As likely as not, "epidemiologic" knowledge was based on highly biased convenience samples of patients attending a clinic. No matter how good the microbiology, the results were of limited value in establishing the epidemiology of infection.

The recent explosion in the amount of epidemiologic information occurred because new strategies were employed. Two of the three strategies discussed in this chapter have produced prevalence data; the third has yielded incidence rates. The distinction between incidence and prevalence is important.[1]

Prevalence refers to the amount of a condition existing in a defined population at a point in time; it is like having a snapshot of skin infections in the community. Incidence, on the other hand, is a measure of the number of new cases in a defined population arising over a period of time. This is analogous to a movie rather than a snapshot. It is dynamic and therefore more useful for detecting cause-and-effect relationships. Prevalence rates indicate the burden of a disease on a population; they are especially useful in determining the need for prevention and treatment services.

Mobile Survey Teams

Determining the epidemiology of common skin infections in certain populations of interest has posed enormous technical, administrative, and logistical challenges to dermatologic investigators. These challenges have been met by the development of mobile survey teams. Jungle inhabitants, people in tropical villages, combat soldiers, and Indians on reservations at the farthest remove from centers of learning have been the object of recent investigations.[1,2,5,11] The high prevalence of infections in these populations, coupled with other special circumstances, has made it possible to obtain new insights into the origin of infection, modes of transmission, and the role of such factors as climate, hygiene, and medical practices.

We have been extensively involved in the development and deployment of mobile survey teams since the 1960s. Our experience began as the result of military interest in disabling skin infections among soldiers in the tropics.[1] This led to studies on a global scale with general application to medical and public health concerns about skin infections.

Much of our progress was made possible by such technological developments as rapid air transport, worldwide radio and telephone linkages, portable refrigerators and incubators, miniaturized laboratory paraphernalia, and highly selective culture media. The rest was due to intensive planning and preparation, willingness to travel to remote and sometimes uncomfortable surroundings, constant attention to administrative as well as technical detail, and, of course, serendipity.

This cannot be a review of the mechanics and concepts of operation of mobile dermatologic survey teams because of a lack of published information on the subject. What follows is of necessity a personal account of the difficulties and rewards, the strengths and limitations of our methods. For brevity, we will focus on one survey to illustrate our points.

In 1971 an opportunity arose to go to Colombia, South America, to study common bacterial skin infections (pyoderma) in schoolchildren. Colombia is a mountainous, geographically diverse country that lies on the equator. Year-round climate is chiefly determined by altitude, with hot climates at sea level, cool climates at high altitudes, and mild climates in between. Comparison of schoolchildren living at different altitudes could therefore be expected to provide information concerning the effect of climate on the frequency and severity of pyoderma.

The children were all from the lower socioeconomic classes and were in institutional schools because they had been orphaned or abandoned by their families. Those fresh from the streets were dirty; they had obviously just come from circumstances in which abysmally low standards of hygiene prevailed. Others had been in the schools for several months or more, and their levels of cleanliness reflected it. Comparison of these otherwise similar populations of children would provide a means of showing the effects of hygiene on reducing the frequency of pyoderma.

Surveys involving over 1 200 schoolchildren were conducted in three humid climates—tropical, temperate, and cool.[11] The results (Fig. 21.1) confirmed clinical impressions that both climate and hygiene have a profound effect on the frequency of infection. Climate also affected the rate of recovery of pathogens. The proportion of lesions that yielded streptococci was lowest in the cool region, intermediate in the temperate zone, and highest in the jungle area.

These results have practical as well as scientific implications. They suggest that improving levels of hygiene may often be the most effective preventive measure, and that this may be of more limited value in hot, humid regions where environmental stresses are greater. Antibiotic treatment

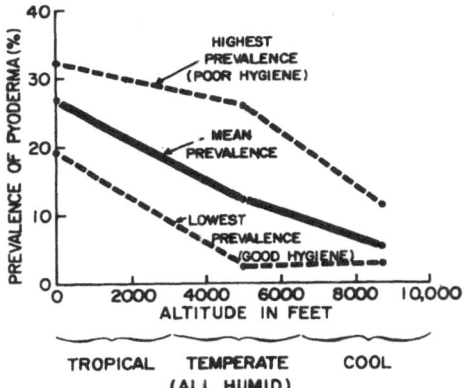

Fig. 21.1. Prevalence of pyoderma in lower-socioeconomic-class Colombian children in relation to altitude (climate) and level of hygiene.

without benefit of culture may be warranted in tropical areas where the overwhelming majority of lesions seem to be streptococcal in origin.

Behind these results were the full-time efforts of three experienced survey team members for a period of 3 months, only one of which was spent on location in Colombia. In addition, there were the efforts of numerous others, ranging from microbiology technicians in the United States to Colombian physicians and scientists. The latter were indispensable because of their assistance in making administrative arrangements, providing facilities, and smoothing the way through difficulties with unfamiliar language, laws, and customs.

The value of the Colombian study was that it exploited a unique set of circumstances to obtain information that probably could not have been obtained in any other way. The costs in terms of time, money, and trained manpower have only been hinted at above. Whether these and similar surveys are worth the effort depends on the urgency of the questions involved and the ability of mobile survey teams to provide valid answers.

The principal strength of mobile survey teams is that they can study infectious disease problems in virtually any population or environment where they might arise. They are not confined to short distances from medical centers or populations of convenience rather than of interest. Properly prepared, they can respond on short notice to urgent needs for information on the causes and means of controlling epidemics.

The chief limitation of mobile survey teams is that they do not have the capability to provide information that is directly referable to large groups of people, such as the inhabitants of a state or a nation. Because of the size and nature of their operation, they are restricted to studying relatively small and nonrandomly selected samples of the population. Extrapolation of the results to the larger population must be based on a judgment of the representativeness of the sample, a process that is subject to considerable bias. It would be extremely risky, for example, to project the results of the Colombian survey to the entire school-aged population of that country.

The kind of data collected by survey teams also has its inherent limitations. Prevalence figures provide only a static picture of disease as it exists in a cross-section of a population at a particular time. This does not have the dynamic quality of incidence rates, which are better for revealing cause-and-effect relationships because they show changes in frequency of disease in the same population over a period of time. Our inference that climate and hygiene contribute heavily to the frequency of streptococcal pyoderma is based on prevalence figures from several populations and therefore cannot be stronger than prevalence data permit.

National Health Survey

An accurate profile of the skin diseases that exist in the U.S. population was compiled for the first time in the mid-1970s. It was based on the dermatological portion of the first Health and Nutrition Examination Survey (HANES I), which was performed during the 3-year period between 1971 and 1974. A comprehensive report on the dermatologic survey findings and methodology was published in 1978.[7]

The report provides data referable to the entire civilian noninstitutionalized population of the United States during the survey period. Among other things, it records the frequency of skin infections by age and sex for persons from 1 to 74 years old (Fig. 21.2). The report furnishes information only on broad-cause categories of disease, which means that for skin infections the listings are limited to dermatophytoses in various body sites as one category and infections of skin and subcutaneous tissue as the other category. However, the data base includes much more detailed information, such as fungal culture results and subcategories of the general classification "infections of the skin and subcutaneous tissue." Consequently, a much more detailed profile of skin infections is potentially available.

The strategy employed to assemble this information was to make use of an existing governmental mechanism for gaining access to a probability sample of the U.S. population. The technical, administrative, and financial difficulties in assembling such a study population would prohibit any other approach. This fact alone confirms the wisdom of having made the considerable expenditure of effort required to organize and conduct the dermatological examinations. Principal credit for seizing this opportunity and organizing the effort on short notice is due the National Program for Dermatology, and particularly the Chairman of the Program's Data Collection Unit, Marie-Louise Johnson.

The advantages to a specialty group of using a government-generated probability sample of the population to arrive at disease prevalence rates for a national population can hardly be overestimated. Not only are there the obvious advantages of cost and convenience, but there is also the consider-

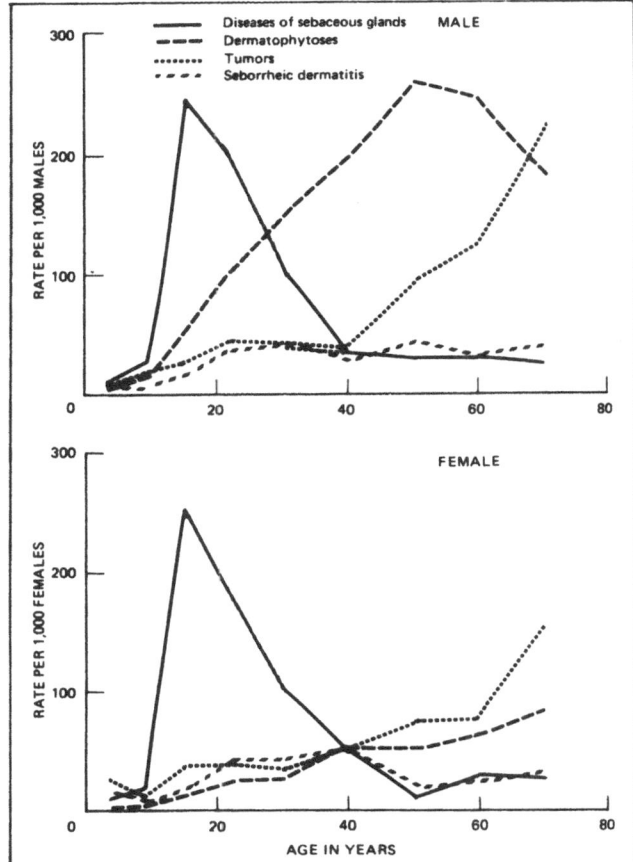

Fig. 21.2. Prevalence rates for the four most frequently occurring types of skin pathology among persons 1–74 years, by age and sex: United States, 1971–1974.

able advantage of gaining access to proper methods of sampling, a technology too often ignored or underrated by medical professionals.

Measurements obtained by probability sampling can differ startlingly from those obtained by more primitive means, such as surveying populations in the convenient form of patients attending a clinic. A case in point is the enormous spread between the estimates of the frequency of psoriasis obtained by the crude and the refined methods. "Convenience sampling" created the impression that as much as 3% of the adult population of the United States had the disease,[6] whereas the National Health Survey indicated that the figure was on the order of one-tenth this amount. It is a virtual certainty that the latter figure is much closer to the true prevalence of psoriasis in the U.S., even though only about .01 of 1% of the total population was examined.

Though a vast improvement over previous methods of determining skin disease prevalence figures for the general population, the Health and Nutrition Examination Survey was not without flaws. Overall, one-fourth of the people in the probability sample could not be examined for various reasons, and in the oldest age group more than one-third were unavailable to the examiners. This creates considerable risk of bias, particularly for the more heavily underrepresented age-sex-income-defined population subgroups. Fortunately, checks into the reasons for nonresponse to the survey revealed that the actual amount of bias was probably far less than suggested above.

A more serious problem involved the quality of the clinical data. There was evidence of considerable variation between the 101 survey dermatologists in the amount of pathology recorded. This occurred despite a standard protocol, frequent review, and constant surveillance. Part of the variation undoubtedly was due to differences between examiners in professional training, personal experience, and degree of motivation. But it is a virtual certainty that the chief reason for the variation was the inherent subjectivity of dermatologic assessments. In the absence of rigidly defined and clearly expressed criteria for abnormal findings, no other result could have been expected.

National Disease-reporting System

Dermatologists and skin microbiologists are seldom afforded the luxury of having incidence data handed to them with negligible effort on their part. In general, disease-reporting systems necessary to generate such data for skin infections simply do not exist. Moreover, even the best disease-reporting systems have little epidemiologic value if the size and demographic characteristics of the population from which the patients have come cannot be known. Such would usually be the case when a dermatologic clinic or hospital keeps count of its cases; it has no way of knowing the size of the potential patient population it serves unless it is part of a highly structured health care system.

There is at least one major exception to this negative pattern, and it comes from a country with a long history of socialized medical care, excellent disease reporting, and complete enumeration of the population. Since 1888 Denmark has had a uniform reporting system for scabies. Christophersen[4] reported the epidemiology of scabies from 1900–1975 using these national data.

The existence of the system makes the quality of the information far superior to that provided by investigators[9,10] who did not have this advantage. The difference is so great that meaningful comparisons cannot be made. Until comparable data can be generated from other countries, the Danish study will stand as the only definitive epidemiologic picture of scabies on a national scale.

The advantages of having information from a population that runs into the millions become apparent when examining the data. In some years, the reported incidence of scabies in Denmark was as low as 0.5 cases per thousand. Reasonably precise rates of this order of magnitude are not possible to obtain unless the populations under study number into the many thousands. The numbers must be greater yet when the data are to be analyzed by such factors as age and sex. The Danish study provides precise figures for 12 different age-sex groups.

The unique and not-soon-to-be duplicated advantage of the Danish study is the length of time for which comparable data were available. Long-term trends were shown that cast doubt on the theory of herd immunity. Also of interest was the extreme feebleness of the recent resurgence of scabies in Denmark as compared to the epidemic years during World Wars I and II (Fig. 21.3). There are no data from other countries capable of demonstrating that the recent worldwide resurgence was stronger anywhere else.

Monthly as well as yearly incidence rates were available also, making it possible to see if there were recurrent seasonal trends. The seasonality was opposite that of gonorrhea, casting suspicion on the venereal transmission of scabies.

Reporting artifacts and deficiencies, the chief limitations of most incidence data based on disease-reporting systems, could have affected the quality of the Danish study. Overreporting was unlikely because relapses of scabies are rare within 1 year. Underreporting was shown not to be a problem: physician reporting was remarkably complete; primary cases routinely entered the medical care system; and the number of secondary cases probably varied in proportion to the number of primary cases.

One objection to this type of study is the lack of ascertainment of cases by qualified specialists. Another is the lack of detailed clinical and environ-

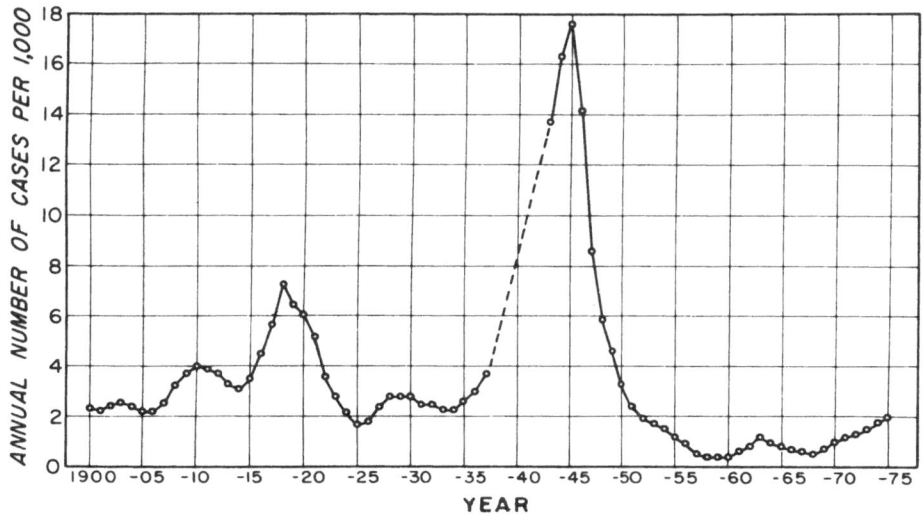

Fig. 21.3. Incidence of scabies in Denmark, 1900–1975.

mental information that might be useful in establishing modes of transmission in various settings. Such objections are understandable when coming from persons whose orientation is heavily clinical, but they are not appropriate when dealing with national incidence rates. The "clinical epidemiology" of diseases such as scabies is best determined by intensive study of small groups.

The use of national disease-reporting systems for epidemiologic studies is limited to situations such as the one for scabies in Denmark. Increasing amounts of disease-reporting data are becoming available each year. With the advent of computers in outpatient clinics, more can be expected. If adequate quality control can be achieved, it may someday be possible to use national disease reports to define the population epidemiology of such entities as dermatophytosis and pyoderma.

A Neglected Strategy

There are a finite number of basic strategies in epidemiology.[8] Some have begun to be exploited with gratifying results in clarifying the epidemiology of skin infections. One is notable by its virtual absence: the case-control study.

Case-control studies, sometimes known as retrospective studies, are generally the sole type of epidemiologic investigations that can be successfully conducted using office, clinic, or hospital patients as the study population. They are useful in testing hypotheses about possible causes of infection. If a contributing factor is identified, the results can be used to establish the degree of risk created by the factor. Case-control studies are not as good as prospective (cohort) studies in determining cause-and-effect relationships, but they are more economical and less time-consuming.

Case-control studies are not of value in determining the incidence or prevalence of infection. Neither are the studies of hospital or clinic populations to determine the proportion that have a particular disease. The percentage figures quoted in these studies are proportional rates, not incidence rates. Though often referred to as "incidence," proportional rates are almost worthless as epidemiologic data. Dermatologists and microbiologists should stop wasting their time on proportional rate studies, and concentrate instead on case-control studies of their patients.

References

1. Allen AM, Taplin D, Lowy JA, Twigg L. (1972) Skin infections in Vietnam. Milit Med 137:295–301
2. Allen AM, Taplin D. (1974) Skin infections in eastern Panama: survey of two representative communities. Am J Trop Med Hyg 23:950–956
3. Allen AM (1978) Epidemiologic methods in dermatology: Part 1, Describing the occurrence of disease in human populations. Int J Dermatol 17:186–193
4. Christophersen J (1978) The epidemiology of scabies in Denmark, 1900 to 1975. Arch Dermatol 114:747–750

 5. Dajani AS, Ferrieri P, Wannamaker L (1973) Endemic superficial pyoderma in children. Arch Dermatol 108:517–522
 6. Farber EM, McClintock RP Jr (1968) A current review of psoriasis. Calif Med 108:440–457
 7. Johnson MLT, Roberts J (1978) Skin conditions and related need for medical care among persons 1–74 years, United States, 1971-1974. Vital and Health Statistics-Series 11-No 212. DHEW Pub No (PHS) 79-1660
 8. MacMahon B, Pugh TF (1970) Epidemiology: Principles and Methods. Little, Brown and Co, Boston, pp 29–46
 9. Orkin M (1971) Resurgence of scabies. JAMA 217:593–597
10. Shrank AB, Alexander SL (1967) Scabies: Another epidemic. Br Med J 1:669–671
11. Taplin D, Lansdell L, Allen AM, Rodriguez R, Cortes A (1973) Prevalence of streptococcal pyoderma in relation to climate and hygiene. Lancet 1:501–503

Chapter 22
The Role of Hands in Nosocomial Gram-negative Infection

M. W. CASEWELL

Ever since 1874 when Semmelweis in Vienna showed that chloride of lime counteracted the agent responsible for the transmission of puerperal fever, there has been an intermittent interest in the transmission of infectious disease by hands. Surprisingly, more than 130 years later, there are still doubts, in theory and practice, about the precise relevance of hand hygiene or disinfection in the transmission of hospital-acquired gram-negative bacilli.

The epidemiology of the classical infections caused by bacteria such as streptococci, staphylococci, shigellae, and *S. typhi* is well understood, and the possibility of transmission of these infections via contaminated hands is universally recognized. But the increasing contribution of the opportunistic gram-negative bacilli to hospital-acquired infection has not always been accompanied by a corresponding clarification of the epidemiology of the various species involved.

There are many subtle traps for the unwary would-be researcher with an interest in the epidemiology of, for example, *Klebsiella, Serratia, Pseudomonas, Enterobacter* or *Acinetobacter* species. These organisms do not produce specific clinical syndromes and the colonization of patients does not declare itself to the clinician. Repeated bacteriological investigation may be needed for its detection. Furthermore, discriminatory typing systems for these species, where these exist, are not universally available. Lastly, these opportunistic gram-negative organisms often form part of the hospital moist environment where it is difficult to assess their significance.

Recently there has been a renewed interest in the hospital-acquired gram-negative bacilli and their epidemiology. It has become clear that these organisms are, above all others, the most resistant to antibiotics that the patient and his physician will have to face. The genetics of this resistance sounds another clarion of alarm.

Perusal of simply the titles of contributions to erudite scientific journals indicates the measure of confusion about the epidemiology of gram-negative bacilli. For example, a letter entitled "Pseudomonas on the chrysanthemums," which appeared in the *Lancet* in 1973, was soon followed by a more ominous article entitled "Flower vases in hospitals as reservoirs of pathogens." A few weeks later there was a third more incisive, and valuable, contribution: "Protecting chrysanthemums from hospital infection." This example illustrates that isolation of an organism does not necessarily implicate causality.

Isolation of Gram-negative Bacilli from Hands

There have been few careful investigations into the role of contaminated hands in the transmission of gram-negative infection. A relatively simple quantitative technique for the bacteriological examination of handwashings was described more the 40 years ago by Price.[17] Many subsequent workers have used neither Price's techniques nor any other means of standardized sampling. Small numbers of hand swabs have sometimes been included as part of a larger epidemiological survey.

In Copenhagen, Ørskov[15] used swabs and hand washings in her investigation into the prevalence of klebsiella capsular type 8 in the urine of male patients attending surgical clinics. Type 8 was demonstrated on the hands of an assistant who cleaned the urinals. In a thorough and useful investigation, Salzman, Clark, and Klemm[20] used a modification of Price's technique and found as many as 65% of 154 staff hands were contaminated with coliforms, 87.5% of which belonged to the *Klebsiella-Aerobacter* group. In addition, 18% of washings yielded 1 000 or more contaminated particles. They did not relate organisms to nearby potential sources or patient infections.

Conversely, Redman and Lockey,[18] working in an intensive care unit, and Farmer,[11] investigating a special care unit, were unable to isolate gram-negative bacilli on swabs from staff hands. By swabbing "the ventral surface of the index finger" Montgomerie et al.[14] found klebsiellae in only 2 of 74 nurses and were unable to relate the strains to recent isolates from infections. One nurse, however, did have the same type as that found in the feces of a patient. Adler et al.,[1] using hand washings and finger-impression plates, found *Klebsiella pneumoniae* type 11 in 30% of nurses working in a nursery for premature infants where umbilical fecal colonization with type 11 was prevalent. Routine hand washing did not eradicate hand-carried klebsiellae.

The hands of 150 patients were examined for gram-negative organisms by Pollack et al.[16] They used a hand-washing technique, and klebsiellae, the commonest organisms, were found in 20% of the patients. Furthermore, colonization was directly related to the length of hospital stay and antibiotic treatment. As patient-to-patient contact is infrequent in, for example, intensive care units, colonization of patients' hands is most likely to represent a source rather than a route of transmission.

Thus it has been established for some time that in certain wards colonization or contamination of hands with klebsiellae can occur in staff and patients. The observed incidence of hand contamination may well depend both on the techniques used and the ward investigated.

Some years ago, with a high incidence of endemic infection in our own intensive care unit at St. Thomas' Hospital, we became interested in clarifying which procedures contaminated the hands, how long klebsiellae survived on dry hands, and what degree of hand washing might be required to remove such organisms. At that time little had been published that combined hand-carriage rates with typing. Apart from the work of Ørskov[15] and Adler et al.[1] there was little epidemiological evidence to correlate hand carriage with potential sources and patient klebsiella types. The hypothesis that hands played a key role in the transmission of klebsiella infection seemed a promising, but unproven, possibility.

The patients in our ten-bed intensive care ward received intensive and continuous bacteriological monitoring for 4 years. Having established capsular typing for the first time in the U.K., we were able to serotype the 986 klebsiella isolates obtained from 2 315 patients admitted during the study period. Our findings[5] showed that individual serotypes of patient isolates clustered—which strongly suggested a changing common source, cross-infection, or both. The distribution in time of the nine commonest types is shown in Fig. 22.1.

Cross-infection, perhaps via hands, would help explain the clusters observed, and by using a sampling and counting technique based upon that of Salzman, Clark, and Klemm[20] we showed that of 28 staff who worked in the intensive care unit, 17% yielded *Klebsiella* spp. in their hand washings.[6] Furthermore, simple supposedly "clean" procedures resulted in the transfer of the patient's serotype to the nurse's hands. The staff thought they had "clean" hands and would not have washed before attending another patient unless a procedure involving an aseptic technique was to be carried out. The most striking feature of these results was the ease with which klebsiellae

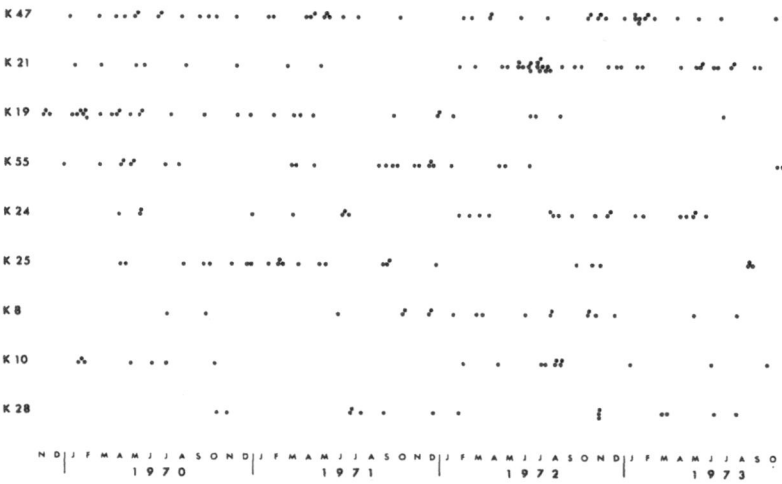

Fig. 22.1. Distribution in time of *Klebsiella* serotypes isolated from intensive care patients: the nine commonest types all show clustering (after Casewell and Phillips 1977a).

were transferred from the colonized patient to the hands of attendant staff. As many as 10^3 detectable viable klebsiellae were transferred following such slight contact as that involved in taking the patient's temperature, lifting the patient into bed, or taking his radial pulse.

Soon after these studies had been completed several reports supporting our findings implicated hands as a route of transmission. In a surgical intensive care unit where hexachlorophene liquid soap was used routinely, "gram-negative organisms" were found in 14 of 20 hand washings taken from nurses.[19] Several serotypes of klebsiellae were found, and the same type was "frequently" isolated from patients and inhalation equipment. The procedures that contaminated staff hands, however, were not defined. It seems likely that the routine use of hexachlorophene in Rosendorf et al.'s study may have exerted a selection pressure that favored gram-negative bacilli.[2] Hexachlorophene was not used by the nursing staff in the St. Thomas' study.

At that time we found no published studies of the survival of klebsiellae on contaminated or artifically inoculated hands, and we were therefore interested to find that we could recover klebsiellae from artificially inoculated hands up to 150 min after inoculation.[6] The reduction in the number of recoverable organisms after such an interval was often less than tenfold. It had often been thought that gram-negative bacilli survived poorly outside moist environments, and the possibility that klebsiellae can both contaminate and survive for periods in excess of 1 hour had not been appreciated either in this study-ward or elsewhere. Such unsuspected contamination provided, in the absence of an effective hand-washing procedure, a highly plausable route of transmission for this species.

Workers in Houston[13] have more recently confirmed the ability of klebsiellae to survive dessication on human skin. By inoculating the forearms (but, unfortunately, not the hands) of ten normal volunteers, viable klebsiellae could be recovered up to 8 hours later. The viable counts of both *Proteus vulgaris* and *Pseudomonas aeruginosa* fell more rapidly.

In order to intercept this route of transmission it was necessary to define an effective procedure for hand antisepsis before passing from one colonized patient to the next. However, most studies investigating the efficacy of hand antiseptics had been concerned with preoperative hand disinfection for surgeons and had thus, understandably, been preoccupied with the reduction in the total bacterial counts of resident flora, especially gram-positive organisms. There were remarkably few studies of the transitory flora of contaminated hands.

We therefore artifically inoculated hands with klebsiellae and compared the effects of rinsing in warm water, washing with chlorhexidine skin cleanser (4% w/v chlorhexidine gluconate), nonmedicated soap, and medicated soap (containing undecyclenic alkalolamide sulphon-succinate 0.5 % w/v, triclosan 0.75% w/v). These studies showed that, of these agents, chlorhexidine skin cleanser most effectively reduced viable counts, reliably giving a 98%–100% reduction of the inoculated organisms.[6]

Patients' Skin As a Source of Contamination for Staff Hands

What then is the relationship of skin colonization of patients, which acts as a potent source of contamination of staff hands, and bowel carriage of *Klebsiella* spp. by patients? The skin of the patients, especially severely ill[21] or antibiotic-treated patients, is often colonized with gram-negative bacilli.

This moves away from the traditional view that the only important source of organisms in klebsiella cross-infection is simply the patients' feces. There is, we now believe, a complex series of combinations of positive sites that include clinical specimens such as urine, as well as the patient's feces and his skin. For effective control of any klebsiella cross-infection or outbreak it is essential that the clinician and, more importantly, his infection-control personnel, are aware of all possibilities.

Tables 22.1–22.3 show the results of a study of 45 patients at St. Thomas' Hospital from whom we isolated various serotypes of gentamicin-resistant klebsiellae. These patients have been arranged into three groups. The first group (Table 22.1) shows 12 patients whose gentamicin-resistant strains declared themselves clinically in specimens of urine or, more unusually, sputum. This group is of interest because they have negative fecal cultures and may, therefore, have acquired their organism by cross-infection.

The second group (Table 22.2) shows 21 patients who have both urine (or sputum) and fecal isolates, and may well represent the traditional concept of autoinfection. But this group is of particular interest because it can be seen that the same klebsiella can also be isolated from various skin sites and from the hands of ten patients. These skin sites may easily contaminate the hands of unsuspecting staff.

The third group (Table 22.3) shows the patients whose klebsiella is absent from their urine (and other clinical specimens) but is present in their feces as

Table 22.1. 12 Patients with Urinary, and 3 with Sputum Isolates. All Have Negative Feces

Patient	Urine	Feces	Groin	Hands	Umbilicus	Throat
1	+	−	+	+	−	+
2	+	−	+	+	−	−
3	+	−	+	+	−	−
4	+	−	+	−	−	−
5	+	−	−	−	−	−
6	+	−	−	−	−	−
7	+	−	−	−	−	−
8	+	−	−	−	−	−
9	+	−	−	−	−	−
10	+	−	−	−	−	−
11	+	−	.	−	−	−
12	+	−	.	.	−	−
	sputum					
13	+	−	−	−	.	+
14	+	−	.	.	.	+
15	+	−	.	.	.	.

Table 22.2. 21 Patients with Urinary and Fecal Isolates. 16 Also Have Skin Colonization

Patient	Urine	Feces	Groin	Hands	Umbilicus	Throat
16	+	+	+	+	+	+
17	+	+	+	+	+	+
18	+	+	+	−	+	+
19	+	+	+	+	+	−
20	+	+	+	+	+	−
21	+	+	+	+	+	−
22	+	+	+	+	+	−
23	+	+	−	+	+	−
24	+	+	+	+	−	−
25	+	+	+	+	−	−
26	+	+	−	+	−	−
27	+	+	+	−	+	−
28	+	+	+	−	·	−
29	+	+	+	−	−	−
30	+	+	+	−	−	−
31	+	+	+	·	−	−
32	+	+	−	−	−	−
33	+	+	−	−	−	−
34	+	+	·	−	·	−
35	+	+	−	·	·	−
36	+	+	·	·	·	·

well as on skin sites. In an outbreak of klebsiella colonization or infection these patients would not be identified by routine clinical specimens and may form an important hidden reservoir for the infection of other patients.

Clinical Importance of Isolation of Gram-negative Bacilli from Skin

What then are the practical implications of the recent findings which show that gram-negative bacilli colonize the skin of patients and may form part of the transitory flora of staff hands? Before answering this question it is worth

Table 22.3 9 Patients with Fecal (and/or Skin) Colonization. ? = Hidden Reservoir for Other Patients

Patient	Urine	Feces	Groin	Hands	Umbilicus	Throat
37	−	+	+	+	+	+
38	−	+	+	+	·	+
39	−	+	·	·	·	+
40	·	+	·	·	·	+
41	−	+	−	+	−	+
42	−	+	+	+	·	·
43	−	+	+	+	−	−
44	−	+	−	−	−	−
45	−	−	+	−	−	−

Table 22.4. Resistance of 108 Strains
of Gentamicin-resistant Klebsiellae to
17 Other Antibiotics

Antibiotic	Per cent strains resistant
gentamicin	100
ampicillin	100
carbenicillin	100
sulfonamide	99
cephaloridine	93
cephalothin	82
tobramycin	79
neomycin	75
chloramphenicol	74
tetracycline	74
cefazolin	57
trimethoprim	44
cephradine	35
cephalexin	14
amikacin	4
cefoxitin	2
cefuroxime	1
cefotaxime	0

reiterating important changes that have recently become most obvious in *Klebsiella* and *Serratia* spp. in the United States and in *Klebsiella* spp. in the U.K. Two developments in *Klebsiella* spp. are of major clinical importance and these may be interrelated.

Firstly, the emergence of gentamicin-resistance has been associated with unprecedented resistance to the majority of other standard antibiotics. In a collection of 1 150 gentamicin-resistant klebsiellae collected from several countries, we selected 108 epidemiologically distinct strains and studied their antibiotic resistance.[7] These klebsiellae are all resistant to the majority of standard antibiotics (Table 22.4). Amikacin, cefuroxime, cefoxitin, and cefotaxime are, at present, the most effective agents in vitro, and should perhaps be reserved for these resistant organisms. Furthermore, in the same collection, 81% of the strains have transmissable plasmids bearing large numbers of R factors. In a few instances, we have good evidence for in vivo transmission of these plasmids from *Klebsiella* spp. to other potentially pathogenic gram-negative bacilli.[4] In one of these patients it was shown[10] that the plasmid in three species was identical in terms of resistances transferred, incompatibility group, and molecular weight. A transposon conferring resistance to trimethoprim and streptomycin was also found.

Secondly, multiply-resistant *Klebsiella* and *Serratia* spp. seem to be distinguished from other enterobactericeae by their ability to give rise to widespread cross-infection from patient to patient, from ward to ward, and even from hospital to hospital. Fig. 22.2 shows the transmission of all isolates of gentamicin-resistant klebsiellae from patient to patient and ward to ward in

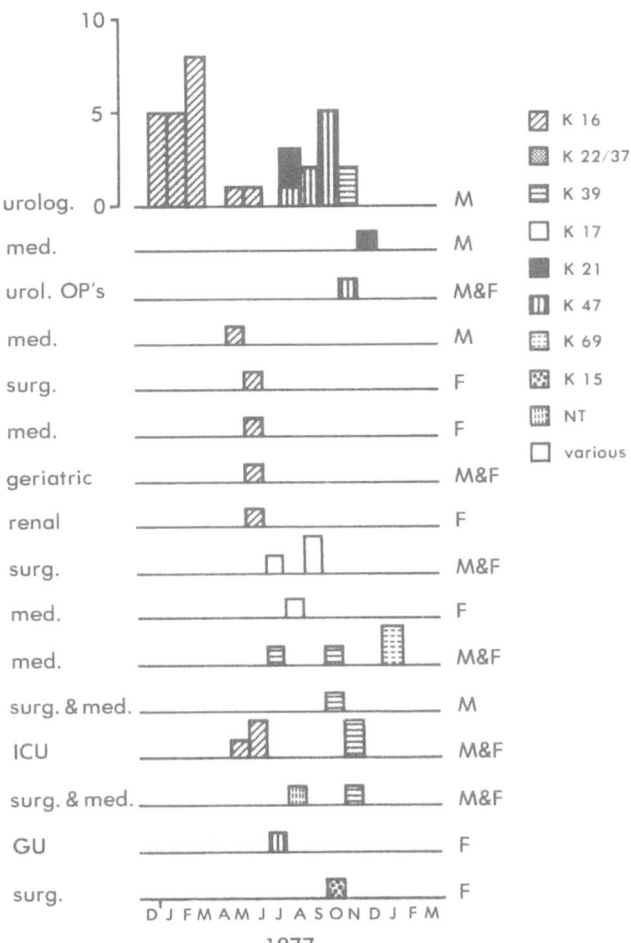

Fig. 22.2. Distribution of gentamicin-resistant klebsiellae among hospital wards.

St. Thomas's Hospital in 1977. With considerable expenditure of energy and resources we contained most of these clusters of infection with individual serotypes to a handful of patients, but our first outbreak involved 19 patients.[3]

Through personal communication and in published accounts we have heard of more widespread outbreaks in the following world centers: Johannesbury, Melbourne, Toronto, Nashville, Munich, and Parma. Fig. 22.3 shows the extent of an outbreak in the U.K. at Bristol[9] where klebsiella capsular type K2 ultimately involved more than 300 patients and spread to three other hospitals. Such outbreaks with a single strain of any organism are uncommon in the 20th century and warrant close scrutiny. Most of the patients were merely colonized, but such outbreaks carry a septicemia rate of about 7% and a mortality of about 2%.

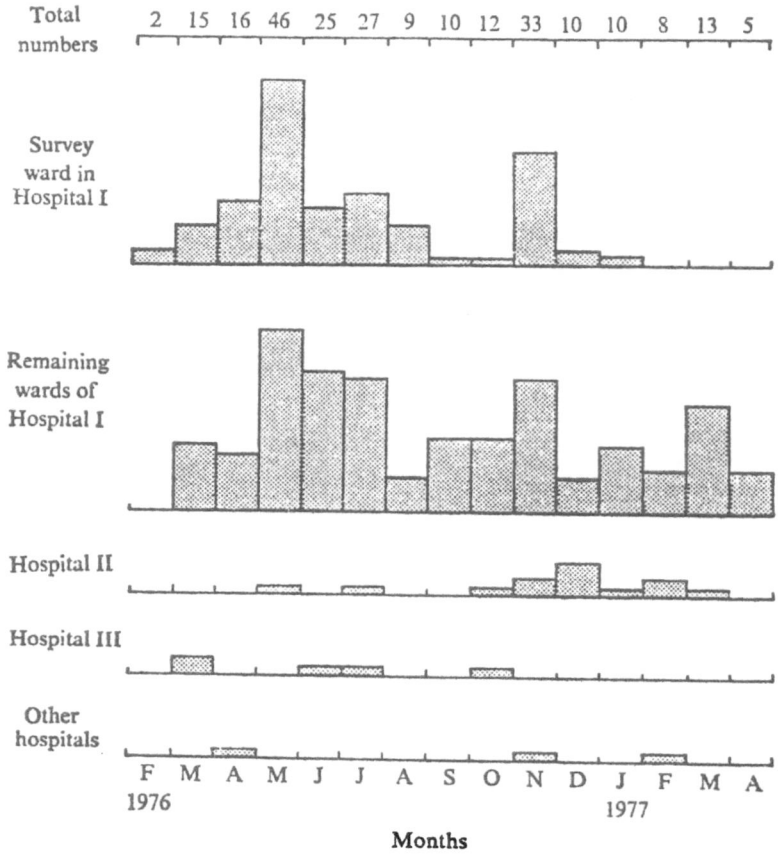

Fig. 22.3. Colonization and infection of 241 patients with gentamicin-resistant multiply-resistant *Klebsiella aerogenes* K2 in Bristol, U.K. (after Curie et al. 1978).

We have already shown[6] that increased attention to staff hand washing can reduce the overall rate of endemic colonization and infection of intensive care patients with sensitive klebsiellae (Fig. 22.4). In an epidemic it is important to define the outbreak fully (including asymptomatic fecal carriers). Our findings on patients' skin colonization and hand transmission have encouraged us to use more rational isolation procedures. We use disposable plastic aprons (not gowns or masks) and pay close attention to hand hygiene and hand washing with chlorhexidine skin cleanser. In this way we have been able to control our own episodes of cross-infection with gentamicin-resistant klebsiellae and have also witnessed the ultimate control of outbreaks in other hospitals.[12]

Finally, I would suggest that the precise epidemiology of the different species of gram-negative bacilli may all have different characteristics, and thus have distinct requirements for successful control. But many questions remain unanswered. Is the ability of klebsiella to survive on staff hands unique to this genus? Is its ability to colonize the patient's skin the key to its success in producing large-scale outbreaks? Does capsular serotype 21, which is particularly common in the U.K.,[7] have a special ability to colonize human skin or contaminate hands and thus bring about outbreaks? Is there an element of virulence that is important in widespread epidemics? How many staff have epidemic gram-negative bacilli as part of their resident, as opposed to transitory, hand flora?

I doubt that we have enough information to understand completely the epidemiology and transmission of all species of gram-negative bacilli. But it seems likely that we shall remain preoccupied with multiply-resistant *Klebsiella* and *Serratis* spp. in particular, and will be making increasing efforts to bring these organisms and their unpleasant plasmids and transposons under complete control.

A more rational approach to our incomplete knowledge of these species' epidemiology and control measures will help, but we may have to depend on

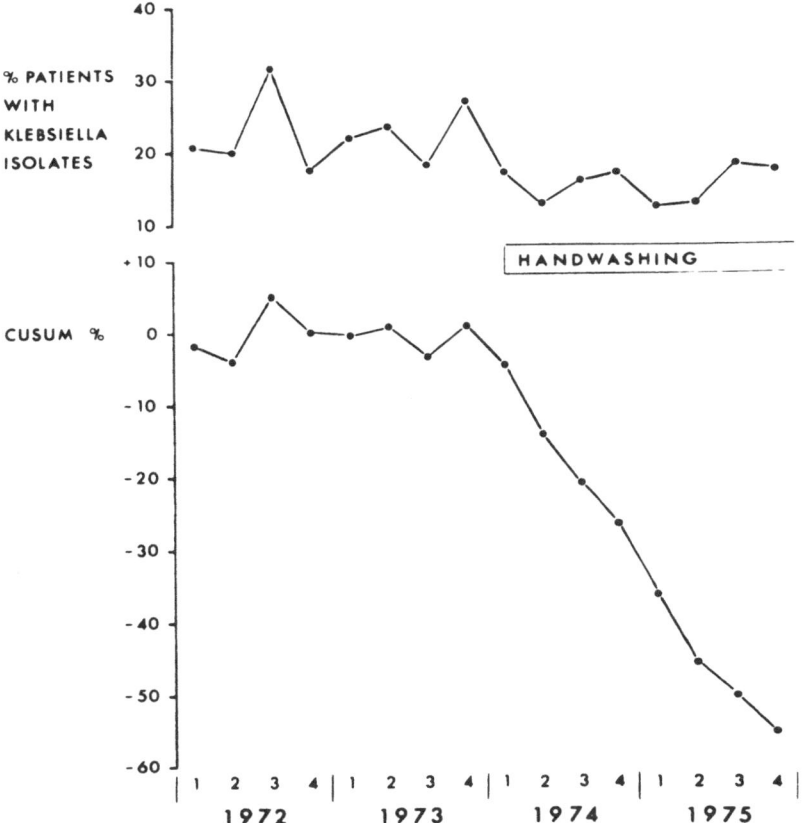

Fig. 22.4. Percentages of patients (per quarter) becoming colonized or infected with klebsiellae, 1972-75.   Also, change in cumulative sum associated with staff handwashing.

the drug companies' ability to keep one step ahead of increasingly antibiotic-resistant clinical isolates that presently are infecting an increasing number of ill patients.

References

1. Adler JL, Shulman JA, Terry PM, Feldman DB, Skalty P (1970) Nosocomial colonisation with kanamycin-resistant *Klebsiella pneumoniae* types 2 and 11 in a premature nursery. J Pediatr 77:376
2. Bruun JN, Solberg CO (1973) Hand carriage of Gram-negative bacilli and *Staphylococcus aureus*. Br Med J 2:580
3. Casewell MW, Dalton MT, Webster M, Phillips I (1977) Gentamicin-resistant *Klebsiella aerogenes* in a urological ward. Lancet II:444
4. Casewell MW, French GL (1979) To be published.
5. Casewell MW, Phillips I (1977a) Epidemiological patterns of klebsiella colonization and infection in an intensive care ward. J Hyg (Camb) 80:295
6. Casewell M, Phillips I (1977b) Hands as a route of transmission of *Klebsiella* species. Br Med J 2:1315
7. Casewell M, Talsania HG (1979b) Predominance of certain klebsiella capsular types in hospitals in the United Kingdom. J Infection 1:77
8. Casewell M, Talsania HG (1979a) To be published.
9. Curie K, Speller DCE, Simpson RA, Stephens M, Cooke DI (1978) A hospital epidemic caused by gentamicin-resistant *Klebsiella aerogenes*. J Hyg (Camb) 80:115
10. Datta N, Hughes VM, Nugent ME, Richards H (1979) Plasmids and transposons and their stability and mutability in bacteria isolated during an outbreak of hospital infection. Plasmid 2:182
11. Farmer K (1968) The influence of hospital environment and antibiotics on the bacterial flora of the upper respiratory tract of the newborn. NZ Med J 67:541
12. Houang ET, Evans MAL, Simpson CN (1979) Control of hospital epidemic of gentamicin-resistant *Klebsiella aerogenes*. Lancet II:205
13. McBride ME, Duncan WC, Knox JM (1975) Physiological environmental control of Gram-negative bacteria on skin. Br J Dermatol 93:191
14. Montgomerie JZ, Doak PB, Taylor DEM, North JDK, Martin WJ (1970) Klebsiella in faecal flora of renal - transplant patients. Lancet II:787
15. Ørskov I (1954) Nosocomial infections with klebsiella in lesions of the urinary tract. II. Acta Pathol Microbiol Scand 35:194
16. Pollack M, Charache P, Nieman RE, Jett MP, Reinhardt JA, Hardy PH (1972) Factors influencing colonisation and antibiotic-resistance patterns of Gram-negative bacteria in hospital patients. Lancet II:668
17. Price PB (1938) The bacteriology of normal skin; a new quantitative test applied to a study of the bacterial flora and the disnfectant action of mechanical cleansing. J Infect Dis 63:301
18. Redman LR Lockey E (1967) Colonisation of the upper respiratory tract with Gram-negative bacilli after operation, endotracheal intubation and prophylactic antibiotic therapy. Anaesthesia 22:220
19. Rosendorf LL, Daicoff G, Baer H (1974) Sources of Gram-negative infection after open-heart surgery. J Thorac Cardiovasc Surg 67:195
20. Salzman TC, Clark JJ, Klemm L (1968) Hand contamination of personnel as a mechansim of cross-infection in nosocomial infection with antibiotic-resistant *Escherichia coli* and *Klebsiella-Aerobacter*. Antimicrob Agents Chemother 1967:97
21. Stratford B, Gallus AS, Matthiesson AM, Dixson S (1968) Alteration of superficial bacterial flora in severely ill patients. Lancet I:68

Chapter 23
The Interaction of Fungi and Bacteria in the Pathogenesis of Athlete's Foot

ALBERT M. KLIGMAN AND JAMES J. LEYDEN

The term "athlete's foot" designates the itchy, scaly, toe-web lesions to which athletic young males are particularly prone. The clinical spectrum ranges from dry, mild, scaling (Fig. 23.1) to a painful, exudative, erosive, inflammatory process with fissuring (Fig. 23.3). The former is mainly asymptomatic and often subclinical. The variety which fits the most familiar image of the disease is characterized by scaly, soggy, whitish hyperkeratotic lesions associated with pruritus and foul odor (Fig. 23.2). Textbooks of dermatology always discuss interdigital athlete's foot in the chapter on superficial fungus infections; various dermatophytes are cited as causative.[5,14] It is established dogma that athlete's foot is a ringworm infection.

This tidy concept is spoiled by the repeated finding that dermatophytic fungi frequently cannot be isolated in culture or demonstrated microscopically. The demonstration of fungi from clinically abnormal interspaces has been surprisingly low; in the majority of studies, fungi could be demonstrated in less than 25% of cases. The exceptions are Marples and Bailey,[11] 58% recovery (3); English and Gibson,[4] 34%; and Goto,[4] who has set a record of 61% by a scrupulous search for fungi. Table 1 summarizes the clinical and mycologic data of different investigators. The contradiction of overt clinical disease in the absence of fungi is highlighted by Gentles and Evans's finding that 40% of interspaces with typical peeling and scaling were mycologically negative.[6] A further complication is that ringworm fungi have been found in up to 9% of normal interspaces (Table 23.1).

Another disquieting feature is that potent antifungal agents do not result in rapid clinical improvement. Tolnaftate, for example, penetrates skin well,[15] but the therapeutic response is disappointing to many patients who want relief in less than two weeks. Likewise, oral griseofulvin, the drug par excellence for ringworm infections, is rather unimpressive in interdigital athlete's foot. Altogether these observations suggest that fungi alone are insufficient to account for the clinical symptomotology.

We have found that resident bacteria collaborate with dermatophytic fungi to produce symptomatic athlete's foot. Until now, bacteria have not been considered except for those instances in which infection by virulent organ-

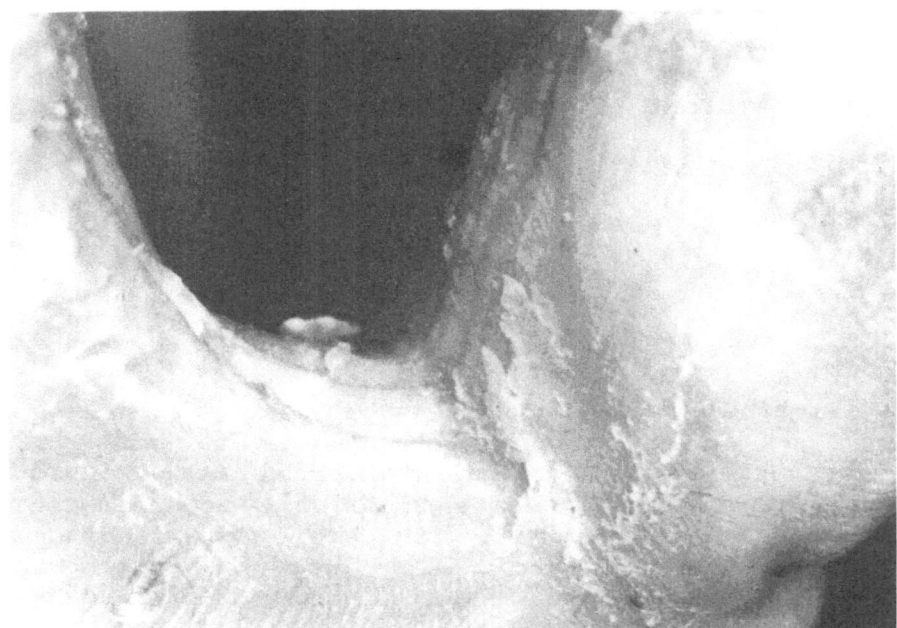

Fig. 23.1. Dermatophytosis simplex. Dry, scaling, relatively symptomatic interspace.

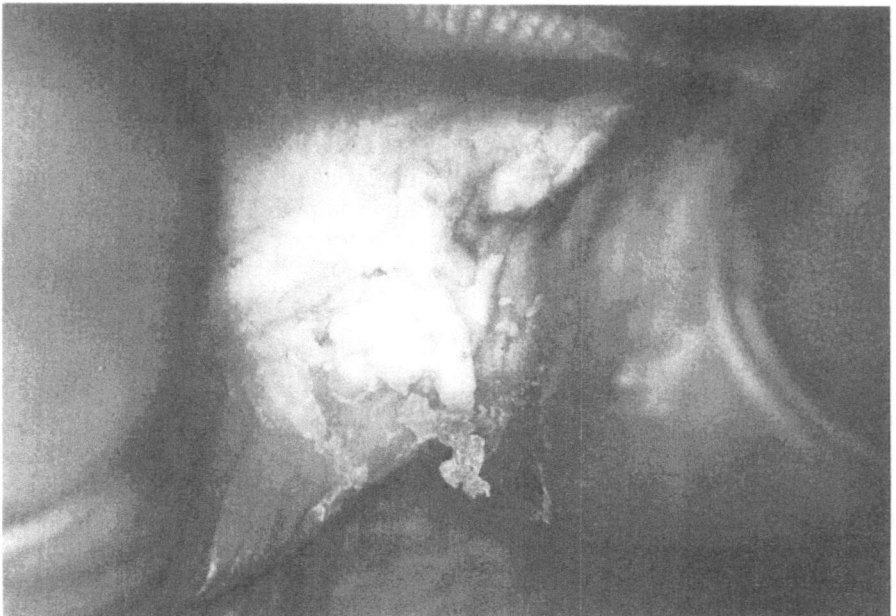

Fig. 23.2. Hyperkeratotic, whitish plaque.   Fungi recovered in 55% of cases. Aerobic diphtheroids begin to predominate.

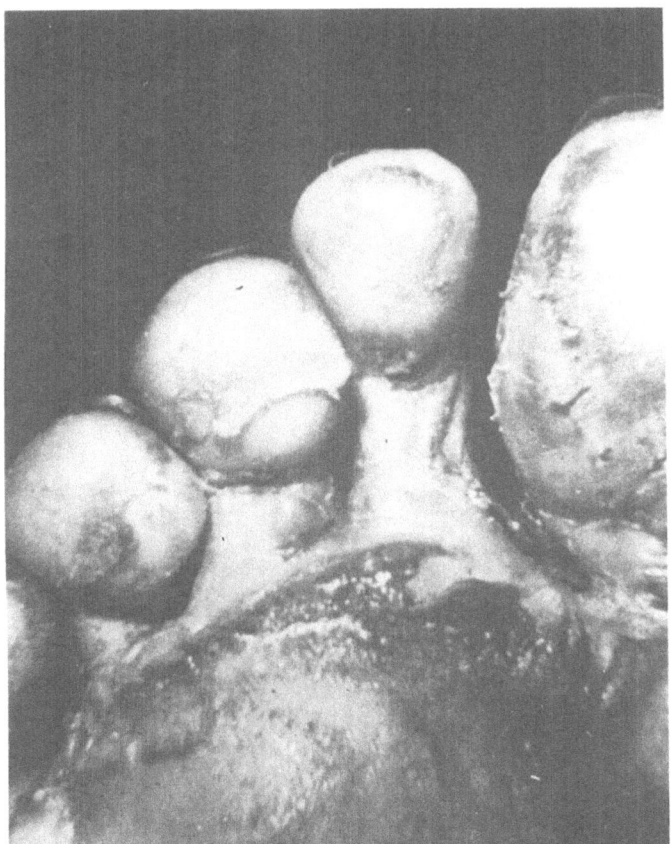

Fig. 23.3. Dermatophytosis complex. Severely inflamed, eroded, symptomatic interspace. Fungi reovered in only 35% of cases, aerobic diphtheroids predominate.

isms has complicated athlete's foot.   Marples and Bailey[11] isolated *Staphylococcus aureus* in 64% of abnormal interspaces; in 47%, both *S. aureus* and a dermatophyte were recovered. While it is recognized that *S. aureus* and β-hemolytic streptococci may occasionally incite a secondary pyoderma, neither we nor others have found such a high prevalence of virulent cocci. In unusual and fortunately rare cases, gram-negative organisms, especially *Proteus* and *Pseudomonas*, may produce a disabling disease with extensive erosions and intense inflammation.[2]

This paper summarizes our concepts regarding the pathogenesis on interdigital athlete's foot. We conceive a continuum from a relatively symptomless scaling lesion due solely to fungi (*dermatophytosis simplex*) to a distressing, wet, pruritic, foul-smelling, macerated condition (*dermatophytosis complex*), representing intense colonization of dermatophyte-infected skin by normal, resident bacteria.

Table 23.1. Reported Frequency of Dermatophytes in Normal and Abnormal Interspaces

Population, ages (no.)	Prevalence of fungi in abnormal interspaces, %	Prevalence of abnormal interspaces, %	Prevalence of fungi in normal interspaces, %	References
Children, 11–14 yr (387)	7.5	70	2.5	12
Adults, male and female (175)	12.5	54	8.9	11
	(asymptomatic subjects)			
	58	19	—	—
	(symptomatic subjects)			
Children, 7–14 yr (4794)	34	34.7	1.5	4
Military recruits (871)	18	60	1.7	1
Military recruits (210)	12.5	42	6.8	3
College students (568)	10.6	63	3.3	13
Adult males (2 101)	38	90	2.5	8
Adult males (201)	61	84	—	7

Clinical-Microbiological Correlations

Dermatophytosis Simplex (Simple Ringworm)

Dry scaling is a common finding in the last interspaces of males (Fig. 23.1). There may be mild, intermittent itching, but otherwise the condition goes unnoticed. In most cases fungi can be demonstrated microscopically or by culture (Table 23.2). When dermatophytes are absent, the diagnosis is usually erythrasma, due to overgrowth of coryneform bacteria which fluoresce brightly under Wood's light. Diphtheroids and cocci (*S. epidermidis*) are always present in simple ringworm. These are the organisms inhabiting normal interspaces but their numbers are greater in dermatophytosis simplex. Gram-negative organisms occur in approximately 20% of cases but only in small numbers. The fungi are either *T. rubrum* or *T. mentagrophytes*.

Dermatophytosis Complex (Bacterially-complicated Ringworm)

The moderate form shows white, hyperkeratotic, macerated, scaling interspaces, associated with itching and an acrid offensive odor (Fig. 23.1). The most severe examples exhibit redness, swelling, sometimes fissurls and erosions, accompanied by foul odor, discomfort, and marked pruritus. Dermatophytes were demonstrated in 55% of moderate cases, falling to 35% in the most severe forms. With increasing symptoms and signs of maceration and hyperkeratosis the numbers of coryneforms increase, especially those which we have designated as large-colony diphtheroids, a group which luxuriates in the cutaneous wetlands. The cocci come to represent a smaller proportion of the microflora (Table 23.2). Gram-negatives (notably Proteus and Pseudomonas) tend to increase pari passu with increasing severity but they comprise only a minority of the flora. Thus, there is a reciprocal relationship between dermatophytes and resident bacteria, particularly coryneforms. As the bacteria expand, the fungi retreat.

Interaction of Fungi and Bacteria in the Production of Symptomatic Athlete's Foot

Role of Aerobic Diphtheroids

Experimental manipulation of the microflora of persons with and without fungi has helped to clarify the relative roles of dermatophytes, aerobic diphtheroids and, to some degree, gram-negative organisms.

When occlusive dressings in the form of several layers of polythylene wrap were applied to the feet of persons with fungus-positive, asymptomatic dermatophytosis simplex, the interspaces became macerated, whitish, itchy and foul-smelling in about a week (Fig. 23.3). Removing the dressings did

Table 23.2. Clinical-Microbiological Features of Interdigital Athlete's Foot

	Clinical features	Dermatophyte	Aerobic diphtheroids		Gram-negative organism, %
			Lipophilic	LCD	
Dermatophytosis simplex (50)	Dry, scaling (Fig. 23.1)	90%	100% (78)	30% (9)	18% (0.06)
Dermatophytosis complex (45)	Hyperkeratotic, leukokeratotic plaques (Fig. 23.2)	55%	90% (50)	70% (33)	19% (0.9%)
Dermatophytosis complex (60)	Macerated, eroded, highly inflamed (Fig. 23.3)	35%	90% (12)	94% (81)	65% (3.0%)
Gram-negative athlete's foot (10)	Painful, hyperkeratotic or erosive with exudation and intense inflammation (Figs. 23.4)	25%	90% (10)	95% (75)	100% (50%)

not result in a rapid return to the original state; symptoms persisted for weeks. Occlusion of normal, fungus-free interspaces produced only asymptomatic, whitish maceration (superhydration). This returned to normal within a couple of days after allowing the feet to dry out. These observations strongly suggest that bacterial overgrowth in an interspace, previously invaded by dermatophytes, results in the hyperkeratotic, macerated, symptomatic type of athlete's foot. Presumably the horny layer is partially digested by the fungi, weakening its capacity to function as a barrier, thus allowing "toxic" products of a huge population of surface bacteria to damage further the viable tissue. Occlusion of dermatophytosis simplex will not result in the transformation to a symptomatic disease if the expansion of the microflora is prevented by topical application of bacteriostatic substances (Table 23.3). Thus, water and increased temperature alone do not account for the worsening that occlusion alone incites.

While the diphtheroids play a central role, gram-negative organisms sometimes become numerous enough to intensify the signs and symptoms. Their presence is suggested by thick, white plaques, intense itching and slow response to treatment. Sometimes a greenish tinge (pyocyanin from Pseudomonas) is detectable along with fluorescence under Wood's light. In rare instances, Pseudomonas takes over more or less completely, eliminating all competitors, including the original dermatophyte. This situation leads to a disabling disorder in which the white horny layer separates in toto, leaving painful, red erosions. Pseudomonas athlete's foot is the most extreme example of the mischief common bacteria can cause when they attain high densities in the interspaces previously invaded by dermatophytes.

Further support for a prominent role of bacteria comes from experiments in which patients with dermatophytosis simplex were treated with either topical antifungal agents or topical antibacterial agents while the feet were occluded.[10] Topical antifungals, such as tolnaftate, produced slow improvement in the symptomatic, macerated type of athlete's foot. Tolnaftate brought about swifter, surer improvement in the dry, scaly variety, for which fungi are primarily responsible. In contrast, topical antibiotics with no antifungal powers at all were capable of decreasing the signs and symptoms to a remarkable degree within about a week. Moreover, as the condition improved, fungus filaments were more readily identified in scrapings; that is to say, the suppression of bacteria allowed the dermatophyte to make a reappearance. In a couple of weeks the picture was typical of dermatophytosis simplex.

These observations make it clear that interdigital athlete's foot fluctuates between the simplex and the complex types, depending on the density of the coryneforms. The symptomatic disease is associated with teaming populations of aerobic bacteria, making the substrate unfavorable for fungus growth. Restraint of the bacteria enables the fungi to reoccupy the site, accompanied by conversion of the symptomatic disease to the simple scaling type.

Table 23.3. Effect of Occlusion With and Without Antimicrobial Treatment

Mycologic status	Treatment	Aerobic flora	Diphtheroids[a]	Cocci[a]	Gram-negative organisms[a]	Mycologic status postocclusion	Clinical appearance
Positive	Occlusion	8.61	8.01	5.47	6.1	8/20	Heavy maceration, hyperkeratosis, intensely symptomatic, lasting 7 to 10 days
Positive	Occlusion and antimicrobial agents[b]	5.51	2.97	4.99	5.11	19/20	Mild peeling, slight maceration; returned to dry, scaling in 2 days
Negative	Occlusion	8.16	7.98	6.82	3.58	0/10	Mild maceration. Normal in 1 to 2 days
Negative	Occlusion and antimicrobial agents**	5.12	3.02	4.94	2.44	0/10	Mild maceration, Normal in 1 to 2 days

[a]Mean expressed as logarithms
[b]Povodine-iodine and chloramphenicol

Role of Gram-negative Organisms

Low numbers of gram-negative organisms can be recovered from about 10% of normal feet.   Their prevalence and numbers increase as interdigital athlete's foot becomes more serious, ending in the thick, macerated plaques (Fig. 23.4).   The disease is most intense when species of Proteus and Pseudomonas are abundant, though not necessarily dominant.   When the interspace becomes essentially a monoculture of Pseudomonas, the clinical picture is frightening and the patient immobilized.   It seems likely that the proteases produced by these organisms can further degrade a horny layer already seriously damaged by dermatophytes and cornyeforms.   In severe dermatophytosis complex, with its mixed population of gram-negatives and gram-positives, it is difficult to estimate how much the former contributes to the symptomology.   This question can be approached experimentally by arranging conditions so that gram-negatives can grow freely while gram-positives cannot.

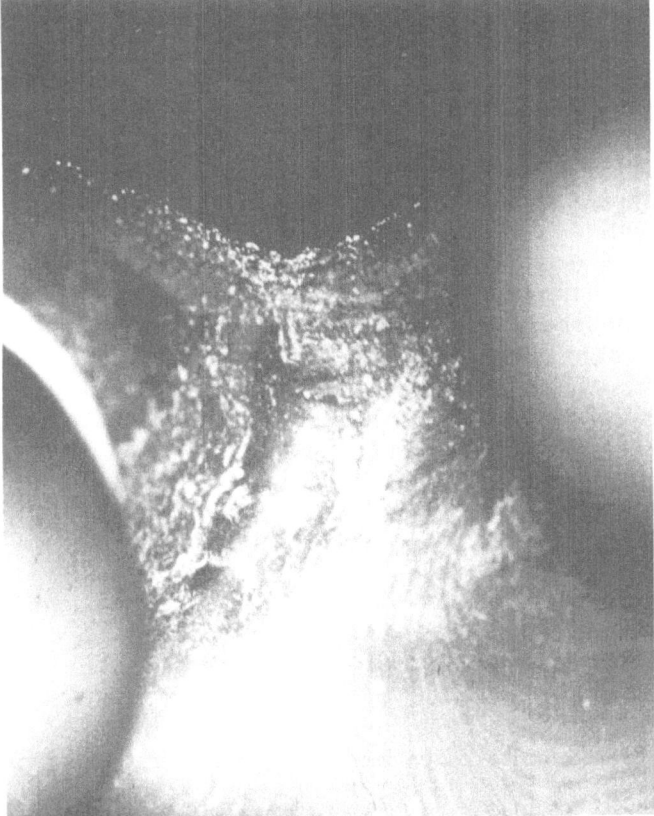

Fig. 23.4. Gram-negative athlete's foot. Eroded, exudative, painful interspace due to overgrowth of Pseudomonas.

Five subjects with clinically normal, mycologically negative interspaces, and five subjects with dermatophytosis simplex had both feet occluded for ten days as described above. The interspaces of one foot were treated daily with 15 g of a 5% hexachlrophene in carbopol gel, while the other foot was treated with the gel alone. Hexachlorophene selectively restrains gram-positives. The results are summarized in Table 23.4.

In the presence of 5% hexachlorophene, dermatophyte-infected interspaces developed a marked hyperkeratosis, maceration and erythema; in two subjects all the interspaces exhibited a severe form of dermatophytosis complex. The dominant organism was Pseudomonas, which attained very high densities. Normal interspaces, on the other hand, showed nothing but a whitish, moist appearance even though there was a comparable overgrowth of Pseudomonas. This regressed swiftly, leaving a normal interspace. High levels of Pseudomonas do not induce a disease. This organism, however, can greatly aggravate a preexisting fungus infaction. Fungus-infected interspaces that did not receive hexachlorophene behaved in the manner already described. They became macerated, hyperkeratotic, foul-smelling, and intensely pruritic—the typical transformation from dermatophytosis simplex to the bacterially complicated dermatophytosis complex.

This experiment demonstrates that gram-negative organisms, even the powerful Pseudomonas species, have limited capacity to induce damage in normal interspaces. The stratum corneum of this region is thick and affords great protection. Probably with longer periods of occlusion or with constant immersion in water, enough damage to the horny layer would have occurred to permit the overgrowth of Pseudomonas to induce clinical signs and symptoms. Taplin et al.[17] have demonstrated such a sequence of events in a military population. Prior to a training expedition in swamplands, gram-negative organisms from toe-web spaces were recovered from only 2 of 76 soldiers. After 19 days in the swamps, nearly 85% had hyperkeratotic, macerated, often painfully fissured interspaces, associated with an abundance of Pseudomonas. Hydration of the foot, especially if prolonged, is the factor above all others that brings forth the worst forms of athlete's foot, culminating in gram-negative intertrigo.

Effect of Partial Occlusion

The above experiments were deliberately extreme. However, the results do not necessarily apply to naturally occurring infections. The following experiment was designed to test whether the experimental findings were relevant to clinical experience.

Seven subjects with dermatophytosis simplex of the fourth interspaces and five subjects with normal interspaces had the fourth and fifth toes taped together for 7 days with the ends of the toes exposed (semiocclusion). Quantitative cultures and clinical assessments were made after the 7-day period. Another five subjects with dermatophytosis simplex had the fourth and fifth toes of both feet taped together for 7 days. Each day, one foot

Table 23.4. Role of Gram-negative Organisms

Mycological status and treatment	Total aerobic count[a]	Cocci[a]	Diphtheroid[a]	Total gram-negative bacteria[a]	Pseudomonas[a]	Clinical appearance
KOH(+) 5% hexachlorophene	8.21	4.8	4.1400	8.010	7.3096	Severe hyperkeratosis, maceration and erythema; persisted 7 to 10 days
KOH(+)	8.06	4.9	8.0200	4.800	0.0000	Marked hyperkeratosis, maceration lasting 7 days
KOH(−) 5% hexachlorophene in gel	8.36	4.3	4.0492	8.056	8.0240	Whitish, slightly macerated appearance; normal in 2 days

[a]Expressed as logarithm mean.

Table 23.5. Effect of Semiocclusion

Myocological status and treatment	Preocclusion			7 days postocclusion		
	Total aerobic count	Fungi	Clinical appearance	Total aerobic count	Fungi	Clinical appearance
KOH(+) occlusion	20.0×10^4	7/7	Scaling	43.3×10^4	6/7	Increased scaling, maceration, erythema and pruritus
KOH(+) magnesium	12.3×10^4	5/5	Scaling	7.8×10^4	1/5	No scaling, no symptoms
KOH(+) 1% tolnaftate	11.1×10^4	5/5	Scaling	28.6×10^4	2/5	Increased scaling and maceration

received 5 g of 1% magnesium pyrithione in hydrophilic ointment and the other foot was treated with 1% tolnaftate. The former is a broad-spectrum antimicrobial with antibacterial and antifungal properties, while the latter has only antifungal properties.

The results were similar to those obtained with more severe and prolonged occlusive dressings (Table 23.5). Fungus-infected interspaces developed maceration, hyperkeratosis, erythema, and increased itching. Interspaces free of fungi showed essentially no clinical changes. Pyrithione-treated dermatophytosis simplex proceeded to improve with a reduction in bacteria and fungi. Interestingly, interspaces treated with tolnaftate showed, as expected, a decrease in fungus content; still, the signs and symptoms actually increased, yielding a mild form of dermatophytotis complex. The expansion of resident coryneforms was sufficient to intensify the disease despite suppression of the dermatophyte.

These experiments are in agreement with the earlier findings. With occlusion, fungus-infected interspaces always show clinical worsening correlated with a huge overgrowth of coryneforms. A pure antifungal medication will not prevent aggravation of the disease by wet conditions. On the other hand, suppressing the overgrowth of the resident bacterial flora and simultaneously restraining fungal growth prevents this transformation.

Comparison of Antifungal, Antibacterial and Combined Therapy

Volunteers, five with dermatophytosis simplex and five with dermatophytosis complex, received one of the following medications twice daily.

1) 1% tolnaftate (solely antifungal);
2) 1% neomycin sulfate in hydrophilic ointment (solely antibacterial); or
3) a combination of the two.

The fourth interspaces were sampled quantitatively for bacteria before and after 1 and 2 weeks of therapy; clinical assessment was made at the same time.

The results (Table 6) demonstrate that a purely antifungal agent, such as 1% tolnaftate, produced satisfactory clinical improvement only in the dry, scaling form of athlete's foot (dermatophytosis complex). The complex form became only a little milder. Neomycin, on the other hand, was beneficial in the wet, macerated form, which became much less symptomatic with complete abolition of odor. Also, it was easier to demonstrate fungi. It had no effect in dermatophytosis simplex. Thus, bacteriostats active against gram-positive organisms will provide appreciable relief but, of course, do not affect the primary fungus infection. Needless to say, the combination of tolnaftate and neomycin was the the most effective mode of treatment. It should be noted that though 2 weeks is adequate for comparisons, it is too short a time period for estimating conclusive efficacy.

Table 23.6. Organism Count Pretreatment and Posttreatment

Clinical type/treatment	Pretreatment		7 days therapy			14 days therapy		
	TA[a]	Fungus	TA	Fungus	Clinical state	TA	Fungus	Clinical state
Dermatophytosis simplex								
1% Tolnaftate	7.6240	5/5	7.46	3/5	50% improved	7.32	1/5	75% improved
1% Neomycin sulfate	7.1070	5/5	5.0792	5/5	No change	4.64	5/5	No change
1% Neomycin sulfate and 1% tolnaftate	7.1816	5/5	4.5052	2/5	50% improved	3.505	1/5	75% to 100% improved
Dermatophytosis complex								
1% Tolnaftate	8.3120	2/5	8.4771	1/5	No change	8.62	2/5	No change
1% Neomycin sulfate	8.7130	4/5	5.505	4/5	20% improved	4.68	3/5	50% improved
1% Neomycin sulfate and 1% tolnaftate	8.0170	2/5	4.3802	2/5	40% improved	2.6990	1/5	75% improved

[a]Total aerobic count per interspace expressed as logarithmic mean.

Comment

The primary event in interdigital athlete's foot is the invasion of the horny layer by dermatophytes.   This state is manifested by mild to moderate peeling, mostly without symptoms. As a result of hot weather, sweating, exercise, or tight shoes enough moisture accumulates to stimulate an overgrowth of bacteria.   In wet toe webs, the number of aerobic bacteria may increase by two to three orders of magnitude. Hydration and an enormous population of resident microbes are not sufficient to produce disease by themselves. When normal interspaces are superhydrated, little change besides a transient soggy whiteness occurs despite a great growth of bacteria. Prior damage to the horny barrier by fungi is a precondition for aggravation.

Athlete's foot is a chronic infection, perhaps a permanent one, and patients are fully aware of the oscillation between the simplex and the complex forms. Flare-ups are common in the summer or after intense exercise and can be experimentally induced by occlusion of fungus-infected feet. As the simple disease converts to a wetter, more symptomatic form, recovery of fungi becomes increasingly difficult.

We have biopsied interspaces in which fungi could not be demonstrated by culture or microscopic examination.   Histologically, sparse filaments could regularly be found in the deepest portion of the horny layer. Hence, fungi do not disappear but rather are driven underground. The stratum corneum in this region may be as thick as $400-500\mu$. If the symptomatic interspace is "dried out," dermatophytosis simplex reappears and fungi again become demonstrable in scrapings. This lability and chronicity of athlete's foot are thoroughly appreciated by clinicians. Merely wearing sandals converts dermatophytosis complex to the simple, scaling form.

The difficulty of demonstrating fungi when dermatophytosis simplex converts to complex was earlier demonstrated by Strauss and Kligman,[16] who kept the feet completely occluded for prolonged periods. We are satisfied that these experiments provide an explanation for the inability to find fungi in typical cases of interdigital athlete's foot. We have not overturned the doctrine that athlete's foot is caused by fungi. However, fungi alone are not sufficient to produce notable clinical disease requiring treatment. Bacteria play a decisive role; without them dermatophytosis cannot develop.

In dermatophytosis complex, a pure antifungal agent like tolnaftate brings about only a slow, modest improvment in a couple of weeks but is effective in the simple scaling form. Topical antibacterial agents alone produce definite clinical benefit of dermatophytosis complex within a short time. As the interspaces objectively and subjectively improve, it becomes increasingly easier to demonstrate fungi. The final picture, weeks later, is typical dermatophytosis simplex. The disease is not cured, merely curbed. A combination of an antibiotic and an antifungal (either by two separate agents or by the new broad-spectrum topical agents, such as miconazole nitrate and clotrimazole) results in swifter and greater resolution of signs and symptoms.

Other maneuvers, such as removal of shoes and separation of interspaces with soft pads, lead to a decrease in the amount of moisture and a corresponding clinical improvement. The bacterial population declines sharply as the interspaces dry.

Treatments that simultaneously suppress bacteria and fungi as well as produce a local drying or astringent effect would be expected to be the most efficacious, since they attack all aspects of the disorder. A 30% concentration of the hexahydrate form of aluminum chloride will produce considerable clinical improvement in symptomatic wet athlete's foot in as short a time as a week.[9] Aluminum chloride suppresses the surface overgrowth of bacteria and is drying in addition. Its penetrating powers through horn are too limited to enable it to reach and kill the deeply situated fungi. Hence, this agent provides immediate symptomatic relief but does not cure the fungus infection.

We see dermatophytosis simplex as a pure fungus infection, a trivial condition by itself. However, whenever the interspaces become hydrated, bacteria grow rapidly and reach densities of millions per interspace. Typically, these organisms are part of the resident microflora, i.e., they are mainly cocci and diphtheroids. Thus, the ordinary form of dermatophytosis complex represents a collaboration between fungi and large-colony diphtheroids. In the more severe cases of dermatophytosis complex occurs when Pseudomonas species take over completely, leading to exfoliation of thick sheets of horny layer and producing an extremely painful, erosive, prurulent interspace that results in incapacitation. In these cases, therapy should include anti-Pseudomonas tactics, such as acetic acid compresses, appropriate topical antibiotics and drying out of the interspaces.

References

1. Ajello L, Keeney EL, Broyles EN (1945) Observations on the incidence of tinea pedis in a group of men entering military life. Johns Hopkins Med J 77:440–447
2. Amonette RA, Rosenberg EW (1973) Infection of toe webs by gram-negative bacteria. Arch Dermatol 107:71–73
3. Davis CM, Garcia RL, Riordan JP et al. (1972) Dermatophytes in military recruits. Arch Dermatol 105:558–560
4. English MP, Gibson MD (1959) Studies in the epidemiology of tinea pedis: 1: Tinea pedis in school children. Br Med J 1:1442–1446
5. Fitzpatrick TB, Arnt KA, Clark WH, et al. (1971) Dermatology in general medicine. McGraw-Hill New York
6. Gentles JC, Evans EGU (1973) Foot infections in swimming baths. Br Med J 3:260–262
7. Goto M (1970) Ecological study of interdigital athlete's foot Jpn J Dermatol 80:130–138
8. Holmes JB, Gentles JC (1956) Diagnosis of foot ringworm. Lancet II:62–63
9. Leyden JJ, Kligman AM (1975) Aluminum chloride in the treatment of symptomatic athlete's foot. Arch Dermatol 111:1004–1010
10. Leyden JJ, Kligman AM (1978) Interdigital athlete's foot. Arch Dermatol 114:1466–1473
11. Marples M, Bailey MJ (1957) A search for the presence of pathogenic bacteria and fungi in the interdigital space of the foot. Br J Dermatol 69:379–388

12. Marples M, Chapman EN (1959) Tinea pedis in a group of school children. Br J Dermatol 71:413–421
13. Marples M, DiMenna ME (1949) A survey of the incidence of interdigital fungus infections in a group of students from the University of Otago. Med J Aust 2:156–161
14. Rook A, Wilkinson DS, Ebling FHG (1972) Textbook of dermatology, 2nd edn. Blackwell Scientific Publications, Oxford, England
15. Stoughton RB (1970) Bioassay of antimicrobials: A method for measuring penetration of agents into human skin. Arch Dermatol 101:160–166
16. Strauss JS, Kligman AM (1957) An experimental study of tinea pedis and onychomycosis of the foot. Arch Dermatol 76:70–79
17. Taplin D, Bassett DC, Mertz PM (1971) Foot lesions associated with *Pseudomonas cepacia*. Lancet II:568–571

Chapter 24
Experimentally Induced Cutaneous Infections in Man

W. Christopher Duncan, Mollie E. McBride,
and John M. Knox

The experimental production of disease often permits a better understanding of etiologic factors than can be obtained by clinical study. The variables encountered in clinical infection and the difficulties encountered with terminology point to the need for experimental models. The development of model infection systems is important not only in elucidating the factors involved in the pathogenesis of spontaneous infection but also because models would be useful in testing antimicrobial formulations.

In the 1880s Garre[15] and Bockhart[5] independently produced experimental cutaneous infections with relative ease. Each produced multiple pustules and subsequently developed furuncles. This type of experimentation was not new to Bockhart who had previously transmitted gonorrhea experimentally.

In 1940, Epstein[12] reviewed efforts to determine the etiology of impetigo and noted that 10 of 13 authors attempting to produce experimental infections succeeded. He produced a large lesion of bullous impetigo with inoculated staphylococci. However, the number of bacteria or method of inoculation were not mentioned.

Three years later Bigger and Hodgson[3] attempted experimental production of impetigo with crusts and isolated staphylococci as the inoculum. Multiple inoculation methods were attempted, but only intradermal injection produced significant reactions. They concluded that streptococci are rarely, if ever, the etiology of impetigo. In the same issue of *Lancet*, Sheehan and Ferguson[23] produced "full impetigo" by inoculating *S. aureus* onto scarified skin in 4 of 13 attempts and in 4 of 18 attempts with impetigo fluid as the inoculum. Inoculation of streptococci failed to produce a lesion "in any way suggestive of impetigo."

In the early 1950s, O'Brien[19,20] applied staphylococci in cups with nutrient agar to intact skin and produced small lesions of bullous impetigo in 4 of 24 experiments. The follicular location of the lesions suggested to him that impetigo may initially be poral in origin.

In the mid-1950s, Elek and Conen[10,11] quantitated the human infections studies.   They determined the minimum pus-producing dose of *Staphylococcus aureus* when inoculated intradermally to be 2–8 million organisms (never less than 1 million), a figure later verified by Maibach.[17] They were unable to demonstrate any difference between staphylococci isolated from healthy nasal carriers and epidemic strains.   Although a variety of adjuvants were tried, only a silk suture significantly reduced the number of staphylococci necessary to produce infection.

In 1960 Foster and Hutt[14] excoriated skin and occluded it following inoculation with 10[6] *S. aureus* organisms.   They were unable to differentiate between the inflammatory response to five types of *S. aureus* or between *S. aureus* and *S. albus*.   They obtained clinical infections with as few as 15 organisms.   Two of their three subjects developed distant furuncles and one became a nasal carrier.

From 1959 to 1963, Maibach[17] amplified work done by previous investigators and introduced several new approaches to gain additional insight into the pathogenesis of cutaneous staphylococcal infections.   Numerous strains of staphylococci, including phage type 80/81, were employed but only strains sensitive to at least one easily administered antibiotic were used. In over 1 000 inoculations, only one untoward infection occurred; however, Maibach termed their success at producing infections "inconstant."

Singh, Marples and Kligman[25] produced a cutaneous infection dependent upon *S. aureus* if they first destroyed the resident flora; however, they likened the reaction to that of toxic contact dermatitis rather than a true infection. Marples and Kligman[18] described a staphylococcal wound infection reminiscent of the method of Foster and Hutt.[14] Stripping skin to the glistening layer both removed its protective microflora and evoked transudation of nutritious serum to establish a site receptive to the infectious overtures of *S. aureus*; however, removing the still-necessary occlusive covering terminated the affair.

In 1966, recognizing as had previous investigators the need for a model of human cutaneous infection, we embarked on a series of experiments directed initially at developing a reproducible cutaneous infection.[8]   We began by simply placing a variety of organisms on the skin and providing them with protection and supplemental nutrients by utilizing small plastic housings filled with nutrient agar.   This device was modified from O'Brien.[19] The organisms were recent isolates from clinical infections, most having been in our laboratory 1 week or less.   *Staphylococcus aureus, Streptococcus pyogenes* group A, *Pseudomonas aeruginosa*, and *Corynebacterium minitumissimum* were used.   These were left in place 24, 48, 72 and 96 h. Few infections were produced by this method; we abandoned *P. aeruginosa* and *Corynebacterium minitumissimum* because they failed to produce any reaction and colonized the skin for only short periods.

Singh[24] also attempted experimental cutaneous infection with *Pseudomonas aeruginosa* but was unable to cause infection on intact skin; however, intradermal inoculation resulted in inflammatory papules, nodules and abscesses in conjunction with systemic symptoms.

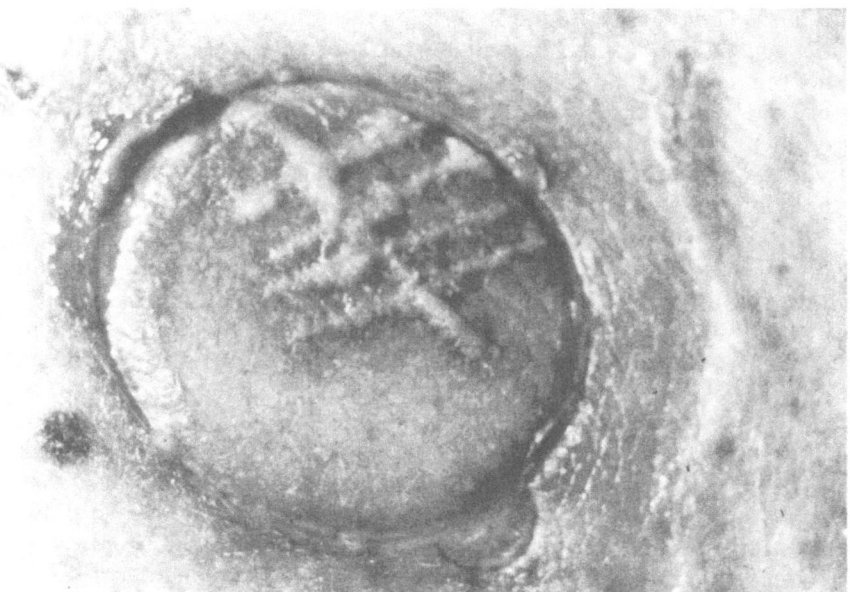

Fig. 24.1. Site superficially scratched prior to inoculating with overnight broth culture containing 2×10^5 streptococci.

Surface Alterations

Following our initial relatively unsuccessful attempts, we slowly added various forms of trauma. This consisted of techniques whereby we felt we were simulating what might occur in day-to-day living. Initially subjects were punctured with a blood lancet to produce a uniform wound 5 mm deep; scratched (Fig. 24.1) sufficiently to produce breeding; or abraded. The inoculum was placed over the traumatized sites and left in place 96 h. This was no more successful then the earlier studies. We reasoned that perhaps the stab wounds were closing before the bacteria could gain entrance and therefore polyethylene granules were rubbed into the stab wounds. This promptly produced six abscesses. Cautiously repeating the studies, however, produced no additional infections in 180 attempts.

Regional Differences

The initial studies utilized only the arms and upper trunk. We then began parallel attempts on the arms and legs using the technique of applying the bacteria, stabbing through the drop of broth, and covering with plastic tape. A remarkable difference was immediately evident. In 290 attempts in 40 subjects, we achieved a success rate of 37% on the leg, as opposed to a 14% success rate on the arms. To further examine regional differences in infectivity, four areas on each subject were inoculated in an identical manner. The back was successfully infected in 15% and the arm in 13% of the attempts. The thigh was successfully infected in 21% of the attempts and the leg in 38%. This difference was significant at the 99% confidence level. We concluded that circulatory factors may account for these differences.[8]

Plucking

During the course of the previous traumatic studies, we had noted that mechanically plucking the hair on the legs followed by application of the bacteria and occlusive plastic patches produced significant pustular reactions in approximately twice as many as those in which we utilized a stab technique. (Maibach[17] plucked hairs, but did not occlude the inoculum.) Further pursuit of this method of producing infections led to the development of a model of human bacterial infection that reliably produced a bacterial folliculitis or impetigo or both in 60%–80% of attempts.[9] The infections were produced only where the hair had been extracted (Figs. 24.2, 24.3). Surrounding intact follicles exposed to the same bacteria and occluded in the same manner were unaffected. The infections resulting from this experimental model could easily be graded on a 1+ to 4+ scale as follows: 1+, one or two follicular pustules; 2+, three or more follicular pustules; 3+, one or

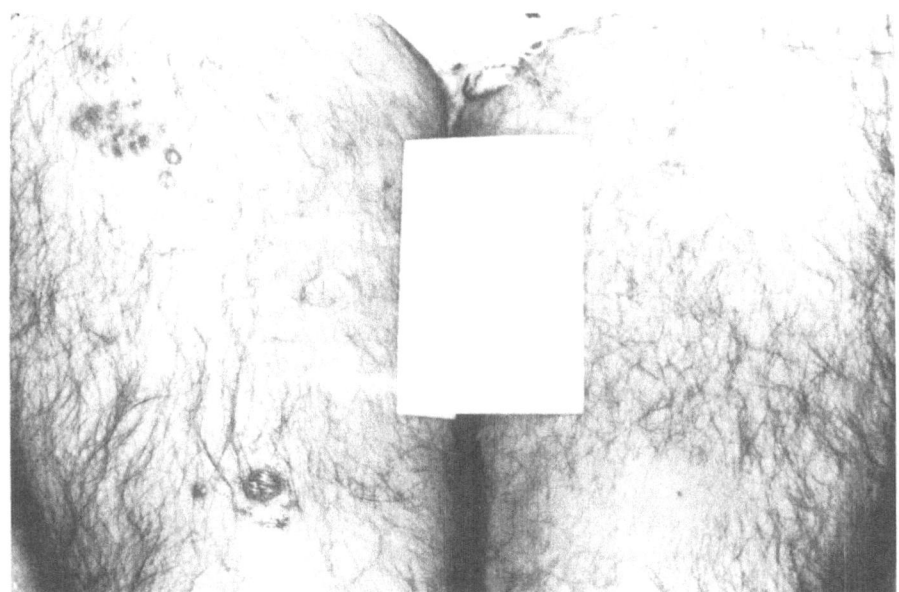

Fig. 24.2. Four sites from which hair has been plucked. Upper left thigh, broth control; lower left, *Streptococcus pyogenes*; upper right, *S. aureus*; lower right, *S. aureus* and *Streptococcus*.

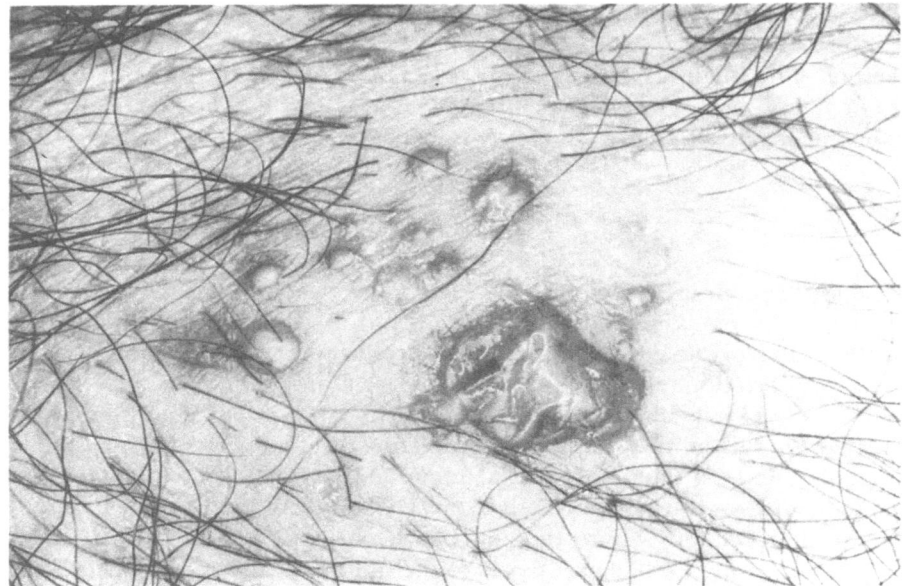

Fig. 24.3. Plucked site inoculated with *S. aureus*; 3+ infection.

more large (3 to 7 mm) extrafollicular pustules; 4+, one or more pustules 8 mm or more in diameter. For purposes of recording successful inoculations, at least a 2+ infection was required. Shaving the hair immediately prior to application of the bacteria was compared with pulling and found ineffective in producing infection.

The interval of time between the extraction of the hair and application of the bacteria was varied to determine its importance in the pathogenesis of this model. No infection occurred at the sites when 24 h elapsed between pulling the hair and application of the bacteria. When the interval between hair extraction and application of bacteria was reduced to 12 h, there was no significant difference in the number of infections produced in the test as compared with the control. Whether it was formation of the keratinous plug or healing of the disrupted follicle wall which prevented entrance of the bacteria was never determined.

Bacteria

Our earliest studies indicated that a combination of *Staphylococcus aureus* and *Streptococcus pyogenes* as the inoculum resulted in more frequent infections than either organism alone. Many different streptococcal strains isolated from impetigo were utilized initially and only rarely produced infections when inoculated alone. When we commenced plucking hairs to initiate infection we utilized *Staphylococcus aureus*, phage 71 or 80/81, and a group

A β-hemolytic streptococcus (T-type 3/13/B3264, M-type 33). The latter was selected as a non-nephritogenic streptococcus; 0.02 ml. of overnight (18-h) cultures in brain–heart infusion broth containing 10^8 staphylococci/ml and 10^7 streptococci/ml resulted in inocula of 10^5 staphylococci and 10^4 streptococci/cm^2. The inoculum was covered with a 4 cm square of polyethylene film and sealed to the skin with impervious plastic adhesive tape. Two sites could be inoculated on each anterior thigh.

Interaction studies utilizing a lawn spotting method between the T-type 3/13/B3264, M-type 33 streptococcus and the phage 71 and the 80/81 staphylococci demonstrated inhibition of the streptococcus; another phage 71 and two group III staphylococci demonstrated less inhibition. Mixed-culture techniques confirmed the results from the lawn spotting method. Despite rather significant differences in vitro, no great differences in recovery could be detected in vivo. The streptococcus survived briefly after the occlusion was removed. whether it had been applied by itself or in combination with a staphylococcus. Log phase streptococcus was compared to stationary phase as the inoculum; no differences in either virulence or recovery were detected.

Histopathology

Biopsies were obtained from 21 infections; all were serially sectioned and alternate sections stained with hematoxylin and eosin. When indicated, the alternate sections were stained with Gram-Weigert.

The pustules were associated with a follicle in virtually every instance. A keratinous plug[26] (Figs. 24.4, 24.5) was present in the roof of all but two pustules. Hairs were never found within any of the pustules, though broken hairs were rarely found in the follicle well below the developing pustule. Rupture of the follicular pustule into the dermis occurred above the level of the sebaceous gland (Fig. 24.6). In many follicles the epithelium was destroyed, leaving only the surrounding connective tissue sheath as the apparent retaining structure. In these instances, the inflammatory reaction in the adjacent dermis was surprisingly scanty. Special stains for the basement membrane were inconclusive; however, electron microscopic studies on a small number of lesions showed an intact basal lamina despite loss of the follicular epithelium. The surrounding collagen fibers also appeared normal except for mild edema between the fibers. The dermal inflammatory response surrounding unruptured follicles consisted almost exclusively of lymphocytes and histiocytes. The sweat glands and ducts were never involved, though occasional polymorphonuclear leukocytes could be seen in the intraepidermal duct lumen and a few lymphocytes and polymorphonuclear leukocytes were found around the dermal duct. In several instances, the uninvolved sweat gland duct was seen coursing through the edge of an intraepidermal pustule. Bacterial stains failed to show any bacteria in the dermis as long as the pustule remained localized to the follicle. Bacteria

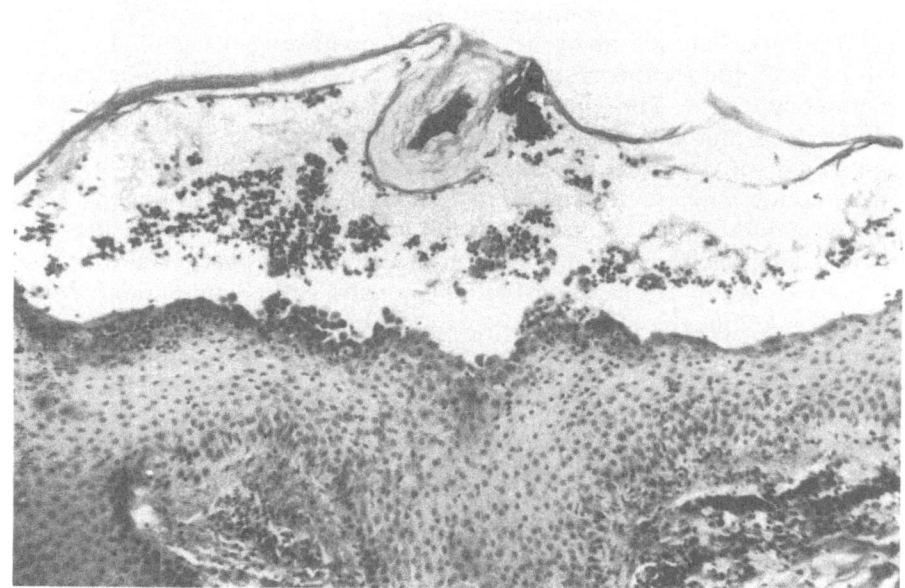

Fig. 24.4. Keratinous plug in subcorneal roof of follicular pustule (*S. aureus*).

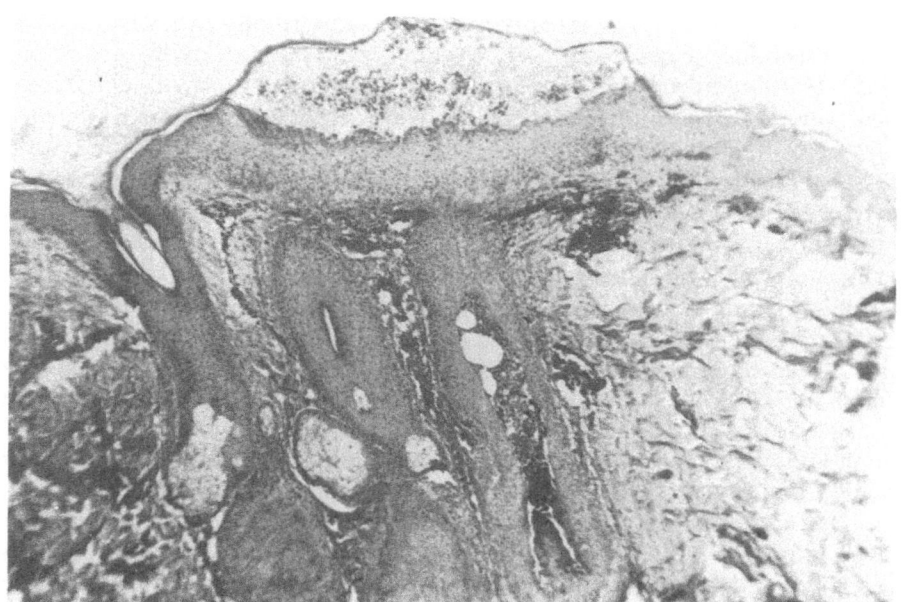

Fig. 24.5. (Same lesion as Fig. 24.4.) Deep follicular component of pustule. There is only minimal inflammatory response in the surrounding dermis (*S. aureus*).

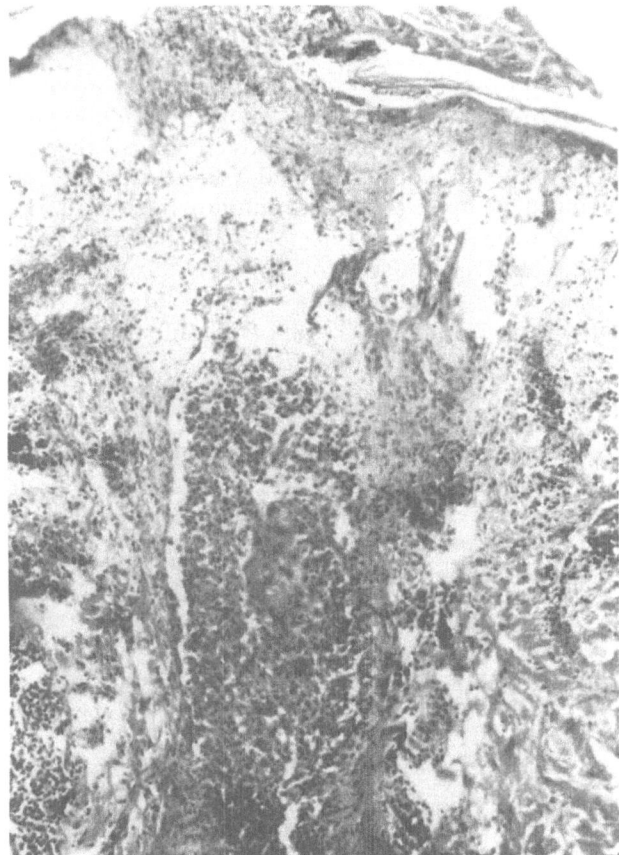

Fig. 24.6. Destruction of follicular epithelium by the infectious process and spread to adjacent dermis (*S. aureus*).

were readily found free in the pustules, within polymorphonuclear leukocytes, and occasionally within the wall of the pustule.

Prophylaxis Studies

Systemic antibiotics were effective in preventing this model of infection only if the staphylococcus utilized was sensitive to the antibiotic. Penicillin and tetracycline, to which the staphylococcus was resistant, did not prevent infection with this model. Erythromycin, on the other hand, to which the staphylococcus was sensitive, completely prevented infection. It would thus appear that only the staphylococcus was essential in initiating infection and that the streptococcus was merely a bystander. This correlated with our finding throughout the series of investigations that the streptococcus by itself was essentially incapable of initiating infection.

Antibody Responses

Antistreptococcal antibody determinations were performed on a group of ten subjects to determine if a correlation between resistance to infection and preexisting antibodies could be made or if an antibody response resulted from the induced infections. Three subjects developed 2+ infection, one developed 3+ infection, and four developed 4+ infection following inoculation with a group A β-hemolytic streptococci T-type 3/13/B3264, M-33 and *Staphylococcus aureus* phage 80/81. Antistreptolysin O (ASO), antideoxyribonuclease B (ADNase B) and antihyaluronidase (AH) determinations were made prior to infection and at 2 and 4 weeks following infection. None of the subjects had abnormal titers prior to the infection nor was there a correlation between the existing level of antibodies and the production of infection. No rises in titers occurred.

The antibody response in patients with streptococcal impetigo to the streptococcal antigens DNAse B and hyaluronidase is reported as vigorous,[4,16] so it is surprising to find no response in these subjects if indeed the streptococcus was an active participant in the infectious process. However, the duration of the antigenic stimulus may have been too brief in this experimental model.

Racial Differences in Infectibility

During the several years in which we worked with the experimental model, there appeared to be a difference in the rate of infections of the white subjects compared to the black subjects. This was examined retrospectively to determine the validity of the observation. Unfortunately, race was not known for every experiment performed, however the following data were obtained: 929 inoculations in 161 white subjects resulted in 231 infections (25%), whereas 319 inoculations in 54 black subjects resulted in only 46 infections (14%). This difference was significant at the 0.05 level and correlates with the finding that tropical pyodermas, mostly streptococcal, were more than 2½ times as prevalent in whites as in blacks.[1]

Summary

What has been learned by these and the earlier investigations, other than that skin is remarkably resistant to bacterial assault? The most frequent causes of infection of the skin are the gram-positive cocci *Staphylococcus aureus* and *Streptococcus pyogenes*, yet all our collective studies would seem to indicate otherwise. Insofar as the pathogenesis of impetigo, our studies, like the earlier ones of Bigger and Hodgson,[3] failed to find a role for the streptococcus, although virtually all epidemiologic studies[6,7,13] find to contrary. There are, however, significant differences in our experimental method and what occurs in real life. All of the experimental infections have been attempted in adults, whereas virtually all the epidemiologic studies of impetigo have been compiled on children; adult contacts of these children are infrequently infected.[7] Allen, Taplin, and Twigg's[1] study of tropical pyodermas is the only modern exception to this.

While streptococci can be recovered from the skin of children and their environment prior to onset of clinical impetigo, we rarely could recover them 24 h after removing the occlusion. Aly et al.[2] found that *S. pyogenes* disappeared completely from the skin within 5 h in 50% of their adult subjects. Why *S. pyogenes* can persist on children's skin but not on adults' has not been elucidated. Although Allen, Taplin, and Twigg[1] implicated *S. pyogenes* as the cause of tropical bacterial pyodermas, *S. aureus* was recovered from 75% of the lesions. They adduced, however, that *S. aureus* was only a secondary wound colonizer.

Plucking hairs would appear to remove a major barrier to infection and this experimental model lucidly demonstrates that follicular openings may represent important portals through which bacteria can gain a foothold and produce infection. It is not difficult to imagine how hairs could be partially or completely pulled from their follicular home as a result of friction from clothing, towels, hands, opposing skin, or razors, with the resultant ingress of bacteria and development of clinical infection.

Quantitative models of experimental human cutaneous infection with *Candida albicans*[21] and *Trichophyton mentagrophytes*[22] have been established and would appear worthwhile for studying pathogenesis, immunity, and treatment. To be noted, however, is that occlusion has been requisite in all methods of inducing infection whether the organism is yeast, dermatophyte, or bacteria. Occlusion to the degree required for successful experimental infection rarely, if ever, occurs naturally so that the conditions of natural infection have yet to be defined.

References

1. Allen AM, Taplin D, Twigg L (1971) Cutaneous streptococcal infections in Vietnam. Arch Dermatol 104:271–80
2. Aly R, Maibach HI, Shinefield HR, Strauss WG (1972) Survival of pathogenic microorganisms on human skin. J Invest Dermatol 58:205–10
3. Bigger JW, Hodgson GA (1943) Impetigo contagiosa, its cause and treatment. Lancet I:544–7
4. Bisno AL, Nelson KE, Waytz P, Brunt J (1973) Factors influencing serum antibody response in streptococcal pyoderma. J Lab Clin Med 81:410–20
5. Bockhart M (1887) Monatsh prakt Dermat 6:450–571
6. Dajani AS, Ferrieri P, Wannamaker LW (1972) Natural history of impetigo. II. Etiologic agents and bacterial interactions. J Clin Invest 51:2863–71
7. Dillon HC (1968) Impetigo contagiosa: Suppurative and nonsuppurative complications. I. Clinical, bacteriologic, and epidemiologic characteristics of impetigo. Am J Dis Child 115:530–41
8. Duncan WC, McBride ME, Knox JM (1970) Experimental production of infection in humans. J Invest Dermatol 54:319–23
9. Duncan WC, McBride ME, Knox JM (1971) Experimental production of cu-

taneous bacterial infections in humans. Antimicrob Agents Chemother, 1970. Amer Soc Microbiol, Washington, D.C. pp. 137–139

10. Elek SD (1956) Experimental staphylococcal infections in the skin of Man. Ann NY Acad Sci 65:85–90
11. Elek SD, Conen PE (1957) The virulence of *Staphylococcus pyogenes* for Man: A study of the problems of wound infection. Br J Exp Pathol 38:573–86
12. Epstein S (1940) Staphylococci. Impetigo contagiosa. Arch Dermatol 42:840–55
13. Ferrieri P, Dajani AS, Wannamaker LW, Chapman SS (1972) Natural history of impetigo. I. Site sequence of acquisitions and familial patterns of spread of cutaneous streptococci. J Clin Invest 51:2851–62
14. Foster WD, Hutt SR (1960) Experimental staphylococcal infections in Man. Lancet 2:1373,1375
15. Garré C (1885) Fortschr Med 3:165–73
16. Kaplan EL, Anthony BF, Chapman SS, Ayoub EM, Wannamaker LW (1970) The influence of the site of infection on the immune response to group A streptococci. J Clin Invest 49:1405–14
17. Maibach HI, Hildick-Smith G (eds) (1965) Skin bacteria and their role in infection. McGraw-Hill, New York
18. Marples RR, Kligman AM (1972) Bacterial infection of superficial wounds: A human model for *Staphylococcus aureus*. In: Maibach HI, Rovee DT (eds) Epidermal wound healing. Yearbook Medical Publishers Chicago
19. O'Brien JP (1950) The etiology of poral closure. II. The role of staphylococcal infection in miliaria rubra and bullous impetigo. J Invest Dermatol 15:102–33
20. O'Brien JP (1952) Experimental staphylococcal folliculitis. Arch Dermatol 65:206–15
21. Rebora AE, Marples RR, Kligman AM (1973) Experimental infection with *Candida albicans*. Arch Dermatol 108:69–73
22. Reinhardt JH, Allen AM, Gunnison D, Akers WA (1974) Experimental human *Trichophyton mentagrophytes* infections. J Invest Dermatol 63:419–22
23. Sheehan HL, Fergusson AG (1943) Impetigo etiology and treatment. Lancet 1:547–50
24. Singh G (1974) Pseudomonas infection of skin: An experimental study. Int J Dermatol 13:90–93
25. Singh G, Marples RR, Kligman AM (1971) Experimental *Staphylococcus aureus* infections in humans. J Invest Dermatol 57:149–63
26. Strauss JS, Kligman AM (1958) Pathologic patterns of the sebaceous gland. J Invest Dermatol 30:51–61

Chapter 25

Exotic Infection: Its Relevance to Clinical Microbiology

W. C. NOBLE

Exotic infection should strictly mean infection from another country or that which is at least strange or bizarre, but is most frequently used among dermatologists to mean infection that is rare. Publishing reports of exotic infection may seem to be a form of clinical gamesmanship, but I believe it has value to others. Rarity may be the result of a genuine lack of cases of infection, there may be local rarity with an abundance of cases elsewhere or infection may appear rare because it is seldom reported.

An example of rarity that may simply reflect a lack of reporting is fish tank granuloma. The first report was in 1962 when Swift and Cohen recognised the connection between lesions, usually of hands, infected with *Mycobacterium marinum* and acquired as a result of cleaning out a tropical aquarium in which fish had recently died, presumably of *M. marinum* infection. Sporadic cases were then reported rising to a peak of ten reported cases in 1977. In 1978 there seems to have been only one case reported in a major journal; this involved a cotrimoxazole-resistant strain,[3] indeed this was the reason for publication. We know of a number of infections with this organism that have not been published, but it seems probable that underreporting of this disease is not a public health problem. The disease is generally minor and self-limiting, human-to-human transmission does not seem to occur, and, unlike the closely related swimming pool granuloma, large outbreaks do not occur.

Rarity as a result of genuine lack of cases seems likely in, for example, anthrax or pseudomonas skin infection. In 1966 Noble and Savin[4] reported infection of the skin with *Pseudomonas aeruginosa* that occurred as a result of a contaminated corticoid cream being applied to skin lesions under plastic occlusion. These lesions were characterized by ecthyma and staining of the skin with pyocyanin produced by the *P. aeruginosa*. We have not seen similar infections in the 13 years that have elapsed since publication, putting them firmly in the "exotic" class. Publication of this particular form of exotica is justified because, as one might have anticipated, others found that corticoid and other creams were similarly contaminated, although no clinically apparent infection had been detected. This particular publication was a warning to others that infection might occur.

A disease rare in Britain but much less so elsewhere is leprosy. Though this particular disease is usually diagnosed fairly promptly, the interest in and awareness of leprosy must mean that few cases slip through the net.

A disease that is a rarity in the U.K., and indeed in most of the U.S.A., is cutaneous diphtheria. Skin infection is usually reported as occurring during the small outbreaks of nasopharyngeal diphtheria that we occasionally experience.[2] There are occasional reports of cutaneous diphtheria in tropical or subtropical areas, e.g., in British troops in Malaysia during the Second World War. However, the value of occasional publications serving to keep memory alive to the possibility of infection is seen in the Seattle outbreak from 1972–1975.[5]

Seattle health authorities had detected only one instance of cutaneous infection with diphtheria in the years 1967–July 1972. From September 1972 to the end of 1975, however, there were 414 cases of skin infection, only 43 of which were also pharyngeal cases and a further 77 pharyngeal cases only. Early in the epidemic, most of the infections were of the pharynx but during 1974–1975 infections were predominantly of the skin; 90% of strains of *C. diphtheriae* isolated were toxigenic. There was a heavy predominance of males ($430/491 = 87.6\%$) many of whom were heavy alcohol drinkers and the epidemic was centered around the Skid Row area.

The importance of skin diphtheria seems only to have been appreciated in recent years. Belsey and LeBlanc[1] have reported that skin lesions persist longer than do pharyngeal lesions, that acquisition of infection is more likely to follow from contact with a patient with skin diphtheria than pharyngeal diphtheria, and they suggest that the number of skin cases may control the progress of the epidemic. If this is so, we can only hope that our clinical colleagues will continue to report their exotica since the rare cases may occasionally herald a period of more frequent disease.

In this paper I have tried to show that, on occasion, publication of exotica or rare infections may be of value in alerting others to the possibility of certain types of infection which may, as a result of undetermined circumstances, become much less rare and prove a major public health problem.

References

1. Belsey MA, Leblanc DR (1975) Skin infections and the epidemiology of diphtheria, acquisition and persistence of *C. diphtheriae* infection. Am J Epidemiol 102:179–184
2. Butterworth A, Simmons LE, Ironside AG, Mandal BK, Fraser-Williams R, Brennard J, Mann NM, Simon S (1974) Diphtheria in the Manchester area 1967–1971. Lancet II:1558–1561
3. Cunningham MJ, White PM, Samman PD (1978) Cotrimoxazole resistant *Mycobacterium marinum*. Br J Dermatol 99:597
4. Noble WC, Savin JA (1966) Steroid cream contaminated with *Pseudomonas aeruginosa*. Lancet 1:347–349
5. Pedersen AHB, Spearman J, Tronca E, Bader M, Harnisch J (1977) Diphtheria on Skid Road, Seattle, Washington 1972–1975. Public Health Rep 92:336–342
6. Swift S, Cohen H (1962) Granulomas of the skin due to *Mycobacterium balnei* after abrasions from a fish tank. N Engl J Med 267:1244–1246

Chapter 26

Dermatitic Skin: Microbiology and Treatment
HOWARD MAIBACH AND RAZA ALY

Microflora of Dermatitic Skin

The microbial flora of dermatitic skin varies from the resident flora of normal skin.[2,3] The normal skin flora consists mainly of coagulase-negative staphylococci and lipophilic and nonlipophilic coryneforms. Lipophilic coryneforms are more common in moist areas (axilla, groin and toe webs), than on dry skin. A small number of gram-negative rods are found in the humid skin regions. The nose and the perineum are the common resident carriage sites of *S. aureus*. The general skin surface yields *S. aureus* in about 5%–10% of the normal population.

Dermatitic skin provides favorable conditions for microbial colonization and multiplications. The overgrowth of the normal flora and the colonization of certain transient pathogens not only present a threat to the individual, but also to those who come into contact with him.

Surprisingly, little is know about the bacterial flora of the dermatitic skin and its role in dermatitis. Previous studies have been concerned with the presence of *S. aureus* or its dispersal in the hospital environment.

We have characterized quantitatively and qualitatively the aerobic microbial flora of inflamed, noninflamed skins, and the anterior nares. Antibiotic resistance of *S. aureus* strains isolated from these patients were determined. Eczematous and psoriatic skin was included as a representative of dermatitic skin.

Microbial Flora of Atopic Dermatitis

Clinically healthy skin areas of atopic patients are more suitable for multiplication of *S. aureus* than normal skin.[3,10] Atopics are often colonized by *S. aureus* to a greater extent than those suffering from psoriasis. In spite of the large number of *S. aureus* in eczematous lesions, tissue damage or inflammatory reaction related to these organisms may not be significant.[3,11] Atopic patients harboring a high density of *S. aureus* contaminate their environment to a greater extent than did those who are nasal carriers only.[4] Due to large number of *S. aureus*, these patients often disperse these organisms in hospitals.[4,12] We examined atopic patients to study their resident flora on inflamed, noninflamed skins (adjacent area), and the anterior nares.

Table 26.1. Average Microbial Counts in Atopic Eczema Patients

Organisms	Anterior nares	Lesions	Normal skin
Staphylococcus aureus	6.5×10^3	7.5×10^4	$7-1 \times 10^3$
Coagulase-negative staphylococci	1.4×10^4	7.1×10^3	1.5×10^4
Micrococci	0.8×10^3	1.6×10^2	9.5×10^2
Streptococci	3.7×10	< 10	0
Nonlipophilic diphtheroids	$2.4 \quad 10^4$	1×10^2	4.4×10^2
Lipophilic diphtheroids	5.4×10^3	0	6.7×10
Bacillus spp.	< 10	< 10	< 10
Gram-negative rods	< 10	< 10	< 10
Yeasts	0	0	< 10

Thirty-nine subjects with an average age of 19 were included (range, 4–52 years; median, 16 years). Patients receiving antibiotics, topically or systemically were not included. Samples were obtained by the detergent scrub method.[18] The nasal samples were collected with a calcium alginate swab.

S. aureus counts were high in the lesions (7.5×10^4/cm²) and on the normal skin (7.1×10^3/cm²). See Table 26.1. *S. aureus* was the predominant organism in the lesion, constituting 91% of the total aerobic flora. On the uninvolved skin, coagulase-negative staphylococci were the predominant organisms (63% of the total flora). Lipophilic coryneforms were not detected in the involved skin; their counts were extremely low on the normal skin (6.7×10/cm²). Lipophilics are part of the resident flora in a normal population.

The incidence of *S. aureus* was high in all the test sites: 93% in the lesions, 76% on the normal skin, and 79% in the nose (Table 26.2). The occurrence of lipophilic coryneforms was 20% in the nose, 2% on uninvolved skin, and none in the lesions. The incidence of micrococci was higher on the normal

Table 26.2. Percent Incidence of Microoorganism (39 Atopics)

Organisms	Anterior nares	Lesions	Normal skin	Normal population (skin) %
Staphylococcus aureus	79	93	76	< 10
Coagulase-negative staphylococci	77	79	82	80
Micrococci	2	13	25	40
Streptococci	2	0	2	0
Nonlipophilic diphtheroids	61	15	18	45
Lipophilic diphtheroids	20	0	2	47
Bacillus spp.	10	15	20	20
Gram-negative rods	2	5	5	20
Yeasts	0	0	2	< 1

skin (25%) than in the lesions (13%). The carriage of coagulase-negative staphylococci was not substantially different in the uninvolved skin and the lesions.

The antibiotic resistance of 140 strains of *S. aureus* was determined. Sixty-three percent were resistant to 2 units and 58% to 4 units of penicillin. Resistance (20%) was noted to 2 μg of tetracycline (14%) and 1 μg of oxacillin (8%).

Phage types of *S. aureus* strains isolated from atopic patients was determined. Thirty-eight percent of the strains belonged to groups 43 and 1; 38% were nontypable. No phages belonging to groups 2 and 4 were detected. Different phage types were noted with the same subjects. It is fortunate that these patients were not colonized by group 2. Toxic epidermal necrolysis and impetigo are most frequently caused by strains belonging to group 2 (type 71).[14]

Microbial Flora in Psoriasis

Most psoriatic patients do not have clinical pyoderma. Their plaques when colonized with *S. aureus* can become a source of hospital cross-infection. Payne[15] reported an epidemic of staphylococcal wound infection traced to an anaesthetist with psoriasis. Nobel and Savin (1968) reported that 44% of 16 inpatients with psoriasis yielded *S. aureus* from plaques or from adjacent normal skin. Selwyn[16] reported a colonization rate of 35%. However, both reported that outpatients with psoriasis seldom yield *S. aureus* on their skin; on the contrary, Marples et al.[7] found high carriage of *S. aureus* (50%) in outpatients.

The resident flora was characterized in the normal skin, in psoriatic plaques, and in the anterior nares of 40 psoriatic outpatients (Table 26.3). The average total aerobic counts were slightly higher in the lesions (7.9×10^3/cm^2) than on the normal skin (3×10^3/cm^2). These findings confirm previous reports[7,13] that the density of microbial flora is higher on the psoriatic plaques than on uninvolved skin. The density of *S. aureus* was higher in the plaques (3.0/cm^2) than in the uninvolved skin (1.5×10^1/cm^2) but this difference was not statistically significant. The density of lipophilic coryneforms was 6.7×10^1/cm^2 in plaques and 1×10^3/cm^2 on normal skin.

Table 26.3. Average Microbial Counts in Psoriatic Patients (40 subjects)

Organisms	Anterior nares	Plaques[a]	Normal skin[a]
S. aureus	1.2×10^4	3.0×10^2	1.5×10^1
Coagulase-negative gram-positive cocci	3.3×10^5	4.1×10^3	1.7×10^3
Nonlipophilic diphtheroids	1.2×10^5	3.0×10^2	2.7×10^2
Lipophilic diphtheroids	9.1×10^4	6.7×10^1	1.0×10^3
Streptococci	2.0×10^2	3.5×10^1	3.8×10^1
Gram-negative rods	2.7×10^3	< 10	< 10

[a]Per cm^2

Coagulase-negative gram-positive cocci were predominant organisms in the plaques and on the normal skin.

The incidence of *S. aureus* was: 30% in the nose, 20% in the plaques, and 13% on uninvolved skin. With the exception of the plaques, the occurrence of *S. aureus* was within the range of the normal population. The carriage rate of *S. aureus* in a normal population is 25%–40% for the nose and less than 10% on the skin. The incidence of lipophilic coryneforms was significantly different in the plaques (4%) as compared with 30% for normal skin ($P = 0.002$). The composition of nasal flora was within the normal range and comparable with normal populations.[1]

Eighty percent of *S. aureus* strains were resistant to 2 units of penicillin; all strains were sensitive to cephalothin, tetracycline, kanamycin, oxacillin, novobiocin, and neomycin.

The following major differences in microbial flora of atopic and psoriatic patients were noted:

1) The incidence and density ($7.5 \times 10^4/cm^2$) of *S. aureus* were much higher (93%) in atopic lesions than in psoriatic plaques (20% incidence and $3 \times 10^2/cm^2$ density).

2) *S. aureus* was the predominant organism in the eczematous lesion, while coagulase-negative gram-positive cocci were the major organisms in psoriatic plaques.

3) The carriage of *S. aureus* was much higher at all the test sites for atopics (76% to 93%) than in psoriatic patients (13 to 30%).

4) The scarcity of lipophilic coryneforms noted in psoriasis was even more pronounced in the eczematous lesion.

Conclusion

The marked differences in qualitative and quantitative flora noted here between psoriasis and atopic dermatitis suggests a complex relationship between the type of cutaneous abnormality and resultant flora. We will not review the differences in histopathology between these disparate clinical entities; nevertheless, histopathology does not explain these microbial changes. Now that we have quantitative information on their microflora, we must look carefully at the many factors believed to control microbial flora in normal skin hopefully their characterization in psoriasis and atopic dermatitis provide insights into the mechanism of the phenomena reported here.

Treatment of Dermatitic Skin

The consensus suggests that the most appropriate treatment for dermatitic skin resides in altering the pathophysiology of the underlying inflammation; corticosteroids are widely used in atopic dermatitis and psoriasis. No experiment indicates a primary microbial etiology in other disease, with the exception of poststreptococcal guttate psoriasis.

Nevertheless, both entities (see discussion of microflora of dermatitic skin, this chapter) support reasonably large numbers of pathogens. What role does this play in the etiology of the primary atopic and psoriatic lesion? No one doubts that both may become grossly infected. The psoriatic not infrequently develops leg cellulitis; and the atopic, a marked pyoderma not infrequently accompanies by adenopathy, leukocytosis, and fever. The role of large numbers of organisms without gross clinical infection is moot. What role is played, and what should be done?

With the availability of active topical antibiotics, the generalist, dermatologist, and pediatrician gravitated to antimicrobial-corticoid combinations. At the time, this appeared a reasonable common sense approach. In recent years, combination therapy has been challenged. The facts, far from complete, support the concept that certain inflammatory cutaneous diseases respond more rapidly and completely to combinations than to individual agents.

For example, the combination of iodochlorhydroxyquin, with its antifungal and antibacterial action, and hydrocortisone, with its anti-inflammatory effects, has been widely used since 1956 for the treatment of fungal and bacterial cutaneous infections. Strains of fungi isolated from superficial skin lesions, including *Candida albicans*, *Trichophyton mentagrophytes*, *Trichophyton rubrum*, *Microsporum canis*, and *Epidermophyton floccosum*, are sensitive to low concentrations of iodochlorhydroxyquin.

In a double-blind controlled study, Carlton Carpenter and associates (unpublished data) compared the combination of iodochlorhydroxyquin and hydrocortisone with each of its components and the cream vehicle in 88 patients with a variety of fungal infections. After 1 week of treatment, a good to excellent response was achieved in 57% of patients given the combination drug compared to 48% given iodochlorhydroxyquin alone, 32% given hydrocortisone alone, and 48% given the vehicle. Response to treatment occurred much earlier in patients receiving the combination than in those receiving the other treatments.

Wachs and Maibach[17] used betamethosone valerate, gentamicin, or a combination of the two to treat 79 patients with impetiginized atopic dermatitis. After 22 days of treatment, results were rated as excellent in 72% and good in 20% of the 25 patients treated with the combination. Excellent results were obtained in 56% and good in 18% of the 27 patients treated with the steroid alone. In the 27 patients given gentamicin alone, 30% achieved an excellent response and 4% a good response.

In a 4-week double-blind trial, Mertens et al.[8] treated 63 patients with inflamed skin infections of bacterial or mycotic origin with a combination of miconazole nitrate and hydrocortisone or one of its constituents alone. In the 21 patients treated with the combination, the infection was cured in 18 and much ameliorated in three at the end of the 4-week treatment period. Similar results were obtained in 70% of the 20 patients treated with miconazole alone and 9% of the 22 patients treated with hydrocortisone alone.

Miller[9] found that 75% of the 16 patients with secondarily infected common dermatoses and primary fungal infection treated with a flumethasone pivalate-iodochlorhydroxyquin combination for 7 days achieved a good to excellent overall response. Flumethasone and iodochlorhydroxyquin alone produced good to excellent results in 56% and 32%, respectively, of patients treated. Similar results were obtained in 31% of patients treated with the cream base.

Konopka et al.[5] reported the results of a double-blind study involving 430 patients with dermatologic conditions complicated by secondary bacterial involvement. Conversion rate and clinical improvement were markedly greater in patients treated with a flumethasone pivalateiodochlorhydroxyquin combination than in those treated with either of the components alone or with the placebo cream.

Carpenter's work was extended in multicenter study investigating the antifungal effects of a combination drug containing the anti-infective agent iodochlorhydroxyquin and the corticoid hydrocortisone.[6] This study was a double-blind, randomized, parallel four-compartment design. Investigators utilized the same protocol and identical patient record forms. To standardize study procedures, methods of recording data, and, by means of clinical slides, diagnosis and severity ratings, the investigators met prior to the study.

Patients selected were male or female ambulatory outpatients, at least 6 years of age, with a cutaneous fungal infection such as tinea cruris, tinea corporis, tinea pedis, or moniliasis. Patients with fungal infections of the scalp and nails were excluded.

Prior to treatment, a patient history was obtained and a microscopic examination of potassium hydroxide-dimethyl sulfoxide-digested samples of the skin and/or hair for the confirmation of the presence of fungal elements was performed in the investigator's office. Before and after the study, scrapings from the affected site were sent to a central laboratory for verification and isolation or identification of the fungal type according to established procedures.

Patients were randomly assigned to one of four treatment groups: iodochlorhydroxyquin 3% and hydrocortisone 1% cream; iodochlorhydroxyquin 3% cream; hydrocortisone 1% cream; and the cream vehicle alone. The cream was applied in a thin layer to the affected area three times daily without occlusive dressing.

Patients returned after 2 or 3 days (visit 2) and 7 days of treatment (visit 3). At the pretreatment visit and at each of the two visits during the drug treatment period, the physician assessed the degree of severity of erythema, scaling, vesiculation, and exudation using the following five-point scale: (1) absent, (2) mild, (3) moderate, (4) severe, and (5) very severe; the patient assessed the severity of itching using the same scale. At visit 2 and 3, the physician evaluated the clinical response according to the following six-point scale: (1) excellent, (2) good, (3) fair, (4) poor, (5) no change, and (6) worse. At these visits, the patient evaluated the change in discomfort since initiation of treatment as completely gone (1), much better (2), somewhat better (3), the same (4), and worse (5).

Twenty-seven dermatologists at 23 centers studied 533 patients; pathogenic fungi were cultured from the specimen obtained at the first visit in 354 patients. Only data from these patients are presented.

These 354 patients were distributed among the four treatment groups as follows: iodochlorhydroxyquin and hydrocortisone, 89 patients; iodochlorhydroxyquin, 83; hydrocortisone, 96; vehicle, 86. Most patients in each group were white, private outpatients. Tinea pedis, tinea cruris, and tinea corporis were the most frequent diagnoses, affecting 121, 105, and 80 patients, respectively. Moniliasis was the diagnosis in 37 patients, while 11 patients had other diagnoses. *Trichophyton rubrum, Trichophyton mentagrophytes, Epidermophytum floccosum,* and *Candida albicans* were the most frequently found fungi.

After visit 2, 18 patients dropped out of the study—12 because of failure to follow the appointment schedule, 4 because of an unsatisfactory therapeutic response, and 2 because of side effects. The visit 2 data became the final visit data for these patients. To obtain a final treatment response from all patients, a "last visit" was created by combining the visit 2 and the visit 3 data from those patients who completed visit 3.

Statistically significant differences between the four treatment groups are summarized in Table 26.4. The iodochlorhydroxyquin-hydrocortisone cream was significantly better than hydrocortisone alone or the vehicle with respect to all five clinical assessments at visit 3 and the last visit. The combination was significantly better than iodocchlorhydroxyquin alone with respect to erythema and physician evaluation at visit 3 and the last visit.

Results of the physician evaluation of overall clinical response and the patient assessment of discomfort at the last visit are shown in Tables 26.5 and 26.6, respectively. The physician evaluation indicated that 65% of patients treated with the iodochlorhydroxyquin-hydrocortisone cream achieved an excellent or good response compared with 46% of those treated with iodochlorhydroxyquin alone, 32% given hydrocortisone alone, and 25% treated with the vehicle.

Discomfort was completely gone or much better in 64% of patients treated with the combination. A similar response was obtained in 55% of patients given iodochlorhydroxyquin alone, 42% treated with hydrocortisone alone, and 30% of those treated with the vehicle.

In the iodochlorhydroxyquin-hydrocortisone group, 49% of the patients who had a positive KOH examination at visit 1 had a negative KOH examination at visit 3. Conversion rate was 56% for those given iodochlorhydroxyquin alone, 19% for those treated with hydrocortisone alone, and 28% for those given only the vehicle. The proportion of patients treated with the combination and iodochlorhydroxyquin alone who changed from a positive KOH examination at visit 1 to a negative KOH examination at visit 3 was significantly greater than that in the hydrocortisone and vehicle groups.

The weighted mean score of the physician evaluation of overall clinical re-

Table 26.4. Statistical Significance of Clinical Assessments [a]

	IHC vs. HC	IHC vs. I	IHC vs. V	I vs. V	I vs. HC	HC vs. V
Erythema						
Visit 2	—	—	—	—	—	—
Visit 3	p < 0.05	p < 0.05	p < 0.01	—	—	—
Last visit	p < 0.01	p < 0.01	p < 0.01	—	—	—
Scaling						
Visit 2	—	—	—	—	—	—
Visit 3	p < 0.05	—	p < 0.01	p < 0.05	—	—
Last visit	p < 0.0	—	p < 0.01	p < 0.05	—	—
Itching						
Visit 2	—	—	—	—	—	—
Visit 3	p < 0.01	—	p < 0.01	p < 0.01	p < 0.05	p < 0.05
Last visit	p < 0.01	—	p < 0.01	p < 0.01	p < 0.05	—
Patient evaluation						
Visit 2	—	—	—	—	—	—
Visit 3	p < 0.01	—	p < 0.01	p < 0.01	p < 0.05	p < 0.05
Last visit	p < 0.01	—	p < 0.01	p < 0.01	p < 0.05	—
Physician evaluation						
Visit 2	—	—	p < 0.01	p < 0.05	—	p < 0.01
Visit 3	p < 0.01	p < 0.01	p < 0.01	p < 0.05	—	—
Last visit	p < 0.01	p < 0.01	p < 0.01	p < 0.05	—	—

[a] IHC = Iodochlorhydroxyquin-hydrocortisone; HC = Hydrocortisone; I = Iodochlorhydroxyquin; V = Vehicle

Table 26.5. Physician Evaluation of Overall Clinical Response at Last Visit

	Treatment			
Response	Iodochlorhydroxyquin-Hydrocortisone	Iodochlor-hydroxyquin	Hydro-cortisone	Vehicle
Excellent	19 (21%)	14 (17%)	7 (7%)	2 (2%)
Good	39 (44%)	24 (29%)	24 (25%)	20 (23%)
Fair	21 (24%)	18 (22%)	33 (34%)	32 (37%)
Poor	7 (8%)	21 (25%)	20 (21%)	20 (23%)
No change	1 (1%)	3 (4%)	9 (9%)	5 (6%)
Worse	2 (2%)	3 (4%)	3 (3%)	7 (8%)
	89	83	96	86
Weight mean[a]	2.30[b]	2.81[c]	3.09	3.31

[a]Scale: (1) excellent, (2) good, (3) fair, (4) poor, (5) no change, (6) worse
[b]Significantly better than iodochlorhydroxyquin alone, hydrocortisone alone, and vehicle
[c]Significantly better than vehicle

Table 26.6. Patient Assessment of Discomfort

	Treatment			
Response	Iodochlorhydroxyquin-hydrocortisone	Iodochlor-hydroxyquin	Hydro-cortisone	Vehicle
Completely gone	21 (24%)	16 (19%)	9 (9%)	3 (3%)
Much better	36 (40%)	30 (36%)	32 (33%)	23 (27%)
Somewhat better	23 (26%)	20 (24%)	23 (24%)	27 (31%)
About the same	5 (6%)	13 (16%)	27 (28%)	25 (29%)
Worse	4 (4%)	4 (5%)	5 (5%)	8 (9%)
	89	83	96	86
Weighted mean[a]	2.27[b]	2.51[b]	2.86	3.14

[a]Scale: (1) excellent, (2) good, (3) fair, (4) poor, (5) no change, (6) worse.
[b]Significantly better than vehicle and hydrocortisone alone.

sponse at the last visit indicated that the combination of iodochlorhydroxyquin and hydrocortisone was consistently better than either of its active components or the vehicle in the treatment of tinea pedis, tinea cruris, and tinea corporis. The combination drug was the best treatment for moniliasis, although the number of patients in the four treatment groups was small.

In summary, a double-blind multicenter study compared the anti-fungal effectiveness of an iodochlorhydroxyquin-hydrocortisone cream with that of its individual components in 354 patients with cutaneous fungal infections. After 7 days of treatment, the combination was significantly better than hydrocortisone or the cream vehicle with respect to erythema, scaling, itching, and patients' and physicians' evaluations. The proportion of patients in the iodochlorhydroxyquin-hydrocortisone and iodochlorhydroxyquin groups who changed from a positive KOH examination at baseline to a negative KOH examination after treatment was significantly greater than that in the hydrocortisone and placebo groups. The conversion rate associated with the iodochlorhydroxyquin-hydrocortisone and the iodochlorhydroxyquin treatments was significantly different from that associated with hydrocortisone alone or placebo treatment.

Neither the efficacy study presented here nor the other double-blind controlled studies quoted should be interpreted out of context. The efficacy must be balanced against potential for toxicity of any added component of a formulation. What they do demonstrate is the fiducial limits of the degree of improvement gained by adding an anti-inflammatory agent to an antimicrobial agent in uncomplicated cutaneous fungal infection.

Furthermore, these studies largely deal with lesions judged "clinically" infected by dermatologists. Further studies must be performed to clarify what role (if any) antimicrobials play in eczematous and psoriatic lesions not considered, on morphologic grounds, infected.

References

1. Aly R, Maibach HI, Shinefield HR, Strauss W (1970) Effect of systemic antibiotic on nasal bacterial ecology in man. Appl Microbiol 20:240–244
2. Aly R, Maibach HI, Mandel A (1976) Bacterial flora in psoriasis. Br J Dermatol 95:603–606
3. Aly R, Maibach HI, Shinefield HR (1977) Microbial flora of atopic dermatitis. Arch Dermatol 113:780–782
4. Hare R, Cooke E (1961) Self-contamination of patients with staphylococcal infections. Br Med J ii:333–336
5. Konopka EA, Kimble EF, Zoganas HC, Heymann H (1975) Antimicrobial effectiveness of Locacorten-Vioform[R] cream in secondary infections of common dermatoses. Dermatologica 151:1–8
6. Maibach H (1978) Iodochlorhydroxyquin treatment of fungal infections: double-blind trial. Arch Dermatol 114:1773–1775
7. Marples RR, Heaton CL, Kligman AM (1973) S. aureus in psoriasis. Arch Dermatol 107:568–571
8. Mertens RLJ, Morias J, Verhamme G (1976) A double-blind study comparing Daktacort[R], miconazole and hydrocortisone in inflammatory skin lesions. Dermatologica 153:228–235

9. Miller RC (1974) Flumethasone pivalate-iodochlorhydroxyquin cream: A new corticosteroid anti-infective combination. Cutis 14:605–609

10. Muller E (1969) Ecology of *Staphylococcus aureus* on the human skin surface vs. *Staphylococcus aureus*, artificially inoculated upon the skin surface after extraction with ether, of patients with dermatoses and during systemic administration of antibiotics. Arch fur klinische und experimentelle Derm 233:376–382

11. Noble WC (1971) The contribution of individual patients to the spread of infection. Br J Dermatol 85:24–29

12. Noble WC, Davies RR (1965) Studies on the dispersal of staphylococci. J Clin Pathol 18:16–19

13. Nobel WC, Savin JA (1968) Carriage of *Staphylococcus aureus* in psoriasis. Br Med J 1:417–418

14. Parker MT, Williams RE (1961) Further observation on the bacteriology of impetigo and pemphigus neonatorum. Acta Pediatr 50:101–112

15. Payne RW (1967) Severe outbreak of surgical sepsis due to *Staphylococcus aureus* of unusual type and origin. Br Med J iv:17–20

16. Selwyn W (1963) Bacterial infections in a skin department. Br J Dermatol 75:26–28

17. Wachs GN, Maibach HI (1976) Cooperative double-blind trial of an antibiotic/corticoid combination in impetiginized atopic dermatitis. Br J Dermatol 95:323–328

18. Williamson and Kligman (1965) A new method for the quantitative investigation of cutaneous flora. J Invest Dermatol 45:498–503

Chapter 27
Staphylococcal Scalded Skin Syndrome: Clinical Features, Biology, and Pathogenesis

PETER M. ELIAS AND PETER FRITSCH

Following the description of an animal model for the staphylococcal scalded skin syndrome (SSSS) by Melish and Glasgow,[30] numerous studies have focused on this fascinating disease entity resulting in a considerable expansion of the knowledge in the microbiological, biochemical, biological and immunological problems involved. Today, little over 100 years after its original description,[37] the concept of SSSS is founded on a new experimental groundwork and appears to be disentangled from the intricacies of scientific controversy so long attached to it. Most important, its nonidentity with toxic epidermal necrolysis—a severe polyetiological reaction pattern of the skin with which SSSS has been lumped for a common resemblance to scalding—has been generally recognized.

Clinical Features

SSSS is an acute, potentially life-threatening disorder caused by an exotoxin (epidermolysin) of phage group 2 staphylococci, which may be harbored in the skin or in extracutaneous foci. SSSS afflicts predominantly infants and children of both sexes and all races; it is characterized by erythema and detachment of the skin vaguely similar to scalding. Two major morphological expressions of SSSS are distinguished which are different in both their clinical symptomatology and prognostic significance as well as their pathogenesis.

Localized SSSS (bullous impetigo). This more common variety is due to invasion of staphylococci into the skin itself and production and centrifugal spread of epidermolysis in loco. Its hallmarks are rapidly enlarging blisters, rimmed by an erythematous halo, which tend to be clustered at first but then spread to distant sites (Fig. 27.1). The originally clear blister fluid turns purulent, followed by confluence of lesions, rupturing, oozing and crust formation. There are no systemic signs. Cocci can be recovered from the lesions.

Systemic SSSS (Ritter's type). In this rare but dangerous manifestations

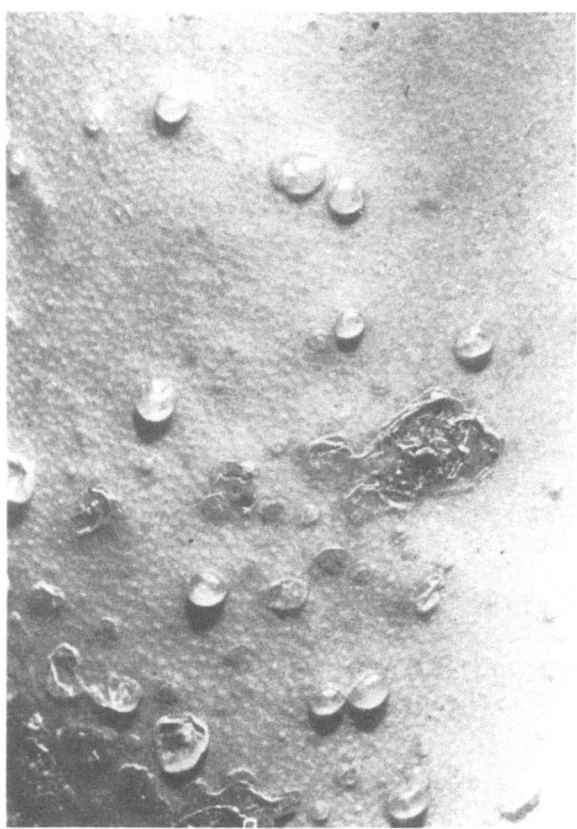

Fig. 27.1. Bullous impetigo represents localized scalded skin syndrome. Painless bullae erupt and rapidly become unroofed leaving grey skin erosions.

of SSSS, epidermolysis is produced by staphylococci in a purulent focal infection and distributed throughout the body by the blood stream. Most often a purulent infection of the upper respiratory tract is present (rhinitis, pharyngitis, tonsillitis), but any other purulent infection, like omphalitis, conjunctivitis and even impetigo, can play this role. The focal infection usually precedes the outbreak of SSSS and may persist throughout its course.

Systemic SSSS follows a distinctive and characteristic time course: at the onset, affected children appear lacrimose and sick without signs of systemic illness; their skin is tender and they are wary about being touched. A distinctive, faint, ill-defined yellow/orange/brick-red rash then erupts, affecting the central portion of the face and the major body folds first, which then extends to the trunk and the extremities. In this erythematous stage, the Nikolski sign (seemingly normal skin wrinkles if gently rubbed) becomes positive. A few hours later, spontaneous detachment of the skin in large sheets takes place (Fig. 27.2), particularly at sites of mechanical stress (shoulders, buttocks, body folds). Large flaccid blisters may arise. Widespread oozing areas of denudation are the characteristic lesions of this epidermolytic stage which quickly progresses into the desquamative (healing) stage, if proper treatment is instituted. Bacterial toxemia and water and electrolyte imbalance may result in hemodynamic shock and death.

Abortive, systemic SSSS may present as scarlatiniform rash without progression to epidermolysis.[31]

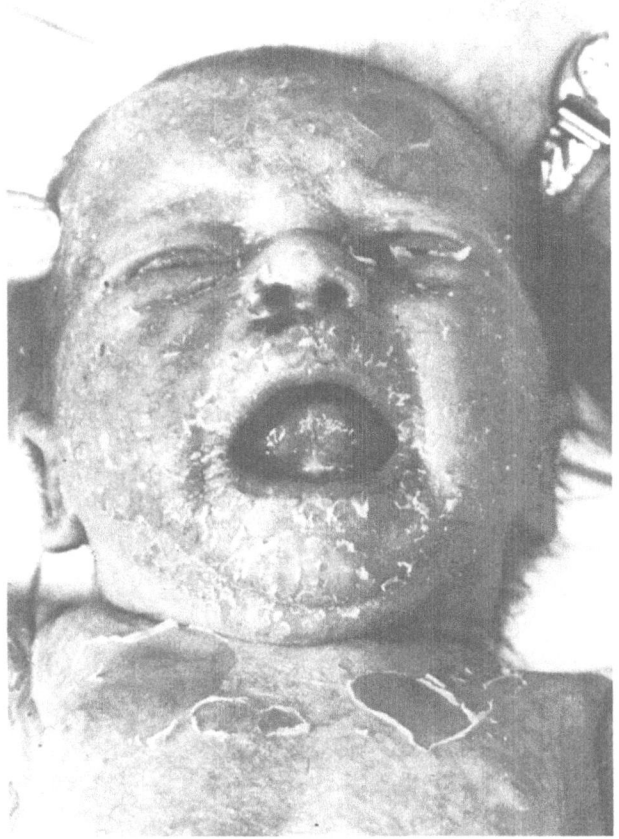

Fig. 27.2. Generalized scalded skin syndrome is a potentially devastating disease that occurs most commonly in the neonatal period. An occult form of infection invariably precedes the syndrome. Epidermolysis begins periorificially and spreads centrifugally, with shedding of skin in sheets and ribbons. Cutaneous tenderness in invariably present, and helps differentiate the *forme fruste*, scarlatiniform eruption of the syndrome from other generalized rashes.

Model Systems

Animal Models

Although an association between SSSS and infection with phage group 2 staphylococci had been noted in the late 1950s,[6,17,18,28,29] it was the development of the neonatal mouse model (Fig. 27.3) of the disease by Melish and Glasgow[30] in 1970 that permitted subsequent isolation and purification of epidermolysis from culture supernatants.[2,7,19,20,23,32] Melish and Glasgow observed that neonatal mice became resistant to both organisms[30] and epidermolysin[32] on around the sixth or seventh postnatal day; they felt that either ill-defined maturational changes or developing hair structures might be responsible for the putative resistance of older animal's skin.[32] Reexamining the susceptibility of adult mouse skin (Fig. 27.4), we found that it is as susceptible to toxin-induced epidermolysis as is neonatal skin,[9,10] but in the adult mouse the intraepidermal rent is less evident mainly because the entire viable epidermis is only three layers thick.[10] Presence or absence of hair did not influence susceptiblity since both glabrous and furry skin was readily split. We confirmed that adult mice resisted injections or organisms, regardless of the size of the inoculum,[10] and systemically administered epidermolysin, even when administered in massive doses.[9]

The parallelism between the neonatal mouse and child was thus further extended to the adult mouse and man (Fig. 27.5). It seemed likely that simi-

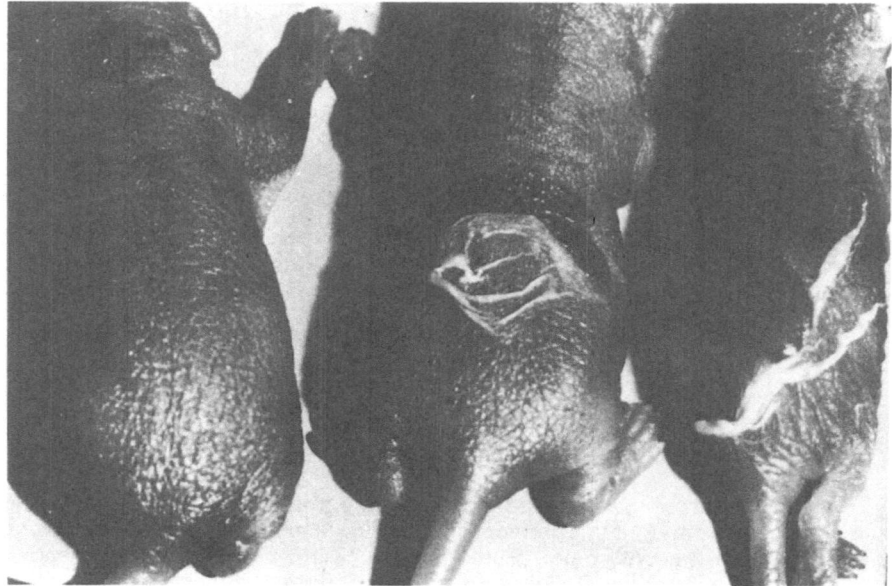

Fig. 27.3. **Neonatal mice injected 2 h earlier with three dilutions of purified epidermolytic toxin. Spontaneous wrinkling and positive Nikolsky sign develop at dilutions of 1:10 and 1:100, but not at 1:1 000 dilutions.**

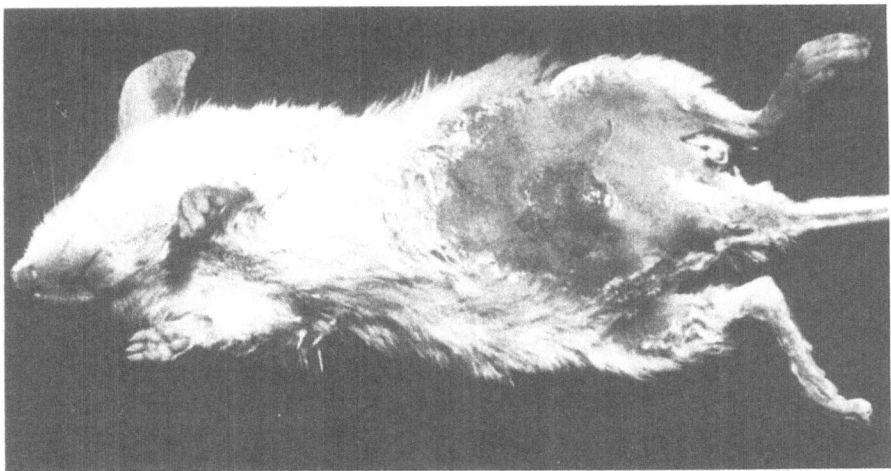

Fig. 27.4. Ten-day-old mouse skin injected 2 h earlier with epidermolytic toxin. Microvesicles form immediately beneath the granular layer and extend into the external root sheath. × 1 000.

lar, intracutaneous factors that limit the full expression of SSSS might be operative in both species.[10] Indeed, although systemic SSSS is rare in adults, occurring primarily in compromised hosts,[14] bullous impetigo is not infrequent in normal adult individuals.[13] Moreover, when purified epidermo-

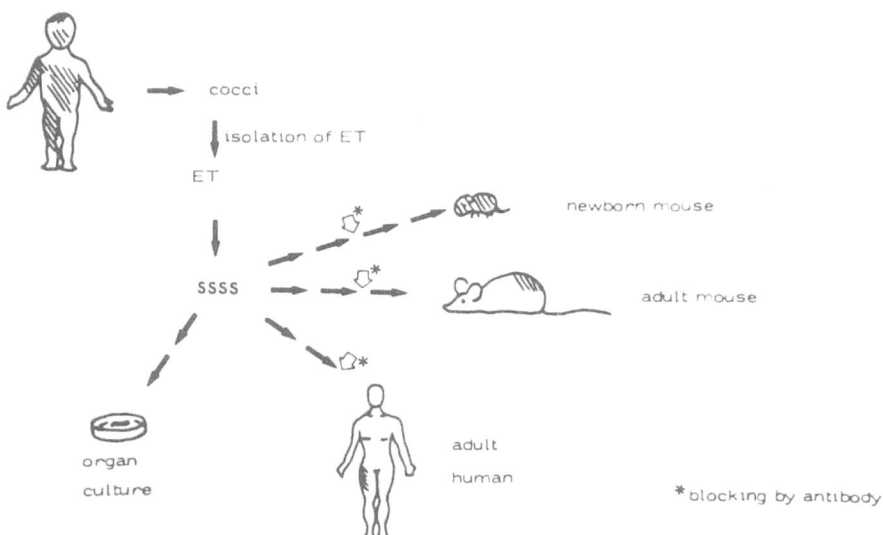

Fig. 27.5. Mouse model of staphylococcal scalded skin syndrome includes not only the original neonatal mouse, but also an adult mouse equivalent, which parallels disease in adults, and an in vitro system which is analogous to the in vivo situation (from Elias et al. (1974) J Invest Derm with permission).

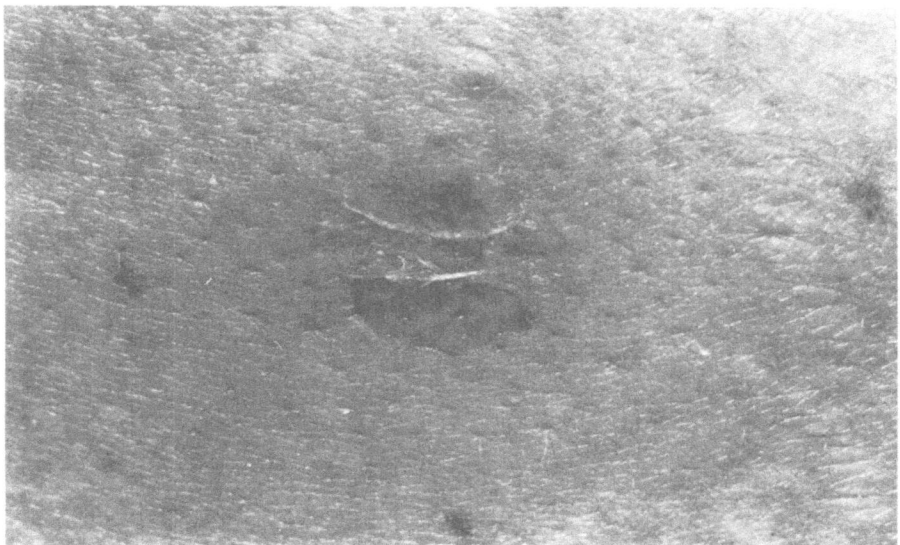

Fig. 27.6. Experimental production of staphylococcal scalded skin syndrome in human volunteer 2 h after intradermal injection of epidermolytic toxin (from Elias et al. (1974) J Lab Clin Med. 84:411–42).

lytic toxin is administered to normal human adults in vivo, bullae develop over injected sites,[9,40] the split occurring at the same intraepidermal level as in patients with SSSS (Fig. 27.6). We concluded that the susceptibility of the adult mouse would prove to be a valuable model for the study of factors that influence the expression of disease in human adults.[10]

In Vitro Model

Epidermolysis can be reproduced in vitro when skin from susceptible species is exposed to epidermolysin (Figs. 27.7, 27.8).[9,29] Both the time and the dose requirements of the reaction are remarkably similar in vitro and in vivo.[9] Epidermolysis is assessed both by peeling and by exfoliative cytology following teasing of tissue samples.[9] Although the in vitro system theoretically should provide unique opportunities to study the biology of the SSSS, the actual yield of new information about the disease has been limited[12]; first, with the in vitro system it has been possible to catalogue and compare the susceptibility and resistance of various species and tissues to the epidermolytic toxin on a quantitative basis (Tables 27.1, 27.2). Thus, to date only epidermis and a few extracutaneous mucosal keratinizing epithelia have been found to be susceptible (Table 27.3). Among subhuman species, only monkeys (Fig. 27.9), mice and, to a lesser extent, hamsters have been identified as susceptible.[12] Surprisingly, several species closely related to the mouse—e.g., including rats and gerbils, as well as many other rodents such as rabbits and guinea pigs[12,16]—are epidermolysin resistant.

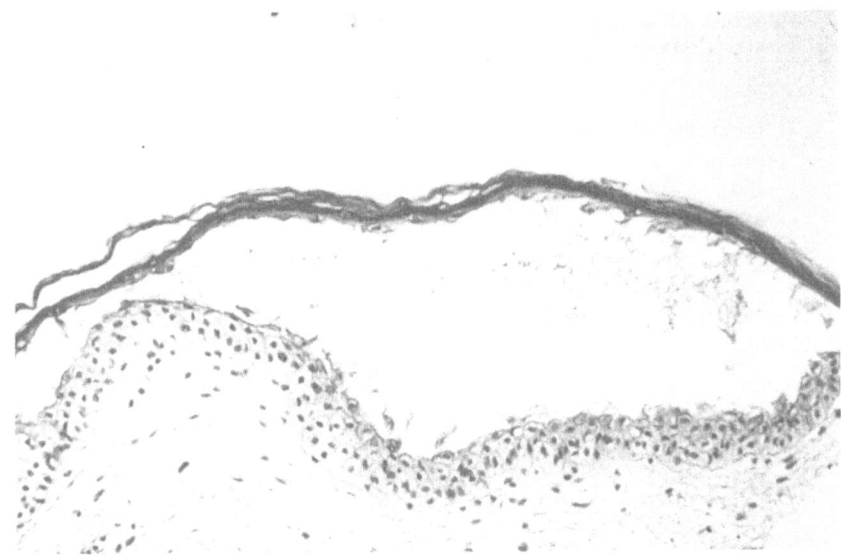

Fig. 27.7. Organ culture of human skin after 4 h in vitro with epidermolytic toxin (10 mg/ml). Upper epidermal sheets, consisting of intact granular and horny layers, can be peeled from underlying epidermis and dermis. Note intact skin markings in detached sheet (from Elias et al. (1974) J Lab Clin Med 84:414–424).

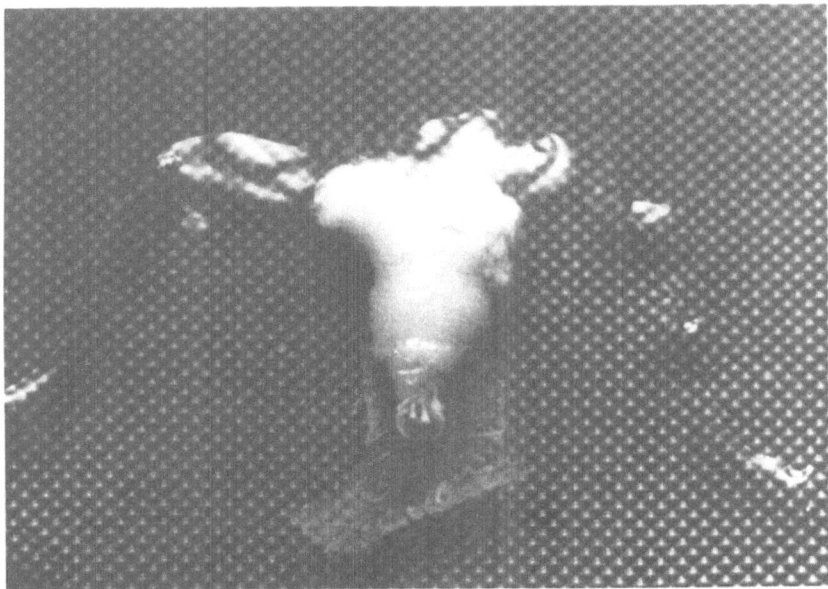

Fig. 27.8. Intraepithelial cleft in specimen of human labial mucosa exposed to epidermolytic toxin for 2 hrs in organ culture. Although this keratinizing epithelium is as sensitive as epidermis, most extracutaneous keratinizing epithelia are refractory to the toxin (from Elias et al. (1976) J Invest Derm 66:80–89).

Table 27.1. Cutaneous Susceptibility or Resistance of Several Species to Epidermolytic Toxin (Elias et al. 1976)

	Species	Age group tested	Sites injected in vivo	Source of skin specimen for incubations
Susceptible	Human	Juvenile	Not tested	Dorsal forearm
		Adult	Forearm, thigh	Multiple regions
	Mouse	Newborn	All routes	Multiple regions
		Adult	All sites	Multiple regions
	Hamster	Newborn	All sites	Multiple regions
		Adult	Ear	Not tested
	Monkey	Adult	Not tested	Eyelid, abdomen
Resistant[a]	Rat	Newborn	All sites	Multiple regions
		Adult	Flank, ear	Ear
	Guinea pig	Newborn	All sites	Flank, ear
		Adult	Flank, ear	Ear
	Rabbit	Newborn	All sites	Flank, ear
		Adult	Ear	Ear
	Dog	Adult	Ear	Ear
	Frog	Adult	Flank	Flank, thigh
	Chicken	Juvenile	Flank	Flank

[a] A negative in vivo response in neonates is based upon the failure of the animal to react to 100 times the epidermolytic toxin dose/gm body weight of minimum required to induce wrinkling in newborn mice (0.01–0.05 mg).

Table 27.2. Adult Human and Mouse Skin of Various Ages: Quantitative Comparison of Response to Epidermolytic Toxin in Vitro[a] (Elias et al. 1976)

Tissue	Exfoliatin concentration	Time exposed to exfoliatin		
		1 hr	2 hr	4 hr
Human (adult)[b]	3 mg/ml[b]	+	+	+
	3×10^{-1}	+/−	+	+
	3×10^{-2}	−	+/−	−
	3×10^{-3}	−	−	−
Mouse (adult)	3 mg/ml	−	+/−	+
	3×10^{-1}	−	−	+
	3×10^{-2}	−	−	−
	3×10^{-3}	−	−	−
Mouse (juvenile)	3 mg/ml	−	+	+
	3×10^{-1}	−	+/−	+
	3×10^{-2}	−	−	−
	3×10^{-3}	−	−	−
Mouse (newborn)	1.5 mg/ml	−	+	+
	1.5×10^{-1}	−	−	+
	1.5×10^{-2}	−	−	+/−
	1.5×10^{-3}	−	−	−

[a] Response was assessed by the following manifestations: peeling, exfoliative cytology, cleavage in histologic sections, and, in some cases, ultrastructural modifications.
[b] The same batch of epidermolytic toxin was used in determining all dose-response curves.

Table 27.3. Susceptibility of Various Human, Monkey, and Murine Epithelia to Epidermolytic Toxin[a,b] (Elias et al. 1976)

Species	Epithelium	In vivo	In vitro
Mouse	Glabrous skin (N,A)	Positive (1,2,3,4)	Positive (1,2,3,4)
	Hairy skin (A)	Positive (1,3,4)	Positive (1,2,3)
	Esophagus (N)	Positive (3,4)	Positive (3,4)
	Vagina (A)	Not tested	Positive (1,2,3)
	Cervix (A)	Not tested	Negative (1,2,3)
	Endometrium (A)	Not tested	Negative (1,2,3)
	Bladder (N)	Negative (3)	Not tested
	Ureter (N)	Negative (3)	Not tested
	Stomach (N)	Negative (3)	Not tested
	Adrenal gland (N)	Not tested	Negative (3,4)
	Epididymis (N)	Not tested	Negative (3,4)
Human	Glabrous skin (A)	Positive (1,2,3,4)	Positive (1,2,3,4)
	Hairy skin (A)	Positive (a,3)	Positive (1,3,4)
	Vulvar mucosa (A)	Not tested	Positive (1,3)
Monkey	Glabrous skin (A)	Not tested	Positive (1,2,3)
	Mucous membrane (A) (oral mucosa)	Not tested	Positive (1,2,3)

[a]N = neonatal; A = adult.
[b]Assessed by: 1 = peeling; 2 = exfoliative cytology; 3 = histology; 4 = electron microscopy.

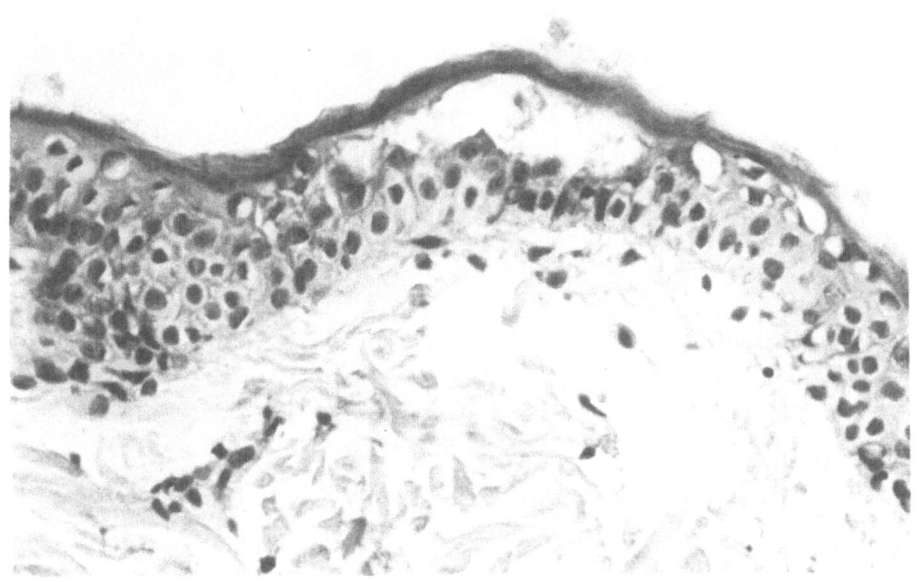

Fig. 27.9. Monkey skin 2 h after incubation with epidermolytic toxin in organ culture. Note appearance of subgranular cleft comparable to mouse and human epidermis (From Elias et al. (1976) J Invest Derm 66:80–89).

Second, by performing recombinant experiments, it has been possible to pinpoint at a tissue level the source of susceptibility and resistance. Neither serum nor mesenchymal factors influence epidermolysis—intraepidermal splitting of susceptible species occurs when epidermal sheets alone are incubated with epidermolysin.[12] Thus, both species resistance and susceptibility can be traced directly to the epidermis itself.[12]

Furthermore, as detailed below, the in vitro system has permitted a detailed search for the metabolic requirements for epidermolysis.

Pathogenesis

Light and Electron Microscopy

All forms of the SSSS in humans and experimental animals are characterized by mid-to-upper epidermal cleavage with splitting beneath and within the stratum granulosum (Fig. 27.10).[22] The cleavage space may contain free-floating or partially attached acantholytic cells, but the remainder of the epidermis appears unremarkable, and the dermis contains no inflammatory cells. In cases of bullous impetigo a dense papillary dermal infiltrate of neutrophils and mononuclear cells may impinge upon the epidermis.[13]

Ultrastructural studies of human skin lesions and of experimental lesions in the mouse analogue have confirmed the intercellular cleft that is produced

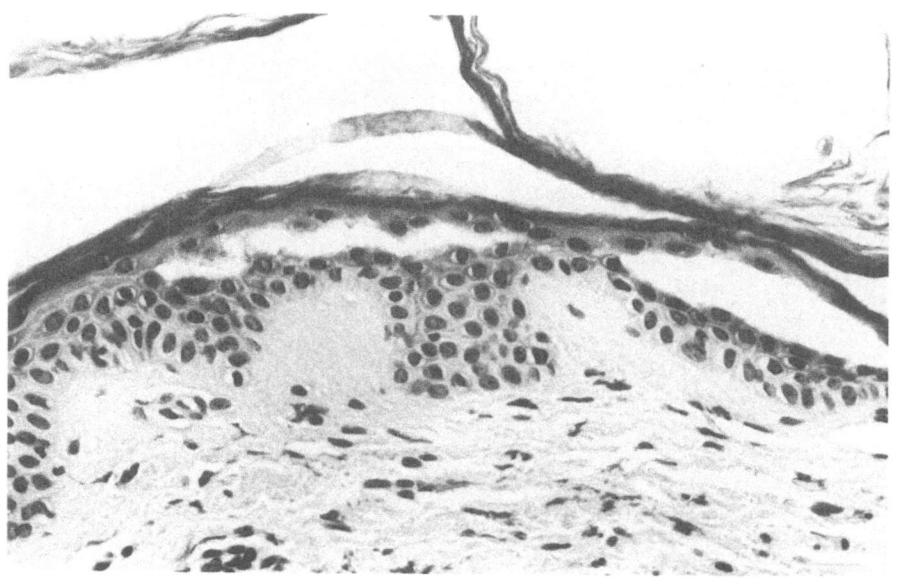

Fig. 27.10. Section of neonatal mouse skin 2 h after injection of epidermolytic toxin. Note intraepidermal cleft and absence of inflammatory cells in dermis. × 240 (from Elias et al. (1974) J Lab Clin Med 84:414–424 with permission).

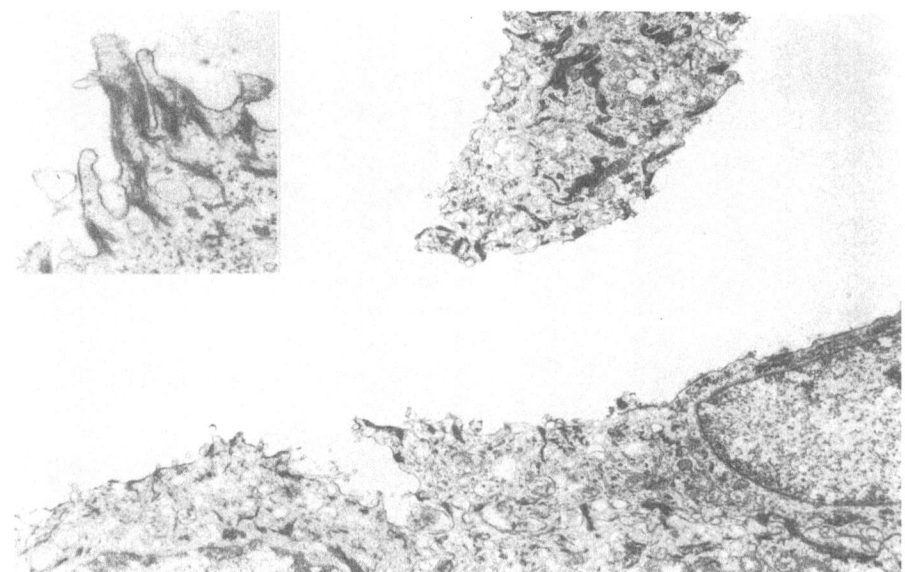

Fig. 27.11. Adult mouse skin incubated for two (2) hours with epidermolytic toxin. Note relatively uninjured, acantholytic keratinocytes lying in cleavage space. Half-desmosomes (insert) are easily seen along split surface. × 12 000.

by epidermolysin.[8,9,25] Both desmosomes and interdesmosomal regions appear to be readily separated (Fig. 27.11), and cytolysis is evident only in advanced lesions. Experimental evidence in favor of a purely intercellular split was provided by studies in which the intercellular spaces were filled with electron dense tracers during the clefting process. Since leakage of tracer into keratinocytes bordering the developing cleavage space was not found, this further verified the purely intercellular nature of the splitting process.[11] Moreover, when cultured mouse keratinocytes were exposed to high concentrations of biologically active epidermolytic toxin for up to 48 h, there was no evidence of cytotoxicity (Fig. 27.12).[11]

The Cell Surface Target

Despite solid evidence for an intercellular split, identification of the cell surface target has proved much more difficult. Lillibridge et al.[25] postulated that the toxin might activate an intercellular proteolytic process directed at desmosomes. But others have shown that tissue protease inhibitors fail to inhibit toxin-induced clefting.[41] Recently, it has been suggested that divalent cations may be involved in the pathogenesis, since ionic calcium reportedly inhibits splitting.[8]

Despite a wealth of evidence that indicates the intercellular nature of the toxin-induced split, the cell surface target of epidermolysin toxin has escaped identification. Stainable cell surface sugars do not appear to be re-

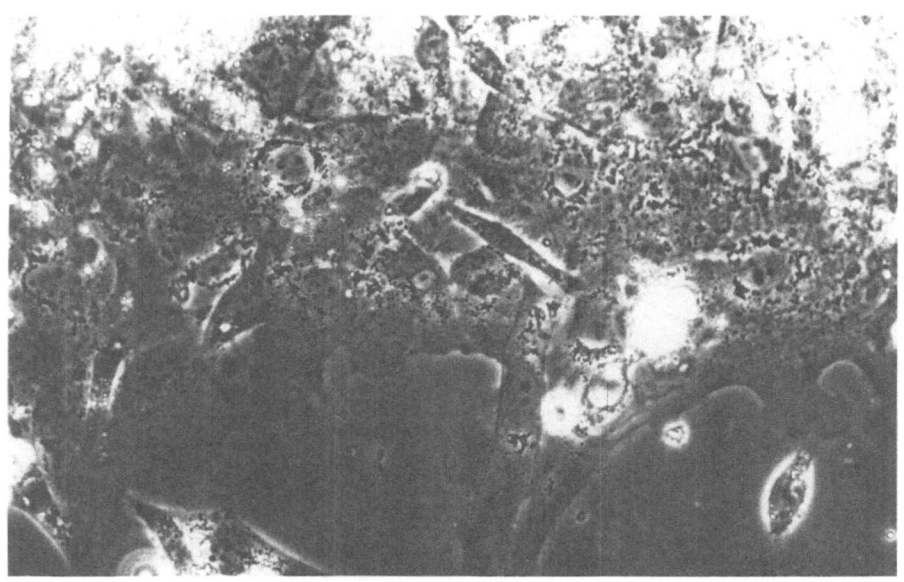

Fig. 27.12. Primary culture of mouse keratinocytes exposed to high concentrations of epidermolytic toxin for 4 h. Note absence of cytolysis. Supernatants retained ability to produce clefting in neonatal mice (from Elias et al. (1975) J Invest Derm 65:501–512 with permission).

moved (Fig. 27.13),[11] nor does the toxin appear to dislodge either pemphigus vulgaris antigen or HL-A surface receptors.[11]

These negative results stimulated investigators to search for a surface-receptor mediated process, but I^{125}-labeled epidermolytic toxin failed to bind preferentially to either epidermis in vivo[15] or to keratinocyte suspensions in vitro.[4] Thus, the identity of the target of the staphylococcal epidermolytic toxin still remains unknown.

Mechanism of Action of Epidermolysin

Since murine epidermis that has been frozen-thawed, formaldehyde-fixed, or heat-treated does not respond to the epidermolytic toxin,[14] we deduced that intracellular metabolic processes must be mobilized as part of the epidermolytic process. In vitro exfoliation occurs somewhat more rapidly at 37°C than at room temperature and effectively ceases at 4°C. Cleavage occurs over a wide pH range (pH 5–9, but not at or below pH 4) and in 100% nitrogen atmosphere. Though treatment with cycloheximide or puromycin produced at 95%–90% decrease in epidermal protein synthesis, these agents did not block in vitro exfoliation. In addition, pretreatment of susceptible tissues with sodium azide, dinitrophenol, 2-deoxyglucose, vinblastine, and soybean trypsin inhibitor did not prevent cleavage.[14] Furthermore, as detailed above, protease inhibitors reportedly likewise did not block exfoliation in vivo.[41]

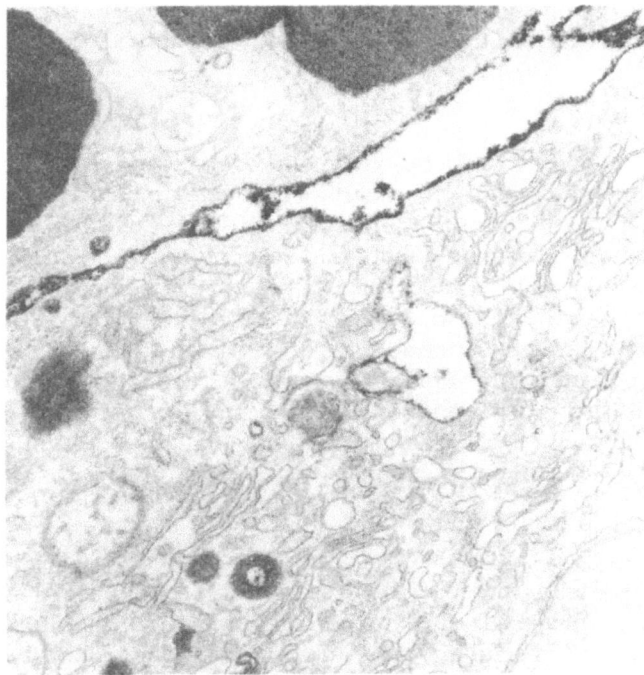

Fig. 27.13. Primary culture of neonatal mouse keratinocytes treated with epidermolytic toxin for six hours, then stained with ruthenium red. Despite cell separation, stainable surface coat is not removed. × 75 000 (from Elias et al. (1977) Arch Derm 113:207–219 with permission).

However, these negative findings do not rule out the possibility of metabolic intracellular participation. Still other mechanisms (or additive mechanisms) may be involved. Alternatively, our assay system may not detect changes which are physiologically relevant. As with the earlier studies on the cell surface target, the question of metabolic participation remains unsolved.

The Problem of the Age-Specificity of the SSSS

A number of hypothetic explanations have been advanced for the age-related susceptibility of human and experimental animals to the SSSS.

Maturation of Skin

Melish and Glasgow's original speculation that the skin might become resistant as it ages has now been disproven. Intracutaneous inoculations of epidermolysin produce clefting in adult humans[9,40] and mice[9] as readily as in neonates. Moreover, differences in susceptibility can be quantitated with the in vitro system, and age-related changes have been found to be inconsequential.[12] Thus, it appears exceedingly unlikely that maturational changes account for the variations in the age distribution of the SSSS.

Pilosebaceous Structures

The eruption of the furry mantle of 5–7-day-old mice coincides with their observed resistance to injected organisms. Therefore, developing hair structures have been considered important for emerging adult resistance.[30] This hypothesis was fortified by the seemingly increased susceptibility of hairless mice to toxin-induced epidermolysis.[3,21] However, evidence from both in vivo inoculations,[10] and from the in vitro system[9,12] have failed to reveal differences in susceptibility. Though epidermolysis may be more difficult to visualize through a furry pelage, microscopic vesiculation is readily apparent, and even extends downward into the external root sheath (Fig. 27.14).[10,12] Thus, available data effectively discount the possibility that pilosebaceous structures confer protection to older individuals.

Protective Antibodies

Still warranting consideration is the possibility that subclinical exposure to epidermolysin elaborating phage group 2 organisms early in life confers lifetime resistance against subsequent exposures. Antibodies can be elicited readily in rabbits treated with epidermolysin,[7,10,34,42] and recent evidence indicates that clinical infections are accompanied by a detectable antibody response against the epidermolysin, measurable by radioimmunoas-

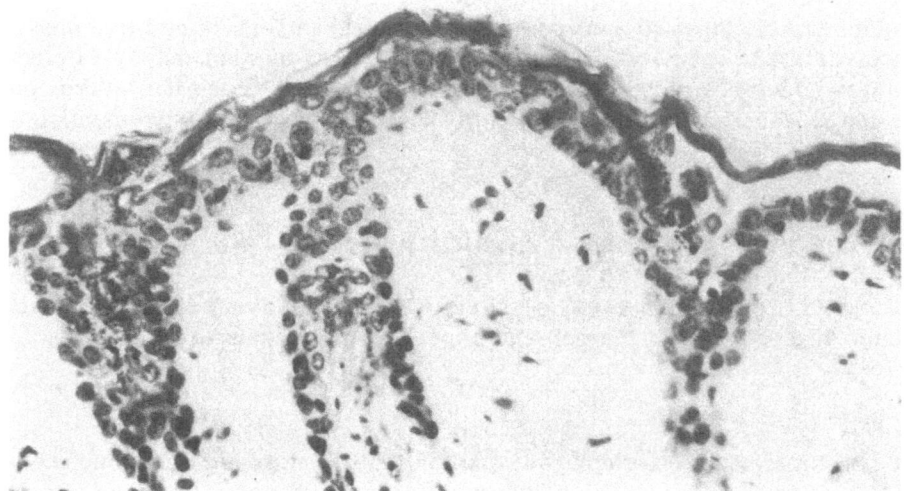

Fig. 27.14. Age distribution of staphylococcal scalded skin syndrome. Note that even in the current era, most cases of SSSS still occur in the first 6 months of life.

Table 27.4. Readministration of Epidermolytic Toxin to Mice Surviving Previous Toxin-Induced Ten (Elias et al. 1974)

Group	Age	Dose (mg protein)	No. of Animals peeling
—	Newborn (3–5 days)	1	12/12
1	21–28 days	1–10	6/6
2	2–3 months	1–10	9/9[a]

[a]Three animals from Group 1 were included in Group 2 for later testing.

say.[5,33] Whether these antibodies persist, and whether they can effectively palliate clinical disease, is not known. We have rechallenged mice previously exposed to epidermolysin later in life and failed to note any diminution in the severity of epidermolysis (Table 27.4).[10] However, this type of experiment does not duplicate actual clinical infections, and the possibility that circulating antibodies play a role in ameliorating or blocking the full expression of the SSSS still warrants consideration.

Antistaphylococcal Immune Factors

Several laboratories have noted that administration of systemic steroids influences the course of experimental SSSS.[10,38,40,41] Specifically, in neonatal mice the threshold dose of pathogenic organisms required to produce generalized epidermolysis is reduced by at least one order of magnitude, and survival is adversely effected.[31] If adult mice are given massive doses of systemic steroids for 2–3 weeks, they also become susceptible to the development of the generalized SSSS following inoculation of organisms.[10] Moreover, retrospective analyses of human cases indicate that steroids can adversely influence the outcome of the SSSS in children.[38] Finally, an ever-increasing number of cases of SSSS have been observed in adults, and many of these patients have been immune-compromised.[1,24,35,36] It appears highly likely, therefore, that immunological factors help regulate disease expression in adults, presumably by inhibiting or localizing the primary infectious process. Since immune function is not known to be blunted in normal neonates and children, it is likely that other factors (see below) underlie the propensity for the disease to occur in normal children.

Metabolism

Although intracutaneous inoculation of epidermolysin readily evokes cleavage in adult skin, systemic administration of toxin, even in massive doses, fails to produce signs of the SSSS (Table 27.5).[9] This observation suggested that extracutaneous factors might be responsible for adult resistance. In particular, we suspected that hepatic detoxification might occur. Two

Table 27.5. Influence of Route of Epidermolytic Toxin Administration on Subsequent Epidermolysis in Adult Mice (Elias et al. 1974)

Route	Dose (mg)	Sites	No. +	Type of Reaction	Comment
Intracutaneous	1	Back, abdomen (epilated); Ear (nontreated)	14/14	Localized	Both chemically epilated and nontreated skin are susceptible.
Systemic	10-25	Intraperitoneal, intramuscular	0/9	None	Local reactions sometimes occurred due to leakage over injection sites.

Table 27.6. Tracer Activity in Various Body Organs (Fritsch et al. 1976)[a]

	Ratio tissue/blood tracer activity	
	Newborn mouse	Adult mouse
Liver	0.5 ± 0.6	0.5 ± 0.24
Bile	——	0.7[a]
Duodenum	——	0.7[a]
Ileum	0.5 ± 0.23	0.4 ± 0.1
Colon	——	0.5[a]
Spleen	0.5[a]	0.4 ± 0.05
Lung	0.7[a]	0.6[a]
Sternum	0.95 ± 0.07	1.4 ± 0.4
Fat	——	0.8 ± 0.3
Muscle	——	0.5[a]
Tongue	0.8[a]	0.95 ± 0.1
Skin (1 h after inj)	0.4 ± 0.05	——
Skin (2–4 h after inj)	1.5 ± 0.4	0.8 ± 0.2
Thyroid	1.2 ± 0.1	1.0[b]
Kidney (1 h)	1.0[b]	5.1[b]
Kidney (24 h)	5.1[b]	1.1[b]

[a]These values represent the mean ratios of tissue/blood tracer activities in two to seven specimens of each organ measured.
[b]Only one measurement done. Specimens were taken 2–4 h after injection of I-ET, except for one measurement each of livers, skin, sternum, and throids which were taken at 24 and 48 h.

approaches were utilized to examine this possibility. First, I[125]-labeled epidermolytic toxin was injected into adult mice and the distribution of label in various organs was studied.[15] Label did not concentrate in liver at any time period (Table 27.6), a finding which mitigated against hepatic catabolism.[15] In the second group of experiments, animals were systemically poisoned with the hepato-toxin carbon tetrachloride and then dosed with epidermolysin. Although these animals did develop generalized scalding, the primary pathology from the carbon tetrachloride was not found in the liver.[15] Instead, severe renal tubular damage was encountered. Thus, although extracutaneous factors clearly contribute to adult resistance, it is not likely that hepatic catabolism plays an important role.

Renal Excretion

The final factor, and perhaps the critical one for the development of disease in children, appears to involve renal excretory mechanisms.[15] Whereas adult mice excrete most of an administered dose of I[125]-labeled epidermolysin by 2 h (Table 27.7), it persists in the blood of neonatal mice for hours and only slowly appears in the urine (Fig. 27.15). Interestingly, neonatal rats display a similar delayed pattern of excretion, yet they do not develop scalded skin syndrome.[15] This further implicates epidermis-specific factors in the resistance of most species to epidermolysin.[12]

Table 27.7. Biological Epidermolytic Toxin Activity in Vitro of Dialysed Adult Mouse Urine (Fritsch et al. 1976)[a]

Concentration of tracer (c/min/ml)	Dialysed urine	Crude ET-'ET mixture (1000 c/min/mg crude ET)
1 000	+	+
500	+	+
200	--	+
100	−	+
50	−	−

[a]Urine was collected for 3 h after injection of crude ET-I-ET mixture (100 c/min/mg crude ET) and processed as described in the Methods section. Newborn mouse skin was incubated 2 h at 37°C and evaluated for exfoliating activity.

If, as suspected, the rate of excretion rather than extrarenal catabolism were the critical factor in the development of the SSSS, then one might make two predictions: First, that biologically active toxin would be excreted by mice inoculated with organisms or injected with epidermolysin; second, that nephrectomy should create a fertile background for epidermolysis. Both of these predictions have been verified.[15]

These studies strongly implicate that renal factors may underlie the age distribution of SSSS.[15] The peak incidence of SSSS is still in the neonatal period, and neonates are known to have immature renal function. Finally, it also is relevant that several adult cases of SSSS have occurred against a setting of renal insufficiency.[24,26,35,36]

At the present time, no definite solution can as yet be offered for the

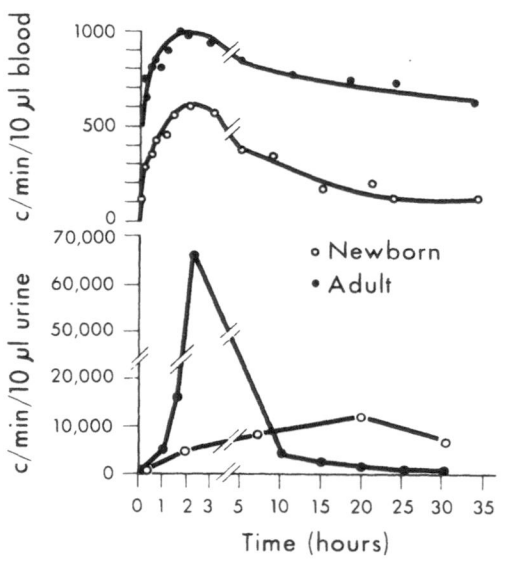

Fig. 27.15. Blood and urine radioactivity in newborn and adult mouse after intraperitoneal injection of 150,000 counts/min of epidermolytic toxin labeled with radioactive iodine (^{125}I) (from Elias et al. (1977) Arch Derm 113:207–219 with permission).

problem of age-related susceptibility to the SSSS; it appears, however, that both immunological factors and renal immaturity may contribute to determine occurrence, expression and severity of SSSS in an affected organism.

References

1. Amon RB, Diamond RL (1975) Toxic epidermal necrolysis: rapid differentiation between staphylococcal-induced disease and drug-induced disease. Arch Dermatol 111:1433–1437
2. Arbuthnott JP, Kent K, Lyell A, et al. (1972) Studies on staphylococcal toxins in relation to toxic epidermal necrolysis (the scalded skin syndrome). Br J Dermatol 86 (suppl 8):35–39
3. Arbuthnott JP, Kent K, Noble WC (1973) The response of hairless mice to staphylococcal epidermolytic toxin. Br J Dermatol 88:481–486
4. Baker DH, Diamond RL, Weupper KD (1978) The epidermolytic toxin of *Staphyloccus aureus:* its failure to bind to cells and its detection in blister fluids of patients with bullous impetigo. J Invest Dermatol 71:274–275
5. Baker DH, Weupper KD, Rasmussen JE (1978) Staphylococcal scalded skin syndrome: detection of antibody to epidermolytic toxin by a primary binding assay. Clin Exp Dermatol 3:17–24
6. Benson PF, Rankin GLS, Ripper JJ (1962) An outbreak of exfoliative dermatitis of the newborn (Ritter's disease) due to *Staphylococus aureus* phage type 55/71. Lancet I:999–1002
7. Diamond RL, Weupper K (1976) Purification and characterization of a staphylococcal epidermolytic toxin. Infect Immun 12:627–633
8. Diamond RL, Wolff HH, Braun-Falco O (1977) The staphylococcal scalded skin syndrome: an experimental histochemical and electron microscopic study. Br J Dermatol 96:483–492
9. Elias PM, Fritsch P, Mittermayer H, et al. (1974) Experimental staphylococcal toxic epidermal necrolysis (TEN) in adult humans and mice. J Lab Clin Med 84:414–424
10. Elias PM, Mittermayer H, Tappeiner G, et al. (1974) Staphylococcal toxic epidermal necrolysis (TEN): The expended mouse model. J Invest Dermatol 63:467–475
11. Elias PM, Fritsch P, Dahl MV, et al. (1975) Staphylococcal exfoliative toxin: Pathogenesis and subcellular site of action. J Invest Derm 65:501–512
12. Elias PM, Fritsch P, Mittermayer G (1976) Staphylococcal toxic epidermal necrolysis: Species and tissue susceptibility and resistance. J Invest Dermatol 66:80–89
13. Elias PM, Levy SW (1976) Bullous impetigo: Occurrence of localized scalded skin syndrome in an adult. Arch Dermatol 112:856–858
14. Elias PM, Fritsch P, Epstein EH Jr (1977) Staphylococcal scalded skin syndrome: Clinical features, pathogenesis, and recent microbiological and biochemical developments. Arch Dermatol 113:207–219
15. Fritsch P, Elias PM, Varga J (1976) The fate of staphylococcal exfoliatin in newborn and adult mice. Br J Dermatol 95:275–284
16. Fritsch PA, Kaaserer G, Elias PM (1979) Action of staphylococcal epidermolysin: further observations on its species specificity. Arch Dermatol Res. 264:287–291
17. Gillespie WA, Pope RC, Simpson K (1957) Pemphigus neonatorum caused by *Staphylococcus aureus* type 71. Br Med J 1:1044–1046
18. Howells CHL, Jones EH (1961) Two outbreaks of neonatal skin sepsis caused by *Staphylococcus aureus* phage type 71. Arch Dis Child 36:214–216

19. Johnson AD, Metzger JF, Spero L (1975) Production, purification, and characterization of *Staphylococcus aureus* exfoliative toxin. Infect Immun 12:1206–1210
20. Kapral FA, Miller MM (1971) Product of *Staphylococcus aureus* responsible for the scalded skin syndrome. Infect Immun 4:541–545
21. Kapral FA, Miller MM (1972) Skin lesions produced by *Staphylococcus aureus* exfoliatin in hairless mice. Infect Immun 6:877–879
22. Koblenzer PJ (1967) Acute epidermal necrolysis (Ritter von Rittershain-Lyell): A clinicopathologic study. Arch Dermatol 95:608–617
23. Kondo I, Sakurai S, Sarai Y (1973) Purification of exfoliatin produced by *Staphylococcus aureus* of bacteriophage group 2 and its physiochemical properties. Infect Immun 8:156–163
24. Levine G, Norden CW (1972) Staphylococcal scalded skin syndrome in an adult. N Engl J Med 287:1339–1340
25. Lillibridge CB, Melish ME, Glasgow LA (1972) Site of action of exfoliative toxin in the staphylococcal scalded skin syndrome. Pediatrics 50:728–738
26. Lyell A (1956) Toxic epidermal necrolysis: An eruption resembling scalding of the skin. Br J Dermatol 68:355–361
27. Lyell A (1967) A review of toxic epidermal necrolysis in Britain. Br J Dermatol 79:662–671
28. Lyell A, Dick HM, Alexander JOD (1969) Outbreak of toxic epidermal necrolysis associated with staphylococci. Lancet 1:787–789
29. McCallum HM (1972) Action of staphylococcal epidermolytic toxin on mouse skin in organ culture. Br J Dermatol (suppl 8):40–41
30. Melish ME, Glasgow LA (1970) The staphylococcal scalded skin syndrome: Development of an experimental mouse model. N Engl J Med 282:1114–1119
31. Melish ME, Glasgow LA (1971) The staphylococcal scalded skin syndrome: The expanded clinical syndrome. J Pediatr 78:958–967
32. Melish ME, Glasgow LA, Turner MD (1972) Staphylococcal scalded skin syndrome: Isolation and partial characterization of the exfoliative toxin. J Infect Dis 125:129–140
33. Melish ME, Chen FS, Murata MS (1979) Epidermolytic toxin (ET) production in human and experimental staph infection. Clin Res 27:114a
34. Miller MM, Kapral FA (1972) Neutralization of *Staphylococcus aureus* exfoliatin by antibody. Infect Immun 6:561–563
35. Norden CW, Mendelow H (1974) Staphylococcal scalded skin syndrome in adults. N Engl J Med 290:577
36. Reid LH, Weston WL, Humbert JR (1974) Staphylococcal scalded skin syndrome. Arch Dermatol 109:239–241
37. Ritter von Rittershain G (1975) Die exfoliative dermatitis jungerer sauglinge. Centralzeit F Kinderheil 2:3
38. Rudolph RI, Schwartz W, Leyden JJ (1974) Treatment of staphylococcal toxic epidermal necrolysis. Arch Dermatol 110:559–562
39. Rycheck RR, Taylor PM, Gezon HM (1963) Epidemic staphylococcal pyoderma associated with Ritter's disease and the appearance of phage type 3B/71. N Engl J Med 269:332–337
40. Wiley BB, Allman S, Rogolsky M, et al. (1974) Staphylococcal scalded skin syndrome: Potentiation by immuno-suppression in mice: Toxin-mediated exfoliation in a healthy adult. Infect Immun 9:636–640
41. Wuepper KD, Diamond RL, Knutson DD (1975) Studies on the mechanism of epidermal injury by a staphylococcal epidermolytic toxin. J Invest Dermatol 65:191–200
42. Wuepper KD, Baker DH, Diamond RL (1976) Measurement of *Staphylococcus aureus* epidermolytic toxin: A comparison of bioassay, radial immunodiffusion, and radioimmunoassay. J Invest Dermatol 67:526–531

Chapter 28
Means of Preventing Burn Wound Infection

JOHN F. BURKE

Prevention of burn wound infection must focus on the basic problems related to a large, open wound, the existence of necrotic tissue, and subsequent decreased host resistance. Antibiotics, both systemic and topical, are an adjunct to treatment and are designed as a supplement to restored normal physiology.

Clinical experience and research have pointed to the following systematic approach for management of the burn wound in respect to bacterial infection:

1) Prompt excision of eschar and immediate wound closure;
2) Protective environmental techniques designed to avoid bacterial cross-infection;
3) Use of preventive antibiotics;
4) Restoration of normal physiology through systemic support measures.

The steps reflect the fact that the burn patient is frequently at risk of serious infection for extended periods due to the time required for complete wound closure. The aim in prevention techniques is to reduce the time that the burn wound is open, prevent wound contamination, and enhance the patient's normal resistance.

Prompt Eschar Excision and Immediate Wound Closure

Prompt eschar excision and immediate wound closure have been found to be fundamental in shortening the time during which metabolic and bacteriologic abnormalities exist.[4] This calls for closing the wound through the use of skin autografts and is therefore limited by the size of the burn area and the amount of skin graft available. Recovery from an extensive burn thus becomes dependent on the availability of normal skin from donor sites if skin graft closure is to be effective in a time period short enough to prevent death from malnutrition and sepsis. If carried out successfully, prompt excision and closure protects host resistance and decreases the time at risk for burn wound sepsis.

In patients with mixed deep dermal and full thickness injuries, the excisional technique of sequential excision[5] is recommended. In cases with large areas of unequivocal full-thickness burns, direct excision to the fascia is the preferred procedure.[5]

In any event, since the consequences of a severe burn are metabolic exhaustion and invasive infection, surgical management must be seen as an early part of an overall plan for preventing infection by restoring normal physiology in the shortest possible time.

Protection from Cross-Infection

While bacterial comtamination of a thermal wound occurs at the time of injury, further contamination continues until the wound is closed or otherwise protected. In the case of the burn patient, wound closure may not be accomplished for weeks after the injury, during which time the wound may have become grossly contaminated if not carefully protected from cross-contamination. In regard to wound closure, the time required to completely close the wound is clearly related to the depth and area of burn. Larger burns requiring staged excision and allograft closure are at risk of infection for long periods of time. In addition, fresh donor sites and superficial burns that are not excised present additional areas which require protection from cross-comtamination.

Cross-infection is most effectively avoided by adopting effective protective precautions specific to the needs of the burn patient. This is a difficult problem because the patient is at serious risk of infection for long periods of time until the wound is closed, and the burn team must maintain complete protection from bacterial cross-contamination 24 hours a day for this period of time. In order to solve this problem, a system of isolation has been developed that is localized to the area around the patient's bed. This Bacteria Controlled Nursing Unit (BCNU)[7] is effective in preventing cross-infection even in massively burned patients receiving immunosuppressive therapy to prolong all graft skin survival.[6] The BCNU is compatible with the intensive care required by acutely ill burned patients cared for in an open intensive care ward.

The BCNU is comprised of thin, transparent plastic walls that surround the patient's bed. The plastic barrier provides conventional isolation, but does not create a visual or sound barrier, enabling the patient to be housed in an open ward without bacterial hazard. In addition, the unit includes a constant, unidirectional flow of sterile air from ceiling to floor, maintaining constant elimination of bacteria shed by the patient and reducing the risk of autoinfection. Access walls are built into the thin plastic walls enabling all medical, nursing care, and monitoring to be carried out without entering the protective environment.

Use of Preventive Antibiotics

The use of preventive antibiotics in burn patients is particularly important because of the increased risk of bacterial spread during debridement of infected or contaminated burn wounds in a patient who already has a weakened resistance to bacterial invasion. The concepts of preventive antibiotics that have been established[1,2,3] indicate that they are useful only during short periods of decreased host resistance. During those brief decreases in the patient's ability to defend himself against bacterial invasion, host resistance can be supplemented by short-term administration of antibiotics.

The most important period of high risk is that immediately surrounding operation. The combination of increased risk of bacterial contamination and lowered resistance is particularly pronounced during the complicated and time-consuming procedures associated with extensive excisions and skin grafting. In these patients, intravenous administration of the appropriate preventive antibiotic (chosen on evidence gained from culture and sensitivity studies of the burn wound) is begun immediately before operation and is continued through the surgical procedure. Antibiotic administration is stopped when the patient recovers from anesthesia and the cardiovascular system becomes stabilized in the postoperative period.

The use of preventive antibiotics as part of a systematic approach to burn management and a means of preventing infection is clearly designed as a supplement to the patient's natural resistance. Consequently, antibiotic therapy is only effective during the period immediately surrounding potential bacterial contamination. The use of preventive antibiotics for prolonged periods is ineffective and, in the long run, self-defeating since it is likely to generate allergic reactions, alterations in the normal bacterial flora of the G.I. and respiratory tract, and increase the likelihood of colonization of the wound with strains of resistant bacteria.

In considering overall therapy for the burn patient, topical antibacterial therapy can also be most effective in preventing colonization of the burn wound by cross-infection or autoinfection. Topical antibacterial agents serve as an environmental control measure in relation to the burn wound. Such agents as 0.5% aqueous silver nitrate solution, 10% mafenide acetate cream, and 1% silver sulfadiazine cream reduce the risk of burn wound infection and are important in overall burn care.

Restoration of Normal Physiology

In the treatment of the burn patient all measures must be designed to bring the patient to the state of near normal physiology. Excision, isolation, antibiotics will accomplish relatively little without the simultaneous resumption of normal physiologic function. In addition to seriously compromising the functions of the brain, heart, and kidneys, low cardiac output and systemic hypoperfusion produce other more subtle but nevertheless potentially lethal effects. These effects are seen, in particular, on the bacterially contaminated burn wound. In fact, immediate correction of circulatory failure is not only essential to prevent death from CNS, cardiac or renal failure, it is also an essential ingredient in the prevention of burn wound sepsis. Restoration of circulation, therefore, includes not only the restoration of the central but also the restoration of the peripheral circulation, if wound infection is to be avoided. In this context, the peripheral hypoperfusion produced by vasoconstrictive agents provides a further reason to avoid their use in burn shock.

In addition to the problems of circulatory volume and level of vascular perfusion, there are further systemic abnormalities that must be corrected before physiologic equilibrium and near-normal antibacterial defenses can be expected. Normal respiratory function and adequate gas exchange are vital; acid-base equilibrium, electrolyte concentration, and hydration are important areas to bring to balance. Further, nutritional abnormalities cause defects in host resistance. Malnutrition and, in particular, protein starvation seriously compromise the ability to mount a substantial antibacterial effort. Vitamin deficiencies act in a similar way. The properly functioning immunological mechanisms form a vital link in the complex chain of defense against bacterial invasion.

For the burn patient, adequate nutrition becomes the critical factor for resistance to infection, wound healing, and survival. Supplemental intravenous hyperalimentation is employed when gastrointestinal function will not satisfy nutritional requirements. The overall metabolic balance of the severely burned patient must be maintained throughout the period of open wound by means of oral and/or intravenous methods.

In the successful prevention of burn wound infection, resistance against bacterial invasion is a cornerstone. The patient's own bacterial defenses are essential and most important in preventing sepsis. Not only must the dead tissues be debrided and the wound closed, but during this period host defense must be maintained if infection is to be prevented.

References

1. Bernard HR, Cole WR (1964) The prophylaxis of surgical infection: The effect of prophylactic antimicrobial drugs on the incidence of infection following potentially contaminated operations. Surgery 56:151
2. Burke JF (1961) The effective period of preventive antibiotic action in experimental incisions and dermal lesions. Surgery 50:161–168, 184–185
3. Burke JF (1975) Preventive antibiotics in surgery. Postgrad Med 58:65
4. Burke JF, Bondoc CC, Quinby WC (1974) Primary burn excision and immediate grafting: a method shortening illness. J Trauma 14:389
5. Burke JF, Quinby WC, Bondoc CC (1976) Primary excision and prompt grafting as routine therapy for the treatment of thermal burns in children. Surg Clin North Am 56:477
6. Burke JF, Quinby WC, Bondoc CC, Cosimi AB, Russell PS, Szyfelbein SK (1975) Immunosuppression and temporary skin transplantation in the treatment of massive third degree burns. Ann Surg 182:183
7. Burke JF, Quinby WC, Bondoc CC, Sheehy EM, Moreno HD (1977) The contribution of a bacterially isolated environment to the prevention of infection in seriously burned patients. Ann Surg 186:377

Chapter 29
Topical Antibiotics in the Prophylaxis of Experimental *Staphylococcus Aureus* and *Streptococcus Pyogenes* Infections in Humans

JAMES J. LEYDEN AND ALBERT M. KLIGMAN

Minor cuts and abrasions are common events in the lives of all and are particularly frequent in young children. While a great deal of study has centered on the healing of superficial wounds, the frequency of infection by virulent organisms has received scant attention. There is no agreement and, indeed, much argument as to how minor cuts and abrasions should be treated to prevent infection. The infection rate varies considerably from one population to another, being strongly influenced by personal hygiene and climate.

The most instructive data derived from the epidemiologic studies of Taplin et al.,[9] who examined specified populations at different altitudes in the same tropical region. Skin infections were less common in the higher, drier, colder altitudes and greatest in the steaming, hot, wet lowlands. At each level, the prevalence of infection was highest among the poor who lived in crowded, unhygienic conditions. Personal cleanliness and social advantage were influential factors. In practically all cases group A hemolytic streptococci were isolated from the lesions.

In their studies of disadvantaged children in hot climates, Dillon[2] and Ferrieri et al.[5] found that group A streptococci accounted for practically all superficial wound infection, which give rise to the nasty lesions known as ecthyma. By contrast, in more temperate climates and in more hygienic populations, it is the impression of experienced clinicians that *Staphylococcus aureus* is the dominant cause of pyodermas that follow bites, burns, abrasions, cuts, etc. Information is lacking regarding the frequency with which minor wounds become infected in different populations. There is agreement, however, that *S. aureus* and hemolytic streptococci are responsible for virtually all superficial pyodermas in all clinical settings worldwide.

Clinicians are sharply divided regarding prophylactic treatment of fresh injuries to the skin. Some think that soap and water alone are sufficient. This seems extreme especially since prompt washing of every superficial wound is impossible. There is much debate regarding the safety and efficacy of so-called first-aid creams. As regards topical antibiotics for the prophylaxis of superficial wound infections, the pro and con arguments are even more passionate. The opponents cite emergence of resistant strains and induction of contact allergy as inevitable consequences. The controversy mainly revolves around neomycin because of its extensive use. However, two recent studies conlude that the risk of contact sensitization may have been greatly exaggerated.[7,8] Further, there is no evidence in nonhospitalized patients that pyoderma-causing bacteria have become resistant to neomycin. In any event, both studies illustrate the costly, burdensome work which is required to settle issues about which there is no consensus.

Early studies advocating the benefits of topical antibiotics were criticized on the grounds that they were not sufficiently controlled nor conducted under double-blind design. Subsequently, more rigorous investigations showed that topical antibiotics were effective treatments of pyodermas due to hemolytic streptococci, though less so than systemic administration in severe cases with multiple lesions.[1,3,4] The paramount difficulty in all field studies is the variability in the nature of the wounds themselves. These, of course, differ in regard to size, depth, tissue damage, contamination with foreign material, body region, time elapsed before treatment, etc. Confronted with the prodigality of these uncontrollable variables, the investigator resorts to the strategy of laboratory models to reduce the variables to a minimum. Experimental infections have again and again proved to be valuable models when the procedures result in reproducible infections that are sufficiently like their natural counterparts to have clinical relevance.

In this paper we shall describe methods for creating superficial infections with β-hemolytic streptococci and *S. aureus* in human volunteers. We shall demonstrate that these models are useful for assessing the value of topical antibiotics in the prevention of superficial wound infection.

Subjects

Healthy, young, adult college students were screened for evidence of systemic disease by physical examination, complete blood count, urinalysis and fasting blood sugar. Written, informed consent was obtained.

Wounding Techniques

Controllable wounds were produced in two ways. The first utilized the ammonium hydroxide blister procedure described by Frosch and Kligman.[6] A 1-cm cylinder on the volar forearm was filled with 50% NH_4OH until sharp burning was felt or small follicular wheals appeared. This exposure, on the average about 12 min, results in a tense intraepidermal bulla accompanied by death of the epidermis. This wound is the laboratory counterpart of a superficial abrasion and has replaced our former technique of stripping with cellophane tape. The blister is technically easier and the wounds are more uniform. Another use for this model is to identify materials which retard wound healing.

The second method involves scarification of the skin with a 30-gauge sterile needle.   Four parallel scratches in one direction and four at right angles were made without producing bleeding. This wound is the equivalent of the superficial scratch or cut.

Inoculation of Wounds

With ammonium hydroxide wounds, inoculation should be delayed for 16–24 h after removal of the blister roof to avoid systemic reactions, especially likely when several blisters are inoculated.   Pathogens can be applied immediately after scarification without fear of untoward reactions.

We emphasize the importance of proper preparation of the inoculum. *S. aureus* and especially group A streptococci must be freshly subcultured at least twice on agar supplemented with sheep's blood to assure virulence. Trypticase soy agar gives variable results.   Strains become progressively less infective when maintained on media other than blood.

Our *S. aureus* is a nontypable strain recovered from a patient with infected dermatitis. It continues to produce uniform lesions and is susceptible to multiple antibiotics. Our group A streptococcus is an antibiotic-sensitive non-nephritogenic strain (T type 11/12, M untypable). We have found many strains of *S. aureus* and group A streptococci that produce similar lesions.

The inoculum size for each blister is 10^5 organisms and 10^7 for each scarification site.   Fresh 18-h subcultures are prepared in normal saline; after inoculation the wound site is covered with plastic wrap and sealed with impermeable plastic tape to promote growth of the organism and to prevent translocation of antimicrobial agents among the sites.

Culture Technique

Quantitative cultures were obtained by the detergent—scrub technique of Williamson and Kligman. Sampling consists of two 1-min scrubs within a glass cylinder of 3.8 sq cm area with 1 ml of 0.1% Triton X-100. In the case of streptococci, 1% Tween-80 is added to the Triton X-100 fluid.   Triton X-100 is inhibiting to streptococci after an hour. This can be prevented by the addition of 1% Tween-80. The scrub samples are pooled, serially diluted in tenfold steps in 0.05% buffered Triton and then drop plated onto Trypticase Soy Agar (TSA), on TSA with lecithin and polysorbate-80, on TSA with 6% sheep's blood.

Treated sites have to be monitored for the possibility of carry-over into the culture medium (false positive results). First, samples are collected no sooner than 12 h after applying the test agent. Second, a 0.5 cc aliquot of each sample is inoculated with 0.1 ml of a fresh saline suspension of a strain *S. epidermidis* highly sensitive to the test antimicrobial. This mixture is allowed to stand for 4 h, subcultured quantitatively and compared with a similar incubation mixture of 0.1 ml *S. epidermidis* and 0.5 ml of Triton

X-100. The possibility of carry-over is examined prior to undertaking a formal study and may help to determine the correct timing for sampling after application of the test antimicrobial.

Effect of Topical Antibiotics

Panels of ten healthy subjects were used for each infection model.

In the *S. aureus* model three blisters were raised on each forearm. After 24 h, each blister base was inoculated with 2×10^5 cells of an overnight growth of *S. aureus* and occluded for 6 h under plastic wrap and impermeable tape. The following treatments were then randomly applied.

  a) 0.1 ml of bacitracin ointment (500 units per gram);
  b) 0.1 ml of neomycin ointment (5 mg neomycin sulfate per gram);
  c) 0.1 ml of polymyxin ointment (10 000 units per gram);
  d), (e) the ointment base alone;
  f) untreated control.

Each site was immediately redressed with plastic wrap and impermeable tape and left for 24 h. Each wound was assessed clinically for erythema and suppuration on a 0–4 scale and sample[8] quantitatively for *S. aureus*.

In the streptococcal model each scarified site was inoculated with 3×10^7 cells of an overnight culture of β-hemolytic streptococci. The site was covered immediately with an impermeable dressing, as above, and left for 6 h. The same treatments were applied. Clinical assessment and quantitative cultures were made 24 h later.

Results

These are summarized in Table 29.1. In the *S. aureus* model there were no differences clinically or bacteriologically among the untreated control site, the ointment base, and polymyxin ointment. Both bacitracin and neomycin produced highly significant reductions ($p < 0.001$) in the number of organisms with the effect of neomycin being the greater (the log mean count for neomycin was 4.6541 compared with 5.1304 for bacitracin).

Table 29.1. Topical Antibiotics in Experimental *S. aureus* Infections

	Untreated control	Vehicle treated	Neomycin	Bacitracin	Polymyxin
# sites positive for *S. aureus*	10	10	10	10	10
Geometric mean count for positive sites	7.06	7.24	4.25	5.1304	7.3184
p value		NS	< 0.001	< 0.001	NS

Table 29.2. Topical Antibiotics in Experimental Streptococcal Infections

	Untreated control	Vehicle treated	Neomycin	Bacitracin	Polymyxin
# sites positive for *S. pyogenes*	10	10	10	10	10
Geometric mean count for positive sites	5.78	5.75	2.64	3.49	5.79
p value		NS	<0.001	<0.001	NS

In the *S. pyogenes* model there was no difference between the untreated control and the ointment base (Table 29.2). Polymyxin achieved a barely significant reduction ($p < 0.05$). Both bacitracin and neomycin produced a highyl significant reduction ($p < 0.001$).

Clinically, striking differences can be seen between wounds in which *S. aureus* or *S. pyogenes* has been successfully prevented from proliferating. Wounds with counts of *S. aureus* in excess of 10^6 show marked erythema and a yellowish exudate. Lesions with *S. pyogenes* in excess of 10^4 organisms develop a palpable erythema and edema along with pustules and crusting along the scarification line.

Discussion

Field evaluation of the efficacy of an agent in the prevention of spontaneous infection of minor cuts and abrasions is usually unfeasible with the exception of circumstances where there is huge logistical support. The variables are simply too great and arriving at conclusions requires the study of a large population under special risk. Moreover, one cannot be sure of compliance with treatment schedules unless each application is monitored, and even then the quantity applied may vary greatly. The use of experimental infections promotes standarization of all these variables. We would emphasize that these superficial, localized infections can be induced in humans without undue risk or discomfort. Moreover, the lesions produced are clinically comparable to spontaneous infections. We have ascertained this by showing that the clinical picture becomes diagnostic when inoculated wounds are left untreated under occlusion for a few days. Thus the results obtained with topical antibacterials are relevant to human infection.

These experimental infections provide a practical and convenient means for assessing the efficacy of topically applied antibiotics. The results were in accordance with expectations. The broader question of whether antibiotics should be used on minor cuts and abrasions is outside the scope of this research, though we have to be registered as protagonists, especially where nephritogenic strains of streptococci abound. We have reason to believe that the same models can be satisfactorily employed for assessing all types of topical antibacterial substances.

References

1. Burnett JW (1963) The route of antibiotic administration in superficial impetigo. N Engl J Med 268:72–75
2. Dillon HC (1968) Impetigo contagiosa: suppurative and non-suppurative complications. Clinical, bacteriologic and epidemiological characteristics of impetigo. Am J Dis Child 115:530–541
3. Dillon HC (1970) The treatment of streptococcal skin infection. J Ped 76:676–684
4. Esterly NB, Markowwitz M (1970) The treatment of pyoderma in children. JAMA 212:1667–1670
5. Ferrieri P, Dajani AS, Wannamaker LW (1972) Natural history of impetigo. I: Site sequence of acquisition and familial patterns of spread of cutaneous streptococci. J Clin Invest 51(11):2851–2862
6. Frosch PJ, Kligman AM (1977) Rapid blister formation in human skin with ammonium hydroxide. Br J Dermatol 96:461–473
7. Leyden JJ, Kligman AM (1979) Contact dermatitis to neomycin sulfate. JAMA 242(12):1276–1278
8. Prystowsky S, Allen A, Smith RW, Nonomura JH, Odom RB, Akers WA (1979) Allergic contact hypersensitivity to nickel, neomycin, ethylenediamine and benzocaine. Arch Dermatol 115:959–963
9. Taplin D, Lansdell L, Allen AM, Rodriguez R, Cortes A (1973) Prevalence of streptococcal pyoderma in relation to climate and hygiene. Lancet I:501–503
10. Williamson PJ, Kligman AM (1965) A new method for quantitative investigation of cutaneous bacteria. J Invest Dermatol 45:498–503

Chapter 30
Current Concepts in Neonatal Bacterial Colonization

WILLIAM T. SPECK, JANE O'NEILL, JOHN M. DRISCOLL, AND HERBERT S. ROSENKRANZ

Bacterial infections are a major cause of neonatal morbidity and mortality. Unfortunately, the impact of antimicrobial therapy on the natural history of bacterial infections in the newborn has been less dramatic than the effect of these agents on the natural history of bacterial infections in the older child and adult. There is a long tradition of perinatal preventive measures designed to decrease the frequency of such infections. Since most of the bacterial pathogens responsible for neonatal infections are part of the normal flora that routinely colonize the newborn, many of the recently developed preventive measures involve the controlled manipulation of bacterial colonization. What follows is a discussion of current concepts of neonatal bacterial colonization and some of the preventive measures recently introduced into nurseries.

Normal Colonization

Most newborn infants first encounter bacteria at the time of labor and delivery when, during passage through the birth canal, the skin (vernix caseosa) and mucus membranes are contaminated with normal vaginal flora—staphylococci, streptococci, anaerobic bacteria, and, occasionally, Enterobacteriacae (*E. coli, klebsiella*, etc.).[6,23] (The skin and mucus membranes of infants delivered by caesarean section are usually sterile). This initial contamination is characteristically transient, as these contaminating microorganisms rarely colonize the newborn infant. However, the initial bacterial contamination may be modified qualitatively and/or quantitatively such that contamination is followed by colonization and/or tissue invasion. Thus qualitative changes in maternal genitourinary flora may result in neonatal contamination with microorganisms of increased virulence, e.g., *N. gonorrhoeae*, group B streptococcus, *C. trachomatis, L. monocytogenes*, etc. These microorganisms quickly colonize the newborn infant and may be responsible for subsequent systemic illness. Similarly, quantitative changes in newborn bacterial contamination may occur. When labor and/or delivery are complicated (intrauterine monitoring, fetal scalp sampling, asphyxia) or prolonged, the extent of neonatal contamination increases and may involve multiple sites including the gastrointestinal tract, the respiratory tract and/or subcutaneous tissue. Such extensive contamination not uncommonly leads to colonization, tissue invasion, and newborn disease.

Following delivery, many of the contaminating microorganisms are removed along with the vernix caseosa during the first ritual bath. Whether delivery is by the vaginal or abdominal route, cultures of the infants nose, throat, umbilicus, and rectum taken immediately after birth are characteristically sterile. However, during the first few days of life these sites become colonized with nosocomially acquired microorganisms. This initial colonization progresses in a predictable fashion beginning in the perineal and umbilical regions and gradually extending to involve the thorax, the axilla, and finally the anterior nares.[17,24,30] Studies performed in our own nursery demonstrate this differential colonization at two specific target sites, e.g., the umbilicus and the anterior nares.[25]

In Table 30.1, we see that culture-negative (sterile) specimens were observed at the umbilicus and anterior nares in 5% and 62% of the infants on the third day of life. However, by day 14 of life the number of negative cultures at these two sites had decreased significantly (approximately 7%). Moreover, the extent of neonatal colonization at these two sites similarly varied as a function of time. Thus "heavy" colonization ($> 10^7$) at the umbilicus and anterior nares on day 3 of life was 56% and 12% respectively, while on day 14 of life, heavy colonization at these two sites had increased to 57% and 59% (Table 30.2).

Thus far we have discussed neonatal bacterial colonization in general

Table 30.1. Colonization Rates on Days 3, 14, and 42 at the Umbilicus and Nasopharynx of Normal Full-Term Infants Receiving No Cord-Care Regimen

	Day	Control	
		NP[a]	UMB[b]
None	3	62%	5%
	14	7%	7%
	42	8%	4%
S. aureus	3	20%	52%
	14	60%	46%
	42	61%	43%
Group B strept.	3	6%	16%
	14	4%	13%
	42	1%	6%
Gram-negative	3	2%	34%
	14	5%	34%
	42	0	24%

[a]NP = nasopharynx
[b]UMB = umbilicus

Table 30.2. Percentage of Children with Colonization Rates Exceeding 10^7 Colony-Forming Units Per Culture Site (Umbilicus and Nasopharynx)

Treatment regimen	Nasopharynx			Umbilicus		
	Day of life					
	03	14	42	03	14	42
Silver sulfadiazine	4	43	48	35	45	26
Triple dye	6	41	33	28	43	42
Castile soap	12	59	41	56	57	31

terms, let us now concentrate on colonization as it relates to three specific bacterial pathogens, e.g., Staphylococcus aureus, gram-negative microorganisms, and group B streptococcus.

Staphylococcus Aureus

Initial interest in neonatal colonization developed in the 1940s and 1950s when nursery epidemics of staphylococcal pyoderma occurred throughout the world. Epidemiological studies soon established that most infants born in a hospital become skin and nasal carriers of Staphylococcus aureus within the first few days of life; that a relationship exists between the rate of neonatal staphylococcal colonization and staphylococcal disease; that there is a seasonal variation in neonatal staphylococcal colonization and disease; and, finally, that the major route by which newborn infants become colonized with these microorganisms is by transfer of staphylococci from one infant to another on the hands of nursing attendants. Numerous preventive measures were utilized in nurseries for the newborn in an attempt to control staphylococcal colonization and disease. These preventive measures included cohort nursing, rooming-in, air conditioning, ultraviolet lighting, elaborate barrier nursing techniques, artificial colonization with nonpathogenic staphylococci, application of antiseptic agents to the nose and/or umbilicus, and total body bathing with antiseptic emulsions. While all of these preventive measures had some beneficial effect in reducing the rate of neonatal staphylococcal colonization and disease, total body bathing with 3% hexachlorophene emulsions was the most widely utilized preventive measure for staphylococcal prophylaxis and was most often given credit for the decreasing incidence of staphylococcal disease observed in the 1960s.[15]

Questions about the efficacy of routine hexachlorophene bathing first arose in the mid-1960s when the incidence of neonatal gram-negative bacterial infections increased. Hexachlorophene had previously been shown to be more effective in vitro against gram-positive than gram-negative bacteria; there was some clinical evidence that its usage actually encouraged colonization with gram-negative microorgansims, presumably through the phenomenon of bacterial interference.[12,29] Several reports were published in the 1960s which suggested that routine hexachlorophene bathing was responsible for the increased incidence of gram-negative bacterial colonization and infections observed in newborn nurseries in the 1960s.[12] Furthermore, it was suggested that the antiseptic effect of hexachlorophene bathing could be increased by simultaneous application to the newborn infant of a second antiseptic agent with activity against gram-negative microorganisms.[15]

Further concern about the efficacy of routine hexachlorophene bathing arose in the late 1960s when several studies questioned the therapeutic index or margin of safety of hexachlorophene. These studies demonstrated that hexachlorophene was absorbed through the intact skin of newborn infants and that this compound was toxic at very low levels to the developing nervous system.[2,20] Following these reports, the Food and Drug Administration issued an order in December 1971 banning over-the-counter products containing 3% hexachlorophene and routine body-bathing of newborn infants with hexachlorophene was discontinued.[11]

Several reports subsequently appeared in the medical literature suggesting an increased incidence of neonatal staphylococcal disease in those nurseries which had discontinued routine hexachlorophene bathing.[8,9] However, the exact relationship between the reported increase in staphylococcal disease and the discontinuation of hexachlorophene bathing was not established in all cases. Nevertheless, there have been repeated attempts to develop a satisfactory alternative to hexachlorophene bathing to control neonatal staphylococcal colonization and disease.

Control of Staphylococcal Colonization

In the United States, recent attempts to control staphylococcal colonization have involved the postnatal application of an antibacterial agent to the umbilicus since staphylococcal colonization begins at this site and inhibition of umbilical colonization retards subsequent colonization of other body sites.[18] Many of the antibacterial agents utilized for this purpose have limitations that include unkown toxicity to the newborn infant, percutaneous absorption into the systemic circulation, and selective activity against gram-positive microorganisms.

Several recent studies have demonstrated that postnatal application of triple dye (brilliant green, 2.29 gm; proflavin hemisulfate, 1.14 gm; crystal violet, 2.29 gm; and water, q.s. to 1 000 ml) to the umbilical stump inhibits staphylococcal colonization.[22] Table 30.3 shows the results of data obtained in our nursery confirming the effectiveness of this agent in reducing postnatal staphylococcal colonization rates. Staphylococcal colonization on day 3 and 14 of life was significantly less in infants treated with a single application of triple dye to the umbilical stump than in control infants receiving similar cord treatment with soap and water.

Although triple dye application has proven effective in inhibiting staphylococcal colonization, this antiseptic agent has two limitations: its potential toxicity and its effect on colonization with nonstaphylococcal bacterial

Table 30.3. Effectiveness of Triple Dye (Brilliant Green, Proflavin Hemisulfate, and Crystal Violet) in Reducing Postnatal Bacterial Colonization Rate[a]

Treatment regimen	Nasopharynx			Umbilicus		
	Day of life					
	03	14	42	03	14	42
	No. organisms isolated					
Silver sulfadiazine	57	8	2	25	1	2
Triple dye	60	2	5	31	3	5
Castile soap	62	7	8	5	7	4
	Staphylococcus aureus					
Silver sulfadiazine	6	51	56	31	50	33
Triple dye	7	37	48	8	38	38
Castile soap	12	58	61	48	52	43
	Group B streptococci					
Silver sulfadiazine	6	1	4	2	8	5
Triple dye	6	3	0˙	14	16	6
Castile soap	5	3	1	10	11	6
	Gram-negative bacteria					
Silver sulfadiazine	3	8	6	14	24	19
Triple dye	2	13	3	36	57	19
Castile	2	4	0	33	32	24

[a]Percentage of infants colonized.

pathogens. Dyes of the acridine and triphenylamine series (proflavine and brillant green) are capable of altering intracellular DNA in eukaryotic and prokaryotic cells.[14] The effect of such agents on the newborn infant are unknown; however, the risks seem minimal. An additional limitation of the triple dye regimen is that a single postnatal application of triple dye to the umbilicus, although effective in reducing staphylococcal colonization rates, promotes colonization with group B streptococci and gram-negative microorganisms (E. coli and klebsiella).[5,25] However, subsequent studies by our group and by Barret and his associates in Houston, have demonstrated that daily application (for 3 days) of triple dye to the umbilicus inhibits staphylococcal colonization and does not promote colonization with streptococci.[5,26] Thus daily application of this antiseptic agent appears to be the preferred regimen.

Another antibacterial agent that we have studied with a potential usefulness in inhibiting neonatal staphylococcal colonization and/or disease is chlorohexidine. Total body bathing and cord application of this agent appears effective in diminishing neonatal staphylococcal colonization and disease. Chlorohexidine bathing offers certain advantages over hexachlorophene bathing: its broad spectrum of activity (gram-positive and gram-negative microorganisms) and the apparent lack of percutaneous absorption in newborn infants demonstrated by Sprunt and her associates (unpublished results). Although chlorohexidine is widely used as a general skin disinfectant, its routine usage in newborn infants is limited to Canada and Europe.[16] Carefully controlled studies to evaluate the efficacy of chlorohexidine bathing in newborn infants have not yet been conducted in the United States; chlorohexidine is presently classified as an investigational drug and thus cannot be recommended as a preventive measure to modify bacterial colonization in newborn infants.

In summary, the most effective technique presently available for reducing neonatal staphylococcal colonization rates and staphylococcal disease is daily application of triple dye to the umbilical stump. However, this technique is no substitute for good nursing practices (see Appendix).

Gram-negative Colonization

The traditional emphasis on neonatal staphylococcal colonization and disease changed in the early 1960s with the emergence of gram-negative microorganisms as the predominate pathogens in neonatal septicemia (a simultaneous ecological shift from nosocomially acquired gram-positive to gram-negative bacterial infections occurred in adult patient populations).[19,31] The mechanisms for this epidemiological shift in neonatal nosocomial infections has not been established. Possible explanations have included the recognized tendency of staphylococcal colonization and disease to undergo unexplained epidemiological shifts; the widespread use of parenteral antimicrobial agents that selected for multiple drug-resistant gram-negative bacteria; routine use of antistaphylococcal preventive measures, e.g., hexachlorophene bathing, which promote colonization with gram-negative bacteria; and, finally, a change in the neonatal population to include more critically ill low-birthweight infants requiring complicated life-support systems.

Neonatal gram-negative contamination and resultant colonization may occur intrapartum from ascending infection in the vagina following rupture of fetal membranes or following skin and mucous membrane contamination with vaginal flora at the time of labor and delivery. Factors that promote intrapartum contamination with gram-negative microorganisms have not been fully evaluated; however, some evidence suggests that premature rupture of fetal membranes, complicated labor and delivery, neonatal asphyxia and/or antimicrobial treatment of the mother prior to delivery may be important factors in determining the rate and extent of intrauterine contamination with these microrganisms.

Neonatal colonization with gram-negative microorganisms occurs predominately in the postpartum period. Several factors predispose to postpartum colonization with these bacteria. Feeding practices are of some importance since infants entirely breast fed harbor fewer gram-negative bacteria in their nasopharynx than infants who are partially or completely artificially fed.[21] In addition, nursery bathing regimens may predispose to colonization with gram-negative microorganisms.[12] Bathing with antiseptic emulsions that preferentially inhibit colonization with gram-positive microorganisms has been shown to promote postnatal gram-negative bacterial colonization. Similarly, routine administration of broad-spectrum antimicrobial agents for presumed septicemia and utilization of the increasingly complex and invasive technology (umbilical catheters, parenteral alimentation systems, inhalation therapy, and pressure transducers) required to monitor and treat low-birthweight infants have been shown to increase neonatal gram-negative colonization.

Control of Gram-negative Colonization

Techniques for controlling neonatal colonization with gram-negative microorganisms have not been fully evaluated. This lack of information in part relates to the failure to demonstrate an exact correlation between colonization rates with these microorganisms and neonatal disease. There is only limited data to suggest that inhibiting colonization with these microorganisms will have any effect on the incidence of neonatal infections. Nevertheless treatment of the umbilical stump with antimicrobial agents active against gram-negative bacteria inhibits subsequent colonization with these microorganisms.[20,25,26]

For example. we have shown (Table 30.3) that a single application of silver sulfadiazine to the umbilicus of the newborn infant inhibits colonization with gram-negative microorganisms on day 3 and day 14 of life. This observation has been subsequently confirmed by other investigators.[5] Alternative measures to control gram-negative colonization are also available. Sprunt and her associates have recently demonstrated the feasibility of artificially colonizing low-birthweight infants with a drug-resistant nonpathogenic streptococcus viridans and thereby inhibiting gram-negative bacterial colonization.[27] This inhibition presumably involves bacterial interference.

In summary. limited data is available to demonstrate a relationship between neonatal gram-negative colonization rates and neonatal disease. Vigorous attempts to control colonization with these microorganisms seem unwarranted at present.

Group B Streptococcus

The group B streptococcus, often incorrectly referred to as the β-hemolytic streptococcus, was initially defined as an animal pathogen in the 1930s and designated *Streptococcus agalactiae*. Following Lancefield's comprehensive classification of the streptococcus based on cell wall carbohydrate, these microorganisms were classified as group B streptococci. These organisms were often recovered from the upper respiratory tract and from female reproductive tracts of asymptomatic individuals and because of their nonpathogenic characteristics, Brown and his associates, suggested that they be named *S. opportunis*.[7] It was soon appreciated that these microorganisms were capable of causing a wide spectrum of human disease.

The impact of group B streptococci on neonatal populations was first emphasized in 1964 by Eickoff, who reported on group B streptococcal infections in newborn infants at the Boston City Hospital.[10] During the early 1970s the group B streptococcus became a common etiologic agent in neonatal infection and presently is the most frequent cause of life-threatening infection in newborn infants. Franciosi and his associates in Denver[13] and Baker and her co-workers in Houston,[4] after reviewing a large series of neonatal group B streptococcal infections, separated group B streptococcal disease into two clinical syndromes based on age at onset. The early onset, or septicemic, type of infection, resulting from bacterial contamination and tissue invasion of this microorganism at the time of labor and delivery occurs within the first few days of life and presents with fulminating septicemia and/or respiratory distress (Table 30.4). The mortality rate of this early syndrome approaches 50%.

The delayed onset, or meningitic, group B streptococcal syndrome results from postnatal nosocomial acquisition of group B streptococci. This clinical entity is characterized by the appearance of signs and symptoms after the first week of life. The late onset syndrome most often involves the central nervous system and has a lower mortality rate than the early onset form of the disease. Although the separation of neonatal group B streptococcal disease into two separate syndromes is not always possible, such a distinction is useful in discussing the epidemiological aspects of neonatal group B streptococcal colonization and disease.

Neonatal group B streptococcal colonization may occur intrapartum from ascending infection in the vagina following rupture of fetal membranes or following skin and mucus membrane contamination with vaginal flora during labor and delivery. Reported colonization rates for pregnant women vary depending on the patient population studied and the bacteriologic techniques utilized to detect colonization. Recent studies report a maternal colonization rate of 25%–35% (similar colonization rates have been observed in nursing personnel caring for newborn infants).[3] Moreover, of newborn infants born to colonized women, approximately 70% will have group B strep-

Table 30.4. Characteristics of Newborn Infants with Meningitis Due to Group B Streptococcus

| Characteristic | Form of disease | |
	"Early onset"	"Late onset"
Time of onset	≤8 days	11 days–12 weeks
Complications in delivery	92%	19%
Clinical presentation	Severe, fulminant, multisystem illness	Mild to moderate, slowly, progressive illness, usually meningitis
Mortality rate	58%	14%
Serotypes of streptococcus	Variable	Type III predominant

tococci on the skin and/or mucus membranes during the first few days of life; there is almost complete concordance between the serotypes isolated from colonized mothers and their colonized infants.[3] This concordance suggests vertical transmission from mother to infant as the mode by which group B streptococci are transmitted to newborn infants developing the early onset or septicemic form of group B streptococcal disease.

Neonatal group B streptococcal colonization may occur in the postpartum period through nosocomial acquisition of these microorganisms.[1] Several studies have demonstrated that nosocomial acquisition of group B streptococci is the major route by which infants born to culture-negative women acquire this microorganism. Thus newborn colonization has been reported to increase from approximately 20%–25% during the first 24 h of life to 60%–65% at the time of nursery discharge. These studies suggest that the major route of postpartum colonization among newborn infants is cross-colonization in the newborn nursery, presumably via the hands of nursery personnel.

Control of Group B Streptococcal Colonization

Numerous techniques have been utilized to prevent the intrapartum acquisition of group B streptococci. The traditional approach has been to administer antimicrobial agents to all colonized women and their sexual partners in an attempt to eradicate these microorganisms from the female genitourinary tract and thereby prevent vertical transmission. Although several antimicrobial regimens have been utilized for this purpose, there are as yet no random, controlled prospective studies that have consistently demonstrated the efficacy of this method of preventing group B streptococcal colonization and/or neonatal disease. An alternative technique for the prevention of vertical transmission of group B streptococci is antimicrobial treatment, e.g., parenteral penicillin, of all infants born to colonized women. Initial uncontrolled studies on this preventive measure have been encouraging and a collaborative study to evaluate the efficacy of this approach has been initiated.[28] However, pending the results of this study, no method of antimicrobial prophylaxis of pregnant women or newborn infants can be recommended at this time.

Prevention of the postpartum acquisition of group B streptococci involves local treatment of the umbilical cord with antiseptic agents. Although no large collaborative studies have been conducted, small studies from a number of centers, including our own, have shown that a number of agents, e.g., daily application of triple dye and silver sulfadiazine, are effective in decreasing nosocomial acquisition of group B streptococci.[26] Whether or not this technique of diminishing nosocomial acquisition of group B streptococci will decrease the incidence of late onset group B streptococcal disease is unknown. In summary, no effective technique is presently available to control vertical transmission of group B streptococci and early onset disease. However, postnatal nosocomial acquisition of this microorganism and the delayed onset group B streptococcal syndrome can be prevented by daily application of antibacterial agents, e.g., triple dye to the umbilical stump during the period of hospitalization.

Appendix: Control of Infection in the Nursery

Admittance to the nursery is granted only to healthy individuals concerned with the care of the infant, including medical and nursing personnel, clergymen, laboratory and x-ray technicians, and the family of the infant.

1) Personnel and mothers with the following illnesses must be excluded from the nursery or contact with the infants:
   a) Fever of undetermined origin
   b) Respiratory infection
   c) Gastroenteritis
   d) Any communicable disease or infection or open draining skin lesions.

2) Infants admitted from outside nurseries must be isolated in an incubator with gown isolation until results of cultures are available. Cultures of the throat, skin, and urine must be obtained.

3) Infants in the same room with infants with possible infection whould be maintained in a forced-draft incubator.

4) Infants excluded from the nursery and to be isolated in separate rooms are those with toxigenic or invasive *E. coli* gastrointestinal infection, viral infections, and draining lesions.

5) Infants delivered to patients who have a communicable disease must be excluded from the nursery.

6) The incubators and cribs must be washed thoroughly with antiseptic solution at least each week and after each separate occupancy. The water is changed every two days.

7) All equipment must be sterilized before use, and respiratory tubing and humidification areas are to be changed every 24 h while infant is on the respirator.

In accordance with state regulation practices recommended by the American Academy of Pediatrics:

1) All nursing personnel wear short-sleeved scrub gowns in the nursery.

2) Physicians and others who enter the nursery briefly wear a gown over street clothes or a uniform.

3) All persons wash their hands before entering any of the nurseries. Antiseptics most useful for handwashing in the nursery are iodophor preparations and antiseptic preparations containing a 3% concentration of hexachlorophene. The iodophors are superior because of their activity against gram-negative organisms but may cause sensitization. Hence, both prepara-

tions should be available at the sink-side. The containers for these preparations should be periodically sterilized.

4) Hands should be considered contaminated unless they are washed just before and just after handling an infant and after touching contaminated materials.

5) To examine infants in incubators, medical personnel should roll their gown sleeves above their elbows and wash to the elbows before each handling of a baby. Gown sleeves once wet are considered contaminated.

6) If medical personnel are to hold babies in their arms or against their gowns, then they must observe individual gown technique.

7) Nurses are required to use an individual gown for each infant, washing their hands and donning a fresh gown before handling the infant. This applies whether the infant is to be held or to be cared for in an incubator. (The individual gown technique used by nurses not only is an ideal method for clean handling of infants but also prevents undue irritation of the arms from the many dozens of washings the nurse does daily).

8) Proper antiseptic and cord care is observed on all infants, including daily cord cleansing with an antibacterial agent, triple dye, silver sulfadiazine, etc. As the cord dries and shrivels, it recedes below skin margins of the umbilicus, creating a moist skin pocket which can become an excellent culture site for staphylococci. Wash this pocket scrupulously with a cotton-tipped applicator moistened with alcohol.

9) Masks are unnecessary for pediatricians and nurses who know how to behave in a nursery. They are only necessary for surgeons. Everyone must wash his hands if he has touched his face or nose.

10) Anything wet in the nursery, e.g., emergency drain in the sink, siphon in the sink, should be thoroughly cleaned and sterilized as often as possible.

11) Any equipment that becomes wet, e.g., oxygen hoods, tubing, should be cleaned and sterilized every 24 h.

References

1. Aber RC, Allen N, Howell JJ, Wilkinson HW, Facklam RR (1976) Nosocomial transmission of group B streptococci. Pediatrics 58:346
2. American Academy of Pediatrics Committee on the Fetus and Newborn (1972) Hexachlorophene and skin care of newborn infants. Pediatrics 49:625
3. Baker CJ (1977) Summary of the workshop on perinatal infections due to group B streptococcus. J Infect Dis 136:137
4. Baker CJ, Barett FF, Gordon RC, Yow MD (1973) Suppurative meningitis due to streptococci of Lancefield group B: a study of 33 infants. J Pediatr 82:724
5. Barrett FF, Mason EO, Fleming D (1977) The effect of 3 cord regimens on bacterial colonization of normal neonates. Pediatr Res 11:497
6. Brook I, Barrett CT, Brinkman III CR, Martin WJ, Finegold SM (1979) Aerobic and anaerobic bacterial flora of the maternal cervix and newborn gastric fluid and conjunctiva: A prospective study. Pediatrics 63:451–455
7. Brown JH (1949) Bovine mastitis and human disease. In: M Bjorneboe (ed) Proceedings of the fourth international congress of microbiology. Rosenkiide and Bagger, Copenhagen, pp. 198–200

8. Center for Disease Control (1972) Morbidity and Mortality Weekly Report 21:37
9. Center for Disease Control (1972) Morbidity and Mortality Weekly Report 21:253
10. Eickoff TC, Klen JO, Daly AK, Ingall D, Finland M (1964) Neonatal sepsis and other infections due to group B beta-hemolytic streptococci. N Engl J Med 271:1221
11. FDA Bulletin, December 8, 1971
12. Forfar JO, Gould JC, MacCabe AI (1968) Effect of hexachlorophene on the incidence of staphylococcal and gram negative infection in the newborn. Lancet II:177
13. Franciosi R'A, Knostman JD, Zimmerman RA (1973) Group B streptococcol neonatal and infant infections. J Pediatr 82:707
14. Greenblatt CL (1970) Competition of polymetic binding sites between acridine and triphenylmethane dyes. Mol Pharmacol 6:649
15. Hexachlorophene in the Nursery (Annotation) (1968) Lancet II:205
16. Hnatko SI (1977) Alternatives to hexachlorophene bathing of newborn infants. Can Med Assoc J 117:223
17. Hurst X (1960) Transmission of hospital staphylococci among newborn infants. II. Colonization of the skin and mucus membranes of the infants. Pediatrics 25:204
18. Jellard J (1957) Umbilical cord as reservoir of infection in maternity hospital. Br Med J 1:925
19. Johanson WG, Pierce AK, Sanford JP (1969) Changing pharyngeal bacteria flora of hospitalized patients. N Engl J Med 281:1137
20. Kimbrough RD (1973) Review of the toxicity of hexachlorophene including its neurotoxicity. J Clin Pharmacol 13:439
21. McFarlan AM, Crone PB, Tee GH (1949) Variations in bacteriology of throat and rectum of infants in two maternity units. Br Med J 2:1140
22. Pildes RS, Ramamurthy RS, Vidyasagar D (1973) Effect of triple dye on staphylococcal colonization in the newborn infant. J Pediatr 82:907
23. Sarkany I, Gaylarde CC (1967) Skin flora of the newborn. Lancet I:589
24. Smith JW, Bloomfield AL The development of the aerobic bacterial flora of the throat in newborn babies. J Pediatr 36:51 1950
25. Speck WT, Driscoll JM, Polin RA, O'Neill J, Rosenkranz HS (1977) Staphylococcal and Streptococcal colonization of the newborn infant: Effect of antiseptic cord care. Am J Dis Child 131:1005
26. Speck WT, Driscoll JM, O'Neill J, Rosenkranz HS (to be published) The effect of antiseptic cord care on bacterial colonization in the newborn infant. Chemotherapy
27. Sprunt K, Leidy G, Redman W (1978) Abnormal colonization of neonates in an intensive care unit: Means of identifying neonates at risk of infection. Pediatr Res 12:998
28. Steigman AJ, Bottone EJ, Hanna BA (1978) Intramuscular penicillin administration at birth: Prevention of early-onset group B streptococcal disease. Pediatrics 62:842
29. Stratford BC (1963) The investigation and treatment of recurrent superficial staphylococcal infections. Med J Aust 1:309
30. Torrey JC, Reese MK (1945) Initial aerobic flora of newborn infants. Selective tolerance of the upper respiratory tract for bacteria. Am J Dis Child 69:208
31. Watt PJ, Okubadejo OA (1967) Changes in incidence of aetiology of bacteraemia arising in hospital practice. Br Med J 1:210

Chapter 31
Propionibacterium Acnes: Present Status and Role in Acne Vulgaris

S. M. PUHVEL

Propionibacterium acnes is the predominant anaerobic bacterium on normal human skin. As such, it has generally been regarded as a relatively harmless saprophyte of minimal clinical significance. The only pathologic process with which *P. acnes* has commonly been associated is acne vulgaris, the disease from which the organism derives its name.

There have been increasing reports[28,47,48,55] describing the occurrence of *P. acnes* infection as a complication after surgery or debilitating illness. Such infections are relatively rare when compared to infections by another cutaneous saprophyte, *Staphylococcus epidermidis*. It has become clear that given a suitably debilitated host, the innocuous *P. acnes* can cause severe clinical infection. This implies that epidemiological studies as described by R. Marples for *S. epidermidis* infections (see Chapter 2) may be essential to establish whether the *P. acnes* in postsurgical infections are endogenous contaminants from the host's own flora, or whether there may be virulent *P. acnes* which are transmitted to the vulnerable host, as reported with *S. epidermidis*.

P. acnes was described in skin in histologic sections of comedones by Unna in 1896.[52] Subsequently this organism—at first called *Bacillus acnes*, later *Corynebacterium acnes*—was thought to be a pathologic factor in acne vulgaris.[4,6,7,51] In 1946 H.C. Douglas and Shirley E. Gunter[3] pointed out that the bacteria found in acne lesions were the same as those found in normal skin. These same investigators suggested that, based on production of propionic acid, the proper classification of *C. acnes* should be *Propionibacterium acnes*.

The observation that *P. acnes* formed an important part of normal cutaneous flora temporarily suppressed interest in the role of this bacterium in acne, for it seemed improbable that normal flora could cause disease. Rather, *P. acnes* in acne lesions became regarded as secondary contaminants of low clinical significance.

In the late 1950s, with the advent of successful use of antimicrobial therapy for the treatment of acne, the pendulum started to swing in the direction of bacteria as significant pathogenic factors in acne. After considerable research during the past two decades, the general concensus among those investigating acne is that *P. acnes*, the predominant organism of normal pilosebaceous and sebaceous follicles of human skin, is a significant contributing factor in the pathogenesis of acne. Use of antibiotics in the treatment of this disease has been correlated with a reduction in the density of cutaneous *P. acnes*[9] and with clinical improvement in acne.[46,50] Although it appears that reduction of *P. acnes* is the reason for the effectiveness of antibiotic therapy in acne, there have been suggestions that the anti-inflammatory effect of certain antibiotics may also be significant in their effectiveness in acne.[31] Strides have been made in elucidating the role of *P. acnes* in the pathogenesis of acne, but many facets still await clarification.

The classification of the anaerobic coryneforms in acne lesions was clarified with the reduction of strains into two species: *P. acnes* and *Propionibacterium granulosum*.[27] *P. acnes* is numerically, by far, the predominant in normal skin and in acne lesions.[26,59] This species has been further subdivided into two serotypes[1] and several complex phage types.[42,53] Generally, subclassification of *P. acnes* has not been useful in elucidating the organism's role in acne in that subclassified strain types have not shown correlation with prevalence in acne lesions.[56]

P. granulosum strains in skin were orginally differentiated from *P. acnes* by Voss[53] and termed *C. acnes II*. In 1972 Johnson and Cummins[15] in an extensive taxonomic investigation of cell-wall composition and DNA similarities of anaerobic coryneforms reclassified Voss' *C. acnes II* as *P. granulosum*. Although taxonomically distinct, *P. acnes* and *P. granulosum* coexist in skin and in acne lesions and cannot be differentiated on morphologic grounds on ordinary media. For purposes of the present paper, both have been included under the *P. acnes* label. There have been suggestions that *P. granulosum* may have particular significance in the pathogenesis of acne because of its seeming greater lypolytic activity[59] and its greater prevalence and density in the more severe forms of inflammatory acne.[26] Since the ratio of prevalence of *P. acnes* to *P. granulosum* is greater in pustules and comedones than on normal skin and in normal sebaceous follicles, it does not seem probable that *P. granulosum* is of particularly greater clinical importance in acne.

P. avidum, the third species of cutaneous anaerobic diphtheroids, is rare in clinical acne lesions and is more typically isolated from nonacne prone areas of skin, particularly from the axilla.[24]

Corynebacterium parvum, the anaerobic diphtheroid used in immunotherapy of cancer, was named by Mayer in 1926 and classified by the French bacterial taxonomist Prevot[33] as a separate species from *C. acnes*. According to Bergey's *Manual of Determinative Bacteriology*,[27] the bacterial classification scheme used on the North American continent, strains labeled *C. parvum* are synonymous with *P. acnes*, belonging predominantely to serogroup II of that species.

Although it is recognized that *P. acnes* is a pathogenic factor in acne, the exact mechanisms of its involvement are unresolved. This paper critically reviews the recent evidence relating *P. acnes* to acne.

It is important to recognize that the pathogenesis of acne involves distinct phases, and the mechanism of *P. acnes* involvement may differ at each phase. A more extensive discussion of the pathogenesis of acne can be found in several textbooks,[2,5,30] but for purposes of simplification the disease can be divided into comedogenesis and inflammation. Comedogenesis begins as retention hyperkeratosis in ducts of sebaceous follicles and in the infrainfundibular part of pilosebaceous units. This initial pathologic change is characteristic of all types of acne. In acne vulgaris, retention hyperkeratosis can lead to the development of visible open or closed comedones or, alternatively, in inflammatory acne, can progress to the inflammatory phase directly from microcomedonal lesions—in which case the comedonal phase remains microscopic and detectable only at the histologic level.

The causative stimulus for the development of retention hyperkeratosis in acne vulgaris is unknown; all theories concerning the role of bacteria in this phase of the disease are highly speculative.

On clinical evidence it appears that antibiotic and antimicrobial therapy is prophylactic in acne. This implies either that antimicrobial therapy may simply prevent the microcomedo from proceeding to the inflammatory phase, without affecting the development of the microscopic comedonal lesion, or alternatively, that antimicrobials prevent new comedo formation which in turn implicates bacterial factors in comedogenesis.

The prevailing theory holds that free fatty acids (FA) released in sebaceous follicles through the action of bacterial lipases (mainly *P. acnes* lipases) on sebaceous triglycerides are the primary comedogenic substances in genetically predetermined sensitive follicles.[17] The main supporting evidence for this theory are the in vivo studies carried out by Kligman et al.[21,22] using the rabbit external ear canal bioassay for comedogenesis. These demonstrated that FA, particularly saturated and monosaturated FA of chain lengths from 10–14, were among the most comedogenic components of human sebum. Kligman et al. found that other components of sebum, for example squalene, had comedogenic potential and proposed that sebum itself is the comedogenic substance initiating retention hyperkeratosis in the vulnerable follicles in skin.

The evidence arguing against the involvement of fatty acids, and thus indirectly bacterial lipases in comedogenesis, is that the spectrum of FA found in skin of acne subjects does not differ from that found in acnefree controls[19]; the concentrations of FA present in skin surface lipids of subjects with acne are on the average lower than that found in skin surface lipids of acnefree controls[8,16]; and studies which have looked for differences in sensitivity of skin to the irritant effects of FA, have failed to find differences between acne subjects and acne-free controls.[43]

The relevance of findings obtained using the rabbit ear bioassay as a model for comedogenesis occurring in acne vulgaris can be questioned. In the bioassay, purified FA preparations were applied topically, and the amounts needed to induce comedo formation were grossly nonphysiologic, i.e., 30–50 mg daily for 10 consecutive days. The quantities of FA measured in 257 isolated sebaceous follicles in human skin have ranged from 1.39 μg to 8.46 μg amounts per duct, with a mean level of 4.24 ± 2.42 μg/duct.[35]

Skepticism regarding the relevance of FA to comedogenesis does not imply that the role of *P. acnes* in this phase of acne should be negated. The possibility that other as yet undefined metabolic interactions involving *P. acnes* and follicular substrates trigger the comedogenic response in follicular epithelial cells has not been investigated to date; this is an area which warrants further study.

The concept that the normal bacterial flora of skin is incapable of inducing pathogenic changes was important in delaying the acceptance of *P. acnes* as a pathogenic organism in acne.

As we have learned more about the actual morphologic changes occurring in follicles during the process of comedogenesis this dilemma appears to have resolved. The milieu of the closed microcomedones is different from that of normal patent follicles and even from that of mature open comedones. In the latter structures the system is open and bacterial products together with other follicular contents are continually eliminated through the follicular opening to the skin surface.[32] By contrast, in closed comedones, bacteria and bacterial metabolic products are trapped any may accummulate to toxic concentrations. Whereas in normal follicles the pH, nutrient supply, and oxygen tension are relatively constant, in closed comedones, bacteria and bacterial metabolic products are trapped and may which in turn affects the rate of production of bacterial metabolic products.[12,14] Thus the argument that *P. acnes* cannot affect a disease process because it is a normal member of the follicular flora becomes invalid. The fact that saprophytic bacteria can cause disease when introduced into an unnatural environment in sufficiently high concentrations is well established.

The mechanisms of *P. acnes* involvement in the inflammatory phases of acne are thought to be:

1) As a source of chemotactic factors which induce the initial accumulation of PMN to the comedonal wall.

2) As a source of enzymes (proteases, lipases, hyaluronidases, neuraminidases, and others) which may contribute directly to the degenerative changes in the comedonal wall, or which may release products which have inflammatory effects.

3) As direct irritants in the dermis possibly through cytotaxigenic effect activating chemotactic complement factors in the dermis.

4) As antigens reacting with specific serum antibodies to *P. acnes*, and possibly in cell mediated immune reactions to *P. acnes*.

Cytotaxins

Viable propagating *P. acnes* produce extracellular chemotactic substances may accumulate in microcomedones in concentrations which stimulate the initial migration of PMN to the intact microcomedo, a step thought to mark the beginning of the inflammatory phase in acne.

The chemotactic substances described as being produced by *P. acnes* include cytotaxins, are small molecular weight (1500–3500) heat stable, dialysable peptides produced during active bacterial growth phase.[37,38] Purified *P. acnes* lipase which has chemotatic potential for polympho nuclear leukocytes,[23] and substances eluted from *P. acnes* in prolonged aqueous suspension have chemotactic potential for human monocytes.[60]

The chemotactic substances described above are cytotaxins; they induce leukocyte chemotaxis independently of the addition of serum factors. In addition to cytotaxic activity, *P. acnes* whole organisms have strong cytotaxigenic potential—that is, the ability to activate chemotactic factors in serum.[10,57] Webster et al.[57] recently demonstrated that cytotaxigenic activity of *P. acnes* was dependent on the activation of the fifth component of complement. To date there is no evidence of complement being present in comedones,[38] therefore it is doubtful that cytotaxigenic activity of *P. acnes* is involved in the initial part of the inflammatory phase in acne; that is, in the inflammation initiated by the stimuli in the intact comedone. In the latter phases of inflammation when the comedonal wall has been disrupted, the cytotaxigenic potential of *P. acnes* in the dermis contributes to the inflammatory reaction.

Enzymes

Enzymes such as hyaluronidase,[34,49] proteases,[13,53] lecithinase,[58] neuraminidase,[29] and lipases,[18,44] have been demonstrated to be produced by *P. acnes* in vitro; each of these may directly, or through their products, be involved in contributing to the disruption of the comedonal wall.

The lipases of *P. acnes* are recognized as the primary lipolytic enzymes in sebaceous follicles[25] and thus indirectly the primary determinants of FA concentrations in skin surface lipids and in comedones. According to the free fatty acid theory of acne pathogenesis[17] FA have been implicated as the main inflammatory agents in acne lesions. Recently the evidence for the significance of FA as inflammatory agents has been critically reevaluated[35,54] and at the present time mechanisms other than lipase production by *P. acnes* offer alternate, and perhaps more satisfactory explanations of *P. acnes* involvement in acne.

Irritants

The massive inflammatory cell response in acne occurs after the break in the comedonal wall, when the dermal–epidermal barrier has been totally disrupted. At that point histologic sections show that comedonal content appears to have streamed into the dermis.[30] Studies of the comparative inflammatory potential of different purified comedonal components[36] indicated that suspensions of physiologic amounts (i.e., 2×10^5) of either viable or heat-killed *P. acnes* were capable of inducing inflammatory reactions on intradermal injection into human volunteers. The reactions consisted of induration and erythema which peaked at 24 h and at the cellular level consisted of a mixture of PMN and lymphocytes. Generally such reactions were longer lasting and more intense in subjects with acne than in acne-free controls, but over 30% of acne-free controls reacted, indicating that the organism in such concentrations must have direct inflammatory effect in the dermis. Whether this is mediated by cytotaxigenic effect or other mechanisms is not known.

Immune Reactions

It is likely that the more intense inflammatory response induced by *P. acnes* injection in subjects with acne, compared to acne-free controls[20,36] is the result of immunologic sensitization of such patients to *P. acnes* antigens during the disease process. Generally such subjects have elevated antibody titers to *P. acnes*[39] and a greater frequency of immediate hypersensitivity reactions.[40] Skin testing with *P. acnes* antigen failed to demonstrate delayed hypersensitivity in subjects with acne[40]; the reactions induced by the intradermal injection of *P. acnes* differ from classical delayed hypersensitivity reactions in the timing of the reaction (peak at 24 h rather than 48 h) and in the cellular response (mixture of PMN and lymphocytes). Such reactions in acne subjects may represent modified Arthus type phenomena involving the toxic effect of antigen-antibody complexes produced as a result of local excess antigen introduction into immune subjects.

Despite negative skin test reactions for delayed hypersensitivity to *P. acnes*, some subjects with severe forms of inflammatory acne have demonstrated positive cell mediated immunity to this organism in in vitro tests.[11,41] The extent to which such cellular reactions are involved in the inflammatory response in vivo, in acne, is undetermined.

In summary, *P. acnes* is presently considered a significant contributing factor in the pathogenesis of acne vulgaris.

Its involvement in comedogenesis remains unclear, but liberation of comedogenic FA from sebaceous triglycerides through action of bacterial lipase, has been proposed as a possible mode of action.

The role of *P. acnes* in inflammatory acne has been better defined. Here it probably acts as a primary source of comedonal chemotactic factors that initiate the inflammatory cascade. Bacterial enzymes and reaction products from enzymatic activity may contribute to enhance the degenerative changes in the comedonal wall at this stage. After rupture of the comedonal wall, *P. acnes* acts as an inflammatory agent directly in the dermis, possibly through its cytotaxigenic potential activating chemotactic components of complement. In addition, in subjects with inflammatory acne who have elevated serum antibody levels to this organism, reintroduction of large amounts of antigenic material into the dermis may aggrevate the inflammatory response through formation of antigen-antibody complexes in a modified Arthus-type reaction.

References

1. Cummins CS (1975) Identification of *Propionibacterium acnes* and related organisms by precipitin tests with trichloroacetic acid extracts. J Clin Microbiol 104–110
2. Cunliffe WJ, Cotterill JA (1975) The acnes. London, W.B. Saunders Co.
3. Douglas AC, Gunter SE (1946) The taxonomic position of *Corynebacterium acnes*. J Bacteriol 52:15–23
4. Fleming A (1909) On the etiology of acne vulgaris and its treatment by vaccines. Lancet I:1035
5. Frank SB (ed) (1979) Acne: Update for the practitioner. Yorke Medical Books, New York
6. Gilchrist TC (1900) A bacteriological and microscopical study of over 300 vesicular and pustular lesions of the skin, with a research upon the etiology of acne vulgaris. Johns Hopkins Hosptl Report 9:400–430
7. Gilchrist TC (1903) The etiology of acne vulgaris. J Cutan Dis 21:107–120
8. Gloor M, Graumann V, Kionke M, Wiegand I, Friedrich HC (1972) Menge und Zusammensetzung der Hautoberflachendenlipide bei Patienten mit Acne vulgaris und gesunden Vergleichspersonen. I Mitteilung Arch Dermatol Forsch 242:316–322
9. Goltz RW, Kjartansson S (1966) Oral tetracycline treatment on bacterial flora in acne vulgaris. Arch Dermatol 93:92–100
10. Gould DJ, Cunliffe WJ, Holland KT (1977) Chemotaxis and acne. J Invest Dermatol 68:251
11. Gowland G, Ward RM, Holland KT, Cunliffe WJ (1978) Cellular immunity to *P. acnes* in the normal population and patients with acne vulgaris. Br J Dermatol 99:43–47
12. Greenman J, Holland KT, Cunliffe WJ (1979) Growth rates and exoenzyme production by *P. acnes* grown in continuous culture. J Invest Dermatol 72:285–286
13. Hoeffler U (1977) Enzymatic and hemolytic properties of *Propionibacterium acnes* and related bacteria. J Clin Microbiol 6:555–558
14. Holland KT, Cunliffe WJ, Roberts CD (1978) The role of bacteria in acne vulgaris, a new approach. Clin Exp Dermatol 3:253–257
15. Johnson GL Cummins CS (1972) Cell wall composition and deoxyribonucleic acid similarities among the anaerobic coryneforms, classical propionibacteria and strains of *Arachnia propionica*. J Bacteriol 109:1047–1066
16. Kanaar P (1971) Lipolysis of skin surface lipids of acne vulgaris patients and healthy controls. Dermatologica 143:121–129
17. Kellum RE (1979) Free fatty acid hypothesis: In: Frank SB (ed) Acne: Update for the practitioner. Yorke Medical Books, New York, pp 65–73
18. Kellum RE, Strangfeld K (1969) Triglyceride hydrolysis by *Corynebacterium acnes*. J Invest Dermatol 52:255–258
19. Kellum RE, Strangfeld K (1972) Acne vulgaris. Studies in pathogenesis: Fatty acids of human surface triglycerides from patients with or without acne. J Invest Dermatol 58:315–318
20. Kersey PJ, Dahl MGC (1979) Delayed skin test reactivity to *P. acnes* in patients with acne. J Invest Dermatol 72:284
21. Kligman AM, Katz AG (1968) Pathogenesis of acne vulgaris, I. Comedogenic properties of human sebum in the external ear canal of the rabbit. Arch Dermatol 98:53–57
22. Kligman AM, Wheatley VR, Mills OH (1970) Comedogenicity of human sebum. Arch Dermatol 102:267–275
23. Lee WL, Shalita AR (1978) Neutrophil chemotaxis by *P. acnes* and its inhibition by antibiotics. J Invest Dermatol 70:219
24. Marples RR, McGinley KJ (1974) *Corynebacterium acnes* and other anaerobic diptheroids from human skin. J Med Microbiol 7:349–357

25. Marples RR, Downing DT, Kligman AM (1971) Control of free fatty acids in human surface lipids by *Corynebacterium acnes*. J Invest Dermatol 56:127–131

26. Marples RR, McGinley KJ, Mills OJ (1973) Microbiology of comedones in acne vulgaris. J Invest Dermatol 60:80–83

27. Moore WEC, Holdeman RV (1974) Propionibacterium. In: Bergey's Manual of Determinative Bacteriology, Buchanon RE, Gibbons NE (eds) 8th edn. The William and Wilkins Co., Baltimore, pp 633–641

28. Morrey BF, Fitzgerald RH Jr, Kelly PJ, et al. (1977) Diphtheroid osteomyelitis. J Bone Joint Surg (Am) 59:527–530

29. Muller HE (1971) Uber das Vorkommen von Neuraminidase bei *Corynebacterium acnes*. Z Med Mikrobiol Immunol 157:240–249

30. Plewig G Kligman AM (1975) Acne: Morphogenesis and treatment. Springer-Verlag, New York, Heidelberg, Berlin

31. Plewig G, Schopf E (1975) Anti-inflammatory effects of antimicrobial agents: an *in vivo* study. J Invest Dermatol 65:532–536

32. Plewig G, Fulton JE, Kligman AM (1971) Cellular dymanics of comedo formation in acne vulgaris. Arch Dermatol Forsch 242:12–29

33. Prevot AR (1948) Manuel de classification et de détermination des bactéries anaerobies. Masson et Cie, Paris, pp 304–318

34. Puhvel SM, Reisner RM (1972) The production of hyaluronidase (hyaluronate lyase) by *Corynebacterium acnes*. J Invest Dermatol 58:66–70

35. Puhvel SM, Sakamoto M (1977) A re-evaluation of fatty acids as inflammatory agents in acne. J Invest Dermatol 68:93–97

36. Puhvel SM, Sakamoto M (1977) An in vivo evaluation of the effect of purified comedonal components in human skin. J Invest Dermatol 69:401–406

37. Puhvel SM, Sakamoto M (1978) The chemoattractant properties of comedonal components. J Invest Dermatol 71:324–329

38. Puhvel SM, Sakamoto M (1980) Cytotaxin production by comedonal bacteria (*Propionibacterium acnes, Propionibacterium granulosum*, and *Staphylococcus epidermidis*). J Invest Dermatol 74:36–39

39. Puhvel SM, Barfatani M, Warnick M, Sternberg TH (1964) Study of antibody levels to *Corynebacterium acnes*. Arch Dermatol 90:421–427

40. Puhvel SM, Hoffman IK, Reisner RM, Sternberg TH (1967) Dermal hypersensitivity of patients with acne vulgaris to *Corynebacterium acnes*. J Invest Dermatol 49:154–158

41. Puhvel SM, Amirian D, Weintraub J, Reisner RM (1977) Lymphocyte transformation in subjects with nodulo-cystic acne. Br J Dermatol 97:205–211

42. Pulverer G, Sargo W, Ko HL (1973) [Orig A] Bakteriophagen von *Propionibacterium acnes*. Zentralbl Bakteriol, 225:353–363

43. Ray T, Kellum RE (1971) Acne vulgaris; studies in pathogenesis: free fatty acid irritancy in patients with and without acne. J Invest Dermatol 57:6–9

44. Reisner RM, Silver DZ, Puhvel SM, Sternberg TH (1968) Lipolytic activity of *Corynebacterium acnes*. J Invest Dermatol 51:190–199

45. Sabouraud R (1897) La seborrhee grasse et la pelade. Ann Inst Pasteur 11:134–159

46. Savin RC, Turner MC (1966) Antibiotics and the placebo reaction in acne. JAMA 196:365–367

47. Schlesinger JJ, Ross AL (1977) *Propionibacterium acnes* meningitis in a previously normal adult. Arch Intern Med 137:921–923

48. Skinner PR, Taylor AJ, Coakham H (1978) Propionibacteria as a cause of shunt and postneurosurgical infections. J Clin Pathol 31:1085–1090

49. Smith RF, Willett NP (1968) Rapid plate method for screening hyaluronidase and chondroitin sulfatase producing microorganisms. Appl Microbiol 16:1434–1336

50. Stewart WD, Maddin S, Nelson AJ et al. (1963) Therapeutic agents in acne vulgaris I. Tetracycline. Can Med Assoc J 89:1096–1097
51. Sudmerson HJ, Thompson ET (1909) The cultivation and biological characteristics of *Bacillus acnes*. J Pathol Bacteriol 14:224–229
52. Unna PG (1896) Histopathology of the diseases of the skin (from Molesworth EH Br Med J 1:1227–1229)
53. Voss JG (1970) Differentiation of two groups of *Corynebacterium acnes*. J Bacteriol 101:392–397
54. Voss JG (1974) Acne vulgaris and free fatty acids: A review and criticism. Arch Dermatol 109:894–898
55. Waitzkin R (1969) Latent *Corynebacterium acnes* infection of bone marrow. N Engl J Med 281:1404–1405
56. Webster GF, Cummins CS (1978) Use of bacteriophage typing to distinguish *Propionibacterium acnes* Types I and II. J Clin Microbiol 7:84–90
57. Webster GF, Leyden JJ, Norman ME, Nilsson UR (1978) Complement activation in acne vulgaris: In vitro studies with *Propionibacterium acnes* and *Propionibacterium granulosum*. Infect Immun 22:523–529
58. Werner H (1964) Untersuchungen uber die Lipase-und Lecithenase-Aktivitat von aeroben und anaeroben Corynebacterium und von Propionibacterium-Arten. Zentralbl Bakteriol [Orig A] 204:127–138
59. Whiteside JA, Voss JG, (1973) Incidence and lipolytic activity of *Proionibacterium acnes* (*Corynebacterium acnes* I) and *P. granulosum* (*C. acnes* group II) in acne and in normal skin. J Invest Dermatol 60:94–97
60. Wilikinson PC, O'Neill GJ, Wapshaw KG (1973) Role of anaerobic coryneforms in specific and non-specific immunological reactions. II Production of a chemotactic factor specific for macrophages. Immunology 24:997–1006

Chapter 32

A Role for *Propionibacterium acnes* in the Production of Inflammatory Lesions in Acne Vulgaris

GUY F. WEBSTER

Acne vulgaris is a two-stage disease. The first stage involves the formation of a comedo, an impaction and distension of the sebaceous follicle by a horny mass containing sebum, hair, and bacteria. Some comedones progress no farther, while others, mainly closed comedones, evolve into angry, potentially disfiguring inflammatory lesions.[7] It is not known why only some comedones become inflammatory lesions (papules, pustules, and nodules), but the stimulus for the inflammation, when it occurs, is increasingly clear.

Propionibacterium acnes is the dominant organism at the bottom of normal and diseased pilosebaceous units and is by far the most numerous comedonal bacteria.[7] Its phlogistic potential has been the object of much research. *P. acnes* and its products stimulate lymphocyte mitogenesis in vitro[23,13] and trigger production of lymphokines in vitro.[4] Neutrophil chemotactic factors have been identified by numerous workers[3,8,15,17] and chemotactic factors can be retrieved from comedones.[15] *P. acnes* is capable of activating serum complement systems to produce neutrophil chemotactic factors.[9,18] This report presents newer findings regarding *P. acnes'* interactions with serum inflammatory systems and the resulting effects on neutrophils.

The complement system is a group of serum proteins (C1–C9) that react sequentially to generate a wide range of biologically active mediators. Complement may be activated by classical and alternative pathways. The former is triggered by immune complexes, such as antibody-coated bacteria; the latter, by a wide range of macromolecules, many of which are bacterial in origin. Activation results in the generation of anaphylatoxins, leukocyte chemotactic factors, opsonins, cellular adherance promotors, cytolysins.[6] The classical pathway is more rapid than the alternative,[2] but both generate the same range of proinflammatory factors.

In 1976, Dahl and McGibbon[1] reported that inflammatory acne lesions contain immunoglobulin and the complement component C3 bound in the region of the comedo. Bound C3 and immunoglobulin was demonstrated near the the lesions of acne fulminans.[22] This is evidence that complement is activated in acne lesions and suggests that complement activation may be a step in the production of inflammatory acne.

P. acnes is most likely the cause of the complement activation. Massey et al.,[9] using crossed immunoelectrophoresis, demonstrated that in vitro *P. acnes* produces cleavage of C3 by the alternative pathway in normal human serum and in serum deficient in C2. They suggested that complement activation might induce chemotaxis and lysosomal release in the comedonal environment.

In 1978, Puhvel and Sakamoto[15] showed that insoluble comedonal fractions and *P. acnes* itself would generate neutrophil chemotactic factors from human and guinea pig serum, presumably through complement activation. It was confirmed that *P. acnes* serotypes I and II and *P. granulosum* can activate complement by both classical and alternative pathways.[18] The different strains' ability to activate complement in normal serum was proportional to the antibody titer in the serum against each bacterial type (see Table 32.1).

P. acnes generated neutrophil chemotactic activity proportional to the activating dose of bacteria. Heat-inactivated serum did not engender chemotactic factors, indicating that complement activation was required. The primary chemotactic factor was a C5-derived molecule, most likely C5a, as absorption of *P. acnes*-activated serum with antiserum to human C5 produced a 78% inhibition of chemotactic activity. Removal of albumin and C3-derived molecules did not inhibit chemotaxis.

Comedones themselves were found to consume hemolytic activity in normal serum, activating complement by both classical and alternative pathways.[20] Serum from inflammatory acne patients with a high anti-*P. acnes* antibody titer (1:1024) enhanced activation by comedones. Absorption of this serum removed the anti-*P. acnes* antibodies and ablated the enhancement of complement activation.

The structures in *P. acnes* that activate the antibody-independent alternative pathway were studied through the fractionation of *P. acnes* cells. Ability to activate the alternative pathway was localized in the cell wall. Extraction of lipid did not affect the ability of cells or cell walls to activate complement. Enzymatic digestion of RNA, DNA, and protein significantly enhanced the ability of cell walls to activate. Thus, none of these cell wall components activate complement, and their removal enriches the complement-reactive molecules. Extraction of cell walls with 10% trichloracetic acid, sodium m-periodate, or formamide completely removed or destroyed the activity of the complement activator. These methods all remove or alter carbohydrates in bacterial cells, thus the alternative pathway activator in *P.*

Table 32.1. Complement Activation by *P. acnes* and *P. granulosum*

Organism	Complement consumption %[a]	Antibody titer
P. acnes 1	93.7	1:128
P. acnes 2	47.3	1:64
P. granulosum	11.7	1:16

[a]Percentage of complement consumption by a 150 μg/ml dose of bacteria.

acnes is a cell wall carbohydrate that is not associated with nucleic acid, protein, or lipid.

Complement activation can also stimulate lysosomal release, resulting in the extracellular liberation of lysosomal constituents (proteases, lactoferrin, glucuronidase, lysozyme, etc.) from macrophages and neutrophils. It has been implicated in the pathogenesis of rheumatoid arthritis,[10] periodontal disease,[16] and immune complex diseases.[5] This phemomenon does not serve to protect the host, but rather, intensifies and prolongs the inflammatory process through tissue destruction.

In order to see if lysosomal release might occur in acne, the ability of *P. acnes* cells to trigger lysosomal release from human peripheral blood neutrophils in vitro was tested under various conditions. Release of lysozyme into incubation mixtures was used as an index of lysosomal release and was assayed by the method of Osserman and Lawlor.[11] Lactate dehydrogenase (LDH), a cytoplasmic enzyme marker, was assayed as a marker for cell death. LDH was not released under any incubation conditions. In the absence of serum neither *P. acnes* serotype I nor II was capable of triggering lysosomal release.

When complement-inactive, heated normal human serum was added, both serotypes induced lysosomal release proportional to the dose of bacteria tested. Addition of complement-active normal serum enhanced the lysosomal release produced by both serotypes. Complement-inactivated serum from inflammatory acne patients promoted lysosomal release greater than that of normal heat-inactivated serum. A similar *P. acnes* absorption procedure as was employed in the comedonal complement-activation experiments was used to remove anti-*P. acnes* antibody. After this absorption, there was a 93% decrease in the ability of acne serum to stimulate lysosomal release (see Table 32.2).

Electron micrographs of *P. acnes*-neutrophil incubation mixtures revealed that conditions which promoted lysozyme release also triggered phagocytosis of the bacteria. It is probable that phagocytosis of *P. acnes* and lysosomal release occur in vivo. All of the required factors, (*P. acnes*, antibody, and neutrophils) have been detected in clinically uninflamed microcome-

Table 32.2. Lysosomal Release[a] in Response to *P. acnes* Bacteria and Neutrophils Incubated with Various Serums

Bacterial[b] dose		Buffer	Heated serums	Fresh serum	Acne serum	Absorbed acne serum
P. acnes 1	25	0	3.5	7.0	—	—
	50	0	9.0	10.0	—	—
	100	2.5	21.5	23.0	36.0	2.31
P. acnes 2	25	0	1.5	5.5	—	—
	50	0	5.0	9.5	—	—
	100	0	8.0	17.4	—	—

[a]Expressed as per cent release of total lysozyme.
[b]Bacterial dosage is expressed as the number of bacteria per neutrophil.

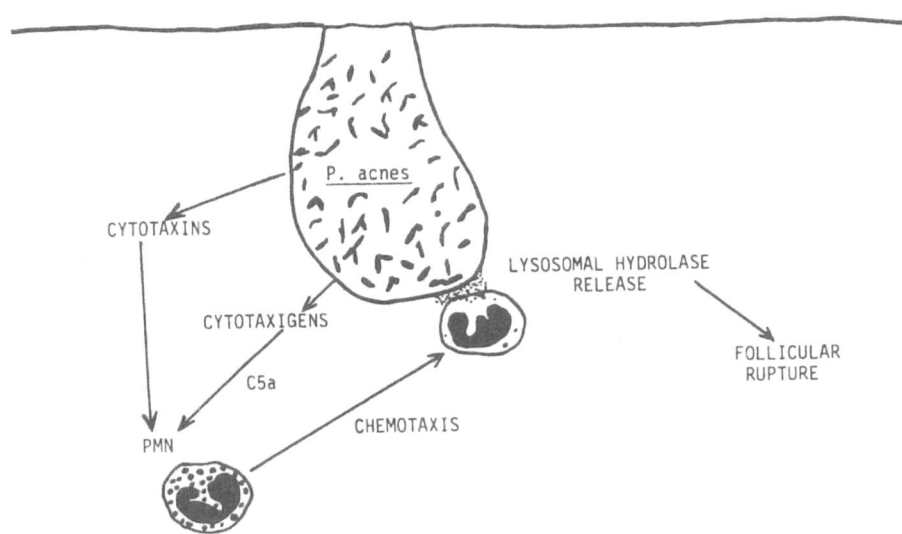

Fig. 32.1. The genesis of inflammatory acne lesions.

dones.[19] This also indicates that arrival of leukocytes at a comedo presages clinical inflammation and confirms previous histological evidence.[7]

P. acnes is thus a multifaceted inflammatory stimulus. It produces neutrophil chemotactic factors, activates complement, and can trigger release of lysosomal hydrolases from human neutrophils in vitro. How may these properties promote inflammation in acne? A possible scenario is presented in Fig. 32.1. The comedo is depicted as a reservoir filled with *P. acnes* and its products. A leakage of this material into surrounding tissue would cause the immigration of neutrophils into the comedo, through direct or complement-derived chemotactic factors. The numerous *P. acnes* would then trigger release of destructive lysosomal enzymes, which might then degrade the comedonal wall, releasing large numbers of *P. acnes* into the viable dermis and inciting a massive inflammatory response. The magnitude of complement activation and lysosomal release is dependent on the anti-*P. acnes* antibody titer of the host.

Puhvel et al.[12,13] found that the anti-*P. acnes* titer of acne patients increases in proportion to the severity of their inflammation. Thus an individual with high levels of anti-*P. acnes* antibody would have a more vigorous inflammatory response to an exposure to *P. acnes* than normal and might be expected to have more severely inflamed lesions. A differing capacity to generate anti-*P. acnes* antibodies could thus be the primary difference between severely inflammatory and largely comedonal acne patients. Conversely, the shift to the inflammatory stage in acne could be determined by greater permeability of "inflammatory" acne follicles (allowing leakage of chemotactic factors), or to a greater capacity of "comedonal" follicles to repair minor damage and thus prevent clinically detectable inflammation.

References

1. Dahl MGC, McGibbon DH (1976) Complement in inflammatory acne vulgaris Br Med J IV:1383
2. Gallin JI, Clark RA, Frank MM (1975) Kinetic analysis of chemotactic factor generation in human serum via activation of the classical and alternative complement pathways. Clin Immunol Immunopathol 3:334–346
3. Gould DJ, Cunliffe WJ Holland KT (1977) Chemotaxis and acne. J Invest Dermatol 68:251
4. Gowland G, Ward RM, Holland KT, Cunliffe WJ (1978) Cellular immunity to P. acnes in the normal population and in patients with acne vulgaris. Br J Dermatol 99:43–47
5. Johnson KJ, Chapman WE, Ward PA (1979) Immunopathology of the lung. Am J Pathol 95:795–839
6. Johnston RB (1977) Biology of the complement system with particular reference to host defense vs. infections a review. In: Suskind RM (ed) Malnutrition and the immune response. Raven Press, New York, pp 295–307
7. Kligman AM (1974) An overview of acne. J Invest Dermatol 62:268–287
8. Lee WL, Shalita AR (1978) Neutrophil chemotaxis by P. acnes and its inhibition by antibiotics. J Invest Dermatol 70:219
9. Massey A, Mowbray JF, Noble WC (1978) Complement activation by Corynebacterium acnes. Br J Dermatol 98:583–584
10. Oronsky AL, Perper RJ (1975) Connective tissue degrading enzymes of human leukocytes. Ann NY Acad Sci 256:233–253
11. Osserman EF, Lawlor DP (1966) Serum and urinary lysozyme (muramidase) in moncytic and monomyelocytic leukemia. J Exp Med 124:921–952
12. Puhvel SM, Barfatani M, Warnick M, Sternberg TH (1964) Study of antibody levels to Corynebacterium acnes. Arch Dermatol 90:421–427
13. Puhvel SM, Hoffman LK, Sternberg TH (1966) Presence of complement fixing antibodies to Corynebacterium acnes in the sera of patients with acne vulgaris. Arch Dermatol 93:364–366
14. Puhvel SM, Amirian D, Weintraub J, Reisner RM (1977) Lymphocyte transformation in subjects with nodulo-cystic acne. Br J Dermatol 97:205–211
15. Puhvel SM, Sakamoto M (1978) The chemoattractant properties of comedonal components. J Invest Dermatol 71:324–329
16. Taichman NS, McArthur WP (1975) Current concepts in periodontal disease. Annual reports in medicinal chemistry 10:228–242
17. Tucker SB, Rodgers RS III, Winklemann RK, Jordan RE (1977) Inflammation in acne vulgaris: mechanisms. J Invest Dermatol 68:237
18. Webster GF, Leyden JJ, Norman ME, Nilsson UR (1978) Complement activation in acne vulgaris: in vitro studies with Propionibacterium acnes and Propionibacterium granulosum. Infect Immun 22:523–529
19. Webster GF, Kligman AM (1979) A method for the assay of inflammatory mediators in follicular casts. J Invest Dermatol 73:266–268
20. Webster GF, Leyden JJ, Nilsson UR (1979a) Complement activation in acne vulgaris: Consumption of complement by comedones. Infect Immun 26:183–186
21. Webster GF, Tsai CC, Leyden JJ (1979b) Neutrophil lysosomal release in response to Propionibacterium acnes. J Invest Dermatol 74:398–401
22. Williamson DM, Cunliffe WJ, Gatecliff M, Scott DG (1977) Acute ulcerative acne conglobata (acne fulminans) with erythema nodosum. Clin Exp Dermatol 2:351–354
23. Zola H (1975) Mitogenicity of Corynebacterium parvun for mouse lymphocytes. Clin Exp Immunol 22:514–521

SKIN INFECTION:
TREATMENT

Chapter 33
Means of Preventing Bacterial Infection

ALEXANDER S. D. SPIERS, SYLVIA F. DIAS
AND JOSÉ A. LOPEZ

In the field of clinical oncology, studies of infection prevention began in the acute leukemias, and the majority of reports still concern the protection of patients with leukemia. This emphasis is explained partly by the disease and partly by the evolution of its treatment. Acute leukemia, more than any other disease, produces neutropenia and thrombocytopenia as a part of its natural history. The ability to control hemorrhage by platelet transfusion led to infection becoming the major cause of death in acute leukemia[4]; despite improvements in antimicrobial therapy, this situation has not altered.[2]

The development of new, intensive chemotherapeutic regimens for acute leukemia has enhanced the risk of serious infection, with particularly high risks attaching to bone marrow transplantation with its attendant supralethal chemotherapy and radiotherapy. As a result, much study has been devoted to the prevention of infection in leukemia patients by use of isolation rooms,[5] laminar air flow (LAF) rooms,[7] and plastic film isolators.[10]

Most studies have combined some form of protective isolation with decontamination of the gastrointestinal tract by orally administered, nonabsorbable antibiotics and attempts to reduce bacteria on the skin with topical application of antiseptics. The overall results of numerous studies indicate that the total number of bacterial infections, the number of febrile days, and the number of infective deaths can all be reduced by these means.

There is now a need to apply these lessons in other, more common conditions. Solid tumors are numerically far more important than leukemias, and modest progress in the treatment of metastatic cancers has led to much interest in more intensive chemotherapy, in the hope of converting partial responses to complete remissions and complete remissions to chemotherapeutic cures. This aggressive approach may be particularly profitable in solid tumors which are moderately or highly responsive to conventional doses of chemotherapy—for example breast cancer and small-cell lung cancer. Intensive treatment with cytotoxic drugs places solid tumor patients at higher risk of infection than was formerly the case, and if the gains to be secured by better drug therapy are not to be offset by an increased number of fatal infections, means for patient protection comparable to those shown to be effective in leukemia patients should be considered.

Economic considerations must receive due attention. It may be possible to provide expensive isolation systems for a few patients with leukemia, but it is unlikely that such facilities could be provided for the much larger number of patients with solid tumors. We therefore set out to evaluate a relatively simple and less expensive system, one using a structurally unmodified hospital room, and to study principally patients with common solid tumors. This pilot study was not randomized and concentrated chiefly upon the microbiological results obtainable and the incidence of documented infections rather than on the results of chemotherapy.

Patients

We performed 21 studies on 18 patients. All patients had malignant disease that could benefit from intensive chemotherapy and gave written, informed consent to the study. Sixteen studies were performed on patients in single rooms equipped with LAF apparatus and 5 studies were carried out in single rooms with conventional reverse precautions (RP) but without LAF. All other components of management were identical in the two groups. Further details of the patients are given in Table 33.1. The cytotoxic chemotherapy administered to these patients varied with their disease but in all instances was intensive and produced significant neutropenia that was monitored by blood counts at least three times a week.

Components of the Protective Regimen

Environment

All patients were nursed in precleaned single rooms with private bathrooms. Doors were kept closed, and staff entering the rooms wore caps, masks, disposable gowns and overshoes, and sterile gloves. In 16 studies, sterile filtered air was supplied by the "Med-Assist" portable LAF apparatus.[11]

Table 33.1. Diagnosis and Nursing Environment of 18 Patients Entered into the Study of Isolation and Decontamination

Diagnosis	Environment		Total
	LAF[a]	RP[b]	
Acute leukemia	4	3	7
Breast cancer	2	0	2
Lung cancer	10	2	12
	16	5	21[c]

[a]laminar air flow.
[b]reverse precautions.
[c]21 studies were performed on 18 patients.

Wheeled LAF cabinets were at the head and foot of the bed; each contained a pump, coarse prefilters, and a High Efficiency Particulate Absolute (HEPA) filter that removed particles as small as 0.3 μ in diam. The cabinet at the head of the bed provided an outflow of sterile air and the cabinet at the bed's foot, an intake: the patient thus is enclosed in a laminar flow of sterile air moving from head to feet. When the cabinets are switched on and the room door is closed, the organism count in the room air is progressively decreased. It is possible for well patients to sit in a chair between the cabinets. Draft and noise from the apparatus is minimal and does not interfere with conversation or sleep.

Topical Antisepsis

The sole antiseptic used was chlorhexidine gluconate, either in 4% solution with detergent (Hibiclens) or as an 0.02% aqueous solution. The regimen used comprised:

  a) a daily tub bath or bed bath using chlorhexidine;
  b) a shower after the bath;
  c) twice-weekly washing of hair and beard with chlorhexidine;
  d) spraying of the nostrils, ears, and throat four times daily with 0.02% chlorhexidine solution from an atomizer;
  e) a mouth rinse four times daily with 20 ml of the same solution;
  f) natural teeth were brushed daily with 0.02% chlorhexidine and dentures were stored in the same solution;
  g) twice daily vaginal douching with 0.02% chlorhexidine.

Although a relatively new product in the United States, chlorhexidine has been used extensively in other countries for many years, and numerous reports attest to its safety and efficacy.[6,10]

Gastrointestinal Decontamination

We used a modification of a regimen previously shown to be effective in leukemia patients[9]:

  nystatin suspension 1 million units by mouth every 6 h
  neomycin 500 mg tablets, one every 6 h
  colistin sulfate 75 mg capsules, one every 6 h

The neomycin and colistin are begun 48 h after the nystatin, and all three medications are continued throughout the period of isolation. Diarrhea occurs in most patients but is either self-limiting or readily controlled with diphenoxylate with atropine (Lomotil). Characteristically the stools are soft and odorless. Occasionally, nausea attributable to the colistin required the subsitution of gentamicin sulfate, 160 mg of the injectable preparation being given every 6 h by mouth, in orange juice.

Nutrition

Most patients received a clean, but not sterile, diet—freshly boiled or baked foods, bottled beverages and juices, fresh tea or coffee, with sterile drinking water and sterile ice cubes. It was necessary to eliminate salads, fresh fruit, uncooked foods, dairy products, candy, pepper, and chocolate. This diet was acceptable but tended to be monotonous. Patients with severe anorexia received total parenteral nutrition via a subclavian venous catheter inserted aseptically, with daily dressing changes and inspection for signs of infection at the skin entry site. Nutrition was monitored by daily weighing and charts of intake and output.

Surveillance

In addition to the routine observations of vital signs, the following surveillance was carried out by the nurse oncologist:

a) daily inspection of skin, mouth, and perianal area for inflammation, abrasions, or dermatitis from antiseptic;
b) weekly cultures of the nostrils, ears, throat, axillae, umbilicus, groins, perianal area, vagina, and any skin lesions;
c) cultures of stool and urine twice weekly;
d) in the event of pyrexia, all cultures were repeated and blood cultures were taken in addition.

The skin cultures were taken before beginning antiseptic treatment and once a week thereafter. Cultures were taken with sterile swabs moistened with sterile distilled water and taken directly to the microbiology laboratory for special processing.

Nursing Implications

The nursing objective was to provide an ultraclean environment in order to protect the patient undergoing aggressive chemotherapy from opportunistic infection during the phase of neutropenia. The nurse provided support while the patient experienced the side effects of the medications and attempted to minimize these effects through drugs, psychological support, and patient education.

The nurse instructed the patient in the use of the LAF or reverse precaution room, the use of topical antiseptics, the purpose of the gut decontamination antibiotics, the methods of nutritional support, and the regimen of cytotoxic chemotherapy. The nurse's goal was to educate the patient for what he or she would experience and thereby to encourage the patient to participate actively in his or her own care. The nurse encouraged the patient to feel a part of the team by familiarizing him or her with the procedures, allowing for greater patient knowledge, understanding, cooperation, and independence.

In patients with solid tumors, eligibility criteria for an aggressive regimen included a good performance status, capacity to participate in self-care, good nutritional status, psychological stability, and freedom from infection.

Before use the room was thoroughly cleaned; the nurse was responsible for the final preparations. These included spraying and wiping all items in the main room and bathroom, all horizontal surfaces, bedtable, phone, television set, and toilet with isopropyl alcohol. In the LAF room the air-flow cabinets were turned on, the door shut, and after two hours the room was ready for patient admission. Preparations for RP rooms were similar, excepting the use of the LAF cabinets. Once the room was prepared, its maintenance became the nurse's responsibility until the patient was discharged. The responsibilities consisted of spraying or wiping all items mentioned previously and mopping the floor with antiseptic three times per week. Mop heads were changed and sterilized after each room cleaning to prevent bacterial growth. The mop was used for the LAF or RP room only to reduce the chance of cross-infection. Cleaning supplies were kept in the room.

It was extremely important that rules be followed by all persons entering the room. To prevent contamination, persons must wear a cap, mask, overshoes, gown, and gloves. Once the room was prepared, the patient entered. Sterile linens, towels, blankets, and pajamas were provided.

Upon admission the patient began the gastrointestinal decontamination regimen described above, together with a specific bathing routine. Before tub bathing the hot and cold taps were run for two minutes each to clean out stagnant water in the pipes. The tub was then filled and 100 ml of chlorhexidine detergent solution was added to the water. The patient was then instructed to first submerge in the water, then stand and rub this detergent solution over the entire body and, finally, resubmerge in the tub water. A shower was allowed after bathing. The hands and perineum were washed with this detergent solution after each bowel movement.

Chlorhexidine gluconate, 0.02% aqueous solution, was provided in an atomizer for spraying the nose, ears, and throat four times daily. The same solution was used as a mouth rinse four times daily—it was expectorated, not swallowed. Female patients also used the solution for a twice-daily vaginal douche. Teeth were brushed with a soft brush and 0.02% chlorhexidine gluconate aqueous solution, and dentures were stored in the same solution, which is changed daily. In the presence of an infected mouth lesion, these antiseptic measures were increased to every 2–3 h. The patient was encouraged to carry out these measures if well enough. If there is painful stomatitis, the chlorhexidine solution may be mixed with 2% lidocaine; this combination was found to provide excellent pain relief coupled with effective antisepsis.

The surveillance measures previously described were carried out daily by the nurse, who also took weekly skin cultures and twice-weekly stool and urine cultures and repeated all the cultures if pyrexia occurred. Adequate nutrition is extremely important for the patient who is receiving aggressive chemotherapy. The nurse supervised the "clean" diet and encouraged the patient to eat. Hyperalimentation is an alternative means of providing adequate nutrition. Each day the nurse changed the dressing and the administration set and inspected the site of insertion for tenderness or inflammation.

Not more than two visitors were permitted at one time and for short visits only. Any person who had symptoms of a cold or viral infection could not enter the room. The patient in the LAF room remained between the LAF cabinets while receiving visitors. The visitors followed the gowning procedure on entering the room and remained at the foot of the bed (i.e., "downstream" in the air flow) while visiting; they avoided physical contact with the patient or the patient's belongings. The LAF and RP rooms are equipped with a television set, radio, and telephone. The patient was encouraged to visit via telephone, thereby reducing the number of people entering the room. Newly purchased items could be brought to the patient, such as books, magazines, newspapers, games, and crafts.

While in the room the patient was encouraged to walk frequently to enhance physiological functions. It was permissible for the patient to leave the area between the LAF cabinets, to walk about or take a bath when there had been no visitors in the room for an hour and the LAF system had therefore had the opportunity substantially to reduce the organism count in the room air. The patient was also instructed in active range of motion, isometric exercises, and deep breathing and coughing exercises.

The patient was discharged from the room when neutropenia was resolving, with the granulocyte count between 500 and 1000/mm³ and documented to be increasing by at least two successive counts. All medications were discontinued and the patient was discharged directly to home with instructions to resume a normal diet. Patients were advised not to eat out or go to public places for at least a week after discharge, in order to lessen the risk of recolonization by potentially pathogenic flora.

Assessment of Microbiological Results

Skin swabs were plated out by the conventional "streak plate" technique and scored as 0 to 4+—0 represented no bacterial growth and 4+ represented growth extending into the fourth set of streaks on the medium. *Contamination scores* for individual sites were made by pooling the score (0–4) for each site sampled and dividing by the total number of samples. The percentage change in total contamination score over a given time inter-

Table 33.2. The Incidence of Neutropenia and the Occurrence of Fatal Infection in Patients Nursed in LAF or RP Rooms

| Room | No. studies | Days with under 1 000 neutrophils | | | Fatal aspergillosis |
		total	mean/study	range	
LAF	16	194	12.1	2–36	1
RP	5	70	14.0	4–28	1
Combined	21	264	12.6	2–36	2

val was then calculated for each site. *Recovery of specific pathogens* at specified intervals was expressed as the number of sites positive for each organism and also as the percentage of all sites sampled which proved to harbor the organism.

Results

Neutropenia and Infection

All patients became neutropenic following the high-dose chemotherapy. They were at enhanced risk of infection (neutrophil count less than $1\,000/mm^3$) for a total of 264 days. The mean duration of neutropenia was 12 days for the whole group. In the solid tumor patients, recovery from neutropenia was quite predictable, usually occurring about day 18 after chemotherapy. The patients with leukemia had longer periods of neutropenia and recovery of the neutrophil count was unpredictable since it depended upon the attainment of hematologic remission of their disease.

For the seven studies carried out in leukemia patients, the mean duration of severe neutropenia (neutrophil count less than $500/mm^3$) was 32 days. In 21 studies there were three documented infections in leukemia patients: one episode of oral candidiasis was successfully treated but two episodes of disseminated aspergillosis were both fatal. The only infection in a solid tumor patient was a *Proteus* urinary tract infection which may have been present before the study. There were several unexplained fevers for which antibiotics were given; all blood cultures were negative and no infection was documented. The only deaths during the study were two patients with aspergillosis. The distribution of neutropenia and infection was comparable between LAF and RP patient groups (see Table 33.2).

Reduction in Topical Flora

The reduction in topical flora achieved by the regimen in the first two weeks of treatment is shown in Table 33.3. Because of the small number of RP patients, it is uncertain whether any difference exists in the efficiency of decontamination between patients nursed in laminar air flow and those exposed to unprocessed room air; the RP patients certainly do not appear to have fared worse. It is apparent from the table that decontamination of the upper respiratory tract is less efficient than that of the vagina and skin. Bacteria-free throat swabs were obtained in only two patients. One of these patients had acute leukemia, severe neutropenia, and severe drug-induced stomatitis. She received local applications of 0.02% chlorhexidine gluconate every 2–3 h, and no bacteria or yeasts were recovered from throat swabs during a 21-day period. This result suggests that intensification of the topical antiseptic regimen might achieve better decontamination in other patients.

The patients with solid tumors had all recovered from neutropenia by day

Table 33.3. The Reduction in Organisms (All Types) at Various Sites, Comparing Day 1 (Before Treatment) with Day 15 of the Decontamination Regimen: Results Are Percentage Reductions

	Patient environment		
Site	LAF	RP	Combined
Ears	48.2	31.7	48.1
Nostrils	35.8	72.0	32.5
Throat	32.1	39.7	32.9
Axillae	69.3	100.0	76.3
Umbilicus	47.4	100.0	58.8
Groins	55.3	97.0	61.8
Vagina	88.6	a	a
Anal margin	82.0	a	a

[a]data insufficient.

22 and were therefore offstudy. In the seven studies carried out in leukemia patients, the day 22 and day 29 cultures showed a percentage reduction in flora remarkably similar to that seen on day 15. It thus appeared that the decontamination could be maintained, without recolonization by new organisms, but was not greatly improved by prolonged exposure to the regimen. Absence of a sterile diet might in part account for this lack of progressive attrition of the flora. In this initial assessment of decontamination, we scored all organisms isolated, since in neutropenic patients with malignant disease it is unwarranted to assume that any organism is definitely nonpathogenic.

Recovery of Pathogens

The recovery of pathogenic microorganisms from all sites is summarized in Table 33.4. It is apparent that our patients were relatively free of pathogens before decontamination—the most common organism, *Staphylococcus aureus*, occurred in only 6.9% of samples. The conspicuous absence of *Pseudomonas aeruginosa* contrasts with the findings of Bodey,[1] who found a high carriage rate for this organism in newly admitted patients with leukemia. The majority of our patients had neither leukemia nor neutropenia before admission, which may explain the difference. Our results show a modest reduction in the rate of staphylococcal carriage during the first two weeks of treatment. There is no evidence of acquisition of pathogens during the period studied, which is encouraging since during this period the patients became neutropenic and also developed areas of traumatized skin (from invasive procedures) and denuded mucous membrane (from cytotoxic drug therapy). Pathogen recovery from day 22 and day 29 samples could not reliably be estimated because of the small numbers involved.

The location of pathogens is shown in Table 33.5. The upper respiratory tract harbors organisms most frequently, accounting for 17/26, 7/14, and 9/16 positive cultures on days 1, 8, and 15, respectively. More effective

Table 33.4. Recovery of Pathogenic Organisms from All Sites in 21 Case Studies, Before Treatment and After 1 and 2 Weeks of the Decontamination Regimen

Pathogen	Day 1: 232 sites		Day 8: 237 sites		Day 15: 169 sites	
	No. +ve.[a]	% +ve.[b]	No. +ve.[a]	% +ve.[b]	No. +ve.[a]	% +ve.[b]
β-hem. Streptococci	1	0.4	3	1.3	1	0.6
S. aureus	16	6.9	6	2.5	7	4.1
E. coli	3	1.3	0	0	0	0
P. aeruginosa	0	0	0	0	1	0.6
Klebsiella spp.	0	0	1	0.4	3	1.8
Proteus mirabilis	1	0.4	1	0.4	2	1.2
Gram -ve.[c] rods[d]	4	1.7	1	0.4	1	0.6
Candida spp.	1	0.4	2	0.8	1	0.6
Total pathogens	26	11.2	14	5.9	16	9.5

[a]No. +ve. = number of gram-positive sites.

[b]% +ve. = percentage of positive sites.

[c]gram −ve. = gram-negative.

[d]speciation undefined

Table 33.5. The Location of Pathogenic Organisms Before and During the Decontamination Regimen

Pathogen	Site		
	Day 1	Day 8	Day 15
β-hem. streptococci	1 groin	3 vagina	1 vagina
S. aureus	11 nose	5 nose	5 nose
	3 throat	1 ear	1 ear
	1 axilla		1 groin
	1 ear		
E. coli	2 anal		
	1 vagina		
P. aeruginosa			1 throat
Klebsiella spp.		1 vagina	2 groin
			1 throat
Proteus mirabilis	1 axilla	1 axilla	2 axilla
Gram -ve.[a] rods	3 anal	1 throat	1 umbilicus
Candida spp.	1 throat	2 groin	1 throat

[a]gram -ve. = gram-negative

decontamination of the upper airways clearly is needed if results are to be improved. Upon examining the incidence of *S. aureus*, the most frequent pathogen, at its most common site, in the nostrils (see Table 33.6), it is seen that decontamination is not very successful, and may even become less effective with the passage of time. Nasal insufflation of chlorhexidine solution does not effectively reach the paranasal sinuses or the hair follicles within the nares, and these may serve as reservoirs of staphylococci.

In contrast to the imperfect decontamination of the upper respiratory tract, serial stool cultures showed regular suppression of flora with many apparently sterile specimens on routine aerobic and anaerobic cultures. These results are in agreement with those we have previously reported with a similar regimen of oral nonabsorbable antibiotics.[3]

Conclusions

We have shown that patients can be isolated in a laminar air flow or reverse precaution environment, fed a restricted diet, and treated with topical antiseptics and oral nonabsorbable antibiotics without major physical or psychological harm. Although patients found the regimen restrictive and sometimes depressing, no patient withdrew from the study, many expressed their

Table 33.6. Nasal Carriage of *Staphlococcus aureus* Before and During Decontamination with Topical 0.02% Chlorhexidine Gluconate

Day	No. sites samples	No. sites positive	% sites positive
1	38	11	29.0
8	37	5	13.5
15	23	5	21.7

willingness to undergo a similar regimen again, and three patients actually did so. The acceptance of this treatment must be attributed to the excellent support given to patients by our nursing staff and the positive relationships the patients enjoyed with their nurses and physicians. Our study demonstrated that chlorhexidine preparations can be applied to the skin and mucous membranes intensively and for prolonged periods without significant toxicity. The only side effect noted was mild drying or defatting of the skin, probably attributable to the detergent solution, and in no case requiring cessation of treatment. We have observed much more severe reactions when topical povidone-iodine preparations were used in a similar manner.

Our microbiological results indicate that the skin can be effectively decontaminated, but the ears, nose, and throat cannot, save in those instances where the antiseptic regimen was greatly intensified. Although we did not bring about a progressive reduction of skin flora to complete sterility, we found no evidence of accumulation of microorganisms during treatment, despite a continued stay in hospital and the predisposing factors of neutropenia and cutaneous trauma. Overall, the recovery of pathogenic organisms from various sites decreased or remained stable despite significant iatrogenic suppression of body defenses. The absence of colonization by *Pseudomonas aeruginosa* was particularly pleasing, since acquisition of this organism is very common in hospital patients.[1,8] Since *Pseudomonas* species are probably transmitted from patient to patient and also by food and medicines,[8] the isolation, clean diet, and gastrointestinal decontamination practiced in our patients might all contribute to the prevention of its acquisition.

The present results do not demonstrate any definite difference between LAF and RP nursing in respect of microbiological findings or the occurrence of infection, but the RP patient group was too small for a meaningful comparison. Since our investigations concentrated upon cutaneous colonization, the absence of an advantage for LAF patients would not be surprising. A much larger study might show an advantage for LAF nursing, perhaps with respect to the acquisition of infections in the respiratory tract.

The failure to eliminate nasal carriage of staphylococci appears to be a weakness of the regimen, and other local measures—for example, nasal application of chlorhexidine cream—should be evaluated. However, it appears that the persistent nasal staphylococci do not colonize the remainder of the body, and this may be attributable to the use of chlorhexidine with detergent on the skin.

It has been adequately demonstrated that regimens similar to ours confer significant protection from infection upon patients with leukemia who are undergoing intensive chemotherapy. Our study shows that effective, though incomplete, decontamination can be achieved in patients with solid tumors and that useful protection may be conferred: no solid tumor patient developed systemic infection despite cytotoxic chemotherapy given at twice the conventional doses. This decontamination regimen should be useful for the microbiological protection of patients with immune deficiency from other causes, and also for patients with serious dermatoses that constitute another serious breach in the body's defenses.

Acknowledgments

We wish to pay tribute to our late colleague, A. Alice Jacobs, Ph.D., microbiologist at University Hospital, who contributed much to the planning and implementation of this study, and whose untimely death prevented her from seeing its completion.

Thanks are owed to the Pharmacy, Dietary, and Housekeeping staff, and especially to the nursing staff of Ward F-4 at University Hospital, whose skill, loyalty, and enthusiasm made this work possible.

References

1. Bodey GP (1970) Epidemiological studies of Pseudomonas species in patients with leukemia. Am J Med Sci 260:82–89
2. Chang H-Y, Rodriguez V, Narboni G, Bodey GP, Luna MA, Freireich EJ (1976) Causes of death in adults with acute leukemia. Medicine (Baltimore) 55:259–268
3. Gompertz D, Brooks AP, Gaya H, Spiers ASD (1973) Volatile fatty acids in the faeces of patients in 'germ-free' isolation. Gut 14:183–186
4. Hersh EM, Bodey GP, Nies BA, Freireich EJ (1965) Causes of death in acute leukemia: a ten-year study of 414 patients from 1954–1963. JAMA 193:105–109
5. Jameson B, Gamble DR, Lynch J. Kay HEM (1971) Five-year analysis of protective isolation. Lancet I:1034–1040
6. Lowbury EJL Lilly HA (1973) Use of 4% chlorhexidine detergent solution (HIBISCRUB) and other methods of skin disinfection. Br Med J I:510–515
7. Schimpff SC, Greene WH, Young VM, Fortner CL, Jepsen L, Cusack N, Block JB, Wiernik PH (1975) Infection prevention in acute nonlymphocytic leukemia. Laminar air flow room reverse isolation with oral, nonabsorbable antibiotic prophylaxis. Ann Intern Med 82:351–358
8. Shooter RA, Cooke EM, Gaya H, Kumar P, Patel N, Parker MT, Thom BT, France DR (1969) Food and medicaments as possible sources of hospital strains of Pseudomonas aeruginosa. Lancet I:1227–1229
9. Storring RA, Jameson B, McElwain TJ, Wiltshaw E, Spiers ASD, Gaya H (1977) Oral non-absorbed antibiotics prevent infection in acute non-lymphocytic leukaemia. Lancet II:838–840
10. Trexler PC, Spiers ASD Gaya H (1975) Plastic isolators for treatment of acute leukaemia patients under "germ-free" conditions. Br Med J II:549–552
11. Young VM, Schimpff SC, Moody MR, Friedman NR, Wiernik PH (1975) Potential utility of a portable air filtration system for prevention of infection in cancer patients. Proceedings Association for Gnotobiotics

Chapter 34
The Topical Treatment of Skin Infections

Sydney Selwyn

Perhaps the most deceptively simple of all therapeutic procedures is the treatment of cutaneous infection with topical medication. Despite the unique accessibility of the skin to scientific investigation, it has for too long been the playground of crude empiricism.

Ancient Remedies

The local treatment of skin disease—including the results of trauma—is the oldest form of therapy. A Sumerian clay tablet dating from 2100 B.C., the earliest extant prescription, lists a topical formulation containing pulverized water snake and bat's dung in an aqueous paste of plant extracts and earths. Irrational pharmacy for internal and external use persisted in orthodox medicine until the last century. Many antiseptic substances have long been available (Table 34.1), and other traditional skin medicaments have useful activity on infected skin due to moderately anti-inflammatory (especially osmotic) and keratolytic properties. Among the most potentially active preparations were the mold extracts widely used in folk medicine, although inevitably their antibiotic content was highly variable.[31]

A much simpler therapeutic approach is described in what is probably the earliest published account of the successful topical treatment of a skin infection. The Old Testament text relates to the ninth century B.C., and illustrates the gap between a patient's expectations and the prescribed treatment.

> Now Naaman, captain of the host of the King of Syria, was a great man and .... he was mighty in valour; but was a leper. So Naaman came with his horses and with his chariot, and stood at the door of the house of Elisha. And Elisha sent a messenger unto him [to avoid cross-infection], saying, Go and wash in Jordan seven times, and thy flesh shall come again to thee, and thou shalt be clean. But Naaman was wroth, and went away, and said, Behold, I thought, He will surely come out to me, and stand, and call on the name of the LORD his God, and strike his hand over the place, and recover the leper. Are not Abouna and Pharpar, rivers of Damascus, better than all the waters of Israel? May I not wash in them, and be clean? So he turned and went away in a rage. And his servants

Table 34.1. Evolution of Topical Agents for Treating Cutaneous Infections

Antiseptics and anti-inflammatory agents	First used[a]	Specific antimicrobial agents[b]	First used[a]
		1. *Antibacterial*	
Earths and salts	in antiquity	(Sulfonamides)	1936
Plant extracts	in antiquity	(Penicillin)	(1871, 1929) 1944
Ethanol solutions	in antiquity	Bacitracin	1940
Mercury compounds	c. 900 A.D.	Gramicidin	1940s
Zinc compounds	c. 1500 A.D.		
Chlorine and compounds	pre-1800	Polymyxins	1950s
Iodine and compounds	c. 1820	Tetracyclines	1948-52
Carbolic acid (phenol)	1860	(Chloramphenicol)	1948
Synthetic organicals	1860s	Aminoglycosides[c]	1945-65
Flavines	1920s	(Fusidic acid)	1962
Cationic detergents	1930s	2. *Antifungal*	
Halogenated quinolines	1940s	Nystatin	1954
Corticosteroids	1949	Amphotericin B	1960
Hexachlorophene	1950s	Imidazoles[d]	1975–78
Chlorhexidine	1950s	3. *Antiviral*	
Triclosan	1960s	Idoxuridine	(1963) 1974

[a] Approximate dates which varied in different countries
[b] Parentheses indicate unsatisfactory for general topical use.
[c] Notably neomycin and framycetin (also gentamicin).
[d] Notably clotrimazole, miconazole and econazole.

came near, and spake unto him and said, My father, if the prophet had bid thee do some great thing, wouldest thou not have done it? How much rather then, when he saith to thee, Wash and be Clean? Then went he down, and dipped himself seven times in Jordan, according to the saying of the man of God: and his flesh came again like unto the flesh of a little child, and he was clean.[11]

Naaman's initial disappointment and anger emphasize the potential therapeutic importance of psychological factors, including the placebo effect. It should be added, however, that his "leprosy" was probably psoriasis, still treated in the Holy Land by exposure to a combination of solar radiation and immersion in waters associated with the River Jordan (mineral waters at its source and the hypertonic waters of its destination in the Dead Sea).

Evaluation of Topical Preparations

Reliable test methods, which must form the basis of rational therapy, have only slowly evolved in the topical field.

In Vitro Tests

The first in vitro evaluation of antiseptics was performed in the 1740s by Sir John Pringle.[32] His semiquantitative procedures, in which the presence or inhibition of putrefaction was the end point, were replaced after 120 years by quantitative serial dilution tests. These procedures, carried out originally in liquid media by Lister and subsequently Koch, have remained virtually unchanged to the present. An essential recent improvement has been the use of neutralizing agents to inactivate residual antiseptic,[34] although this has been neglected in the case of antibiotics. Agar diffusion tests have been recommended as more realistic in vitro models of the activity of antimicrobial agents in skin.[10]

In Vivo Tests

When in vitro tests show satisfactory activity against appropriate microorganisms, and toxicologic studies in animals indicate that a preparation is safe for topical use, its efficacy as an anti-infective agent both in prophylaxis and treatment must then be investigated in vivo.

Animal Studies. Experimental models of cutaneous infections in animals have recently been reviewed by McRipley.[20] He stressed that erroneous conclusions may be drawn if results are extrapolated uncritically from laboratory animals to man. Of particular importance are interspecies differences in the anatomy of the skin and in its susceptibility to various species and strains of microorganisms. Nevertheless, infections in experimental surgical and traumatic wounds—includings burns—have been useful systems in which to test topical antibacterial agents. Of more direct dermatologic relevance are the few well-standardized skin sepsis models, notably streptococcal impetigo in hamsters (omitted by McRipley),[6] and dermatophyte infections in guinea pigs.[2] Vaginal candidiasis in mice or rats has also proved a useful test system.[20]

Human Volunteer Studies. Ideally, the activity of topical antimicrobial agents should be evaluated against microorganisms growing on human skin under a variety of conditions. This approach has been extensively explored by Kligman, Marples, Leyden and their colleagues during the past two decades.[16] Experimental infections with *Staphylococcus aureus*[18] and *Candida albicans*[25] have been studied under polyethylene wrap occlusion, with or without previous trauma. The tendency for severe sepsis to develop led the Philadelphia group to concentrate on commensal bacteria as test organisms.

A sequence of five test procedures is currently recommended for evaluating topical antimicrobials.[16] First an occlusion test is performed. Here an agent's ability to prevent the usually rapid expansion of the normal flora under occlusion is determined over periods of 24 and 48 h on volunteers' forearms. This measures bacteriostatic activity. If results are satisfactory, the bactericidal potency of the topical agent is assessed in the expanded flora test by first boosting the skin microflora under occlusion (e.g., from 10^2 to 10^5 or more organisms per cm^2), and then observing any decline in numbers after contact with the agent for 10 and 30 min, and 6, 24 and 48. A persistence test is carried out to determine whether twice-daily application of a test agent for 4 days produces a persistent antimicrobial effect. This is assessed when the same skin area is subsequently occluded after an interval of 24 h. An ecological shift test is performed to indicate how much qualitative and quantitative disturbance of the skin flora occurs after 6 days of occlusion and three applications of test agent. The ability of the topical agent to function on exudative lesions is assessed in a serum inactivation test. This necessitates the production of ammonium hydroxide blisters, which are opened and inoculated with a red-pigmented strain of *Staphylococcus epidermidis*. After initial occlusion for 6 h, the wounds are treated with the topical agent under occlusion for 24 h, and the antimicrobial effect is determined in comparison with control areas.

Clinical Trails. The ultimate test of an antimicrobial agent is its performance in the prevention or treatment of clinical infection. Unfortunately, clinical trials are fraught with difficulties, particularly in the topical field. For example, studies on the prophylaxis of minor wound infections require extremely large numbers of patients to include an adequate number of equivalent infections in treated and control groups and to deal satisfactorily with the many variables, such as microbial inoculum size and environmental factors.[16]

The treatment of actual clinical infection of the skin is ostensibly easier to study. However, again adequately controlled trials are rare, and assessment of results is difficult due to great variability in the natural history of primary and secondary skin infections. When widespread infection is present, as in many generalized dermatoses, half-body (symmetric paired) comparative studies have been highly regarded. Unfortunately, translocation of small amounts of potent agents, such as neomycin, readily occurs and invalidates this method.[15]

In a detailed review of relevant clinical trials, Leyden and Kligman[15] concluded that properly controlled studies with adequate microbiologic monitoring all confirmed the value of neomycin, bacitracin, and other powerful agents in the topical treatment of cutaneous infection. Moreover, their own work has indicated that even when clinical signs of infection are absent, a *S. aureus* count of $10^6/cm^2$ or greater is a significant factor in perpetuating the disease process and justifies the use of antibiotics. Leyden and Kligman were able to draw some useful conclusions from several uncontrolled trials about the duration of clinical infection in treated and untreated lesions.

Apart from investigating the activity of topical antimicrobial agents—with or without corticosteroids[35]—the influence of different bases or other vehicles is of potential importance. This aspect of topical therapy has been seriously neglected, although it has been found, for example, that a cream base exerted a mildly anti-inflammatory effect in experimental human infections due to *S. aureus* and *Candida albicans*.[15,19] This effect may have been due to the absorption of toxic microbial products by the cream.

Adverse Effects of Topical Antimicrobials

Toxicity

Many topical antimicrobial agents may damage the skin or mucosae locally to a varying degree, and some agents can produce systemic toxicity after absorption from the skin.

Local Damage. Studies in animal models, tissue cultures, and other systems have shown that delayed wound healing and even cell destruction is produced by instilling alcoholic solutions, quaternary ammonium compounds, soaps, hydrogen peroxide, hexachlorophene and various other antiseptic agents into exposed tissues.[1] In a more subtle manner, antibiotics may exert a detrimental effect on local immunity. For example, tetracyclines and aminoglycosides depress leukocyte chemotaxis.[8]

Systemic Toxicity. Although neomycin is no longer used parenterally due to the high risk of damage to the auditory division of the eighth cranial nerve (and, to a lesser extent, nephrotoxicity), a similar result can follow systemic absorption of neomycin from the skin.[17] This is particularly likely in patients with renal impairment and can follow prolonged or widespread use of ointments, creams, lotions, sprays, and even ear or eye preparations. Apart from their potentially deleterious effect on the environment and the highly variable dose delivered,[5] aerosol sprays (still widely used outside the United States) are associated with an unusually high risk of ototoxicity, notably when applied to absorptive surfaces such as burns or surgical wounds during and after operations.[3] In patients with normal renal function, the maximum recommended daily dose is one can, and a course of treatment should not exceed 7 days.

Other topical aminoglycosides, including gentamicin and the neomycin analogue, framycetin, are potentially ototoxic. In addition, if the aminoglycosides and polymyxins are systemically absorbed they can potentiate neuromuscular blocking agents used in surgery. Hexachlorophene is another important example of a topical antibacterial agent that produces systemic toxicity if absorbed to a significant extent. Brain damage has resulted from its use on extensive burns or other damaged skin areas and on neonates in excessive doses or under conditions which favored permeability of the skin.[9] Fortunately the similarly named agent, chlorhexidine, is without such risks. Many other long-established antiseptics, such as mercurials and boric acid, can also produce systemic toxicity if absorbed significantly from the skin. Hydrargaphen appears to be the least hazardous of the topical mercurials.

The well-known adverse effect of chloramphenicol—aplastic anemia—has usually been associated with systemic therapy in a dose-related manner, but there are well-documented cases of the disease following the use of chloramphenicol eye drops or other topical preparations.[7] In some cases, topical therapy has been unreasonably prolonged; occasionally, however, there has been minimal exposure to the antibiotic and a relatively long, variable interval before the onset of aplastic anemia. An immunological mechanism involving hypersensitization seems to underlie this form of hemapoietic paralysis, and the prognosis is extremely poor.[7]

Allergy

Contact sensitization is remarkably uncommon with most topical antimicrobial agents. It was, however, a major problem with sulfonamides and penicillin; hence the topical use of these agents was virtually abandoned except in specific eye infections. Similarly, the topical use of chloramphenicol is now limited almost entirely to the eye. The aminoglycoside antibiotics occupy a controversial position. The topically important examples of the latter are neomycin and its close relative, framycetin, as well as gentamicin, all of which produce cross-sensitization. Although neomycin has been ranked among the ten most frequent contact allergens, Leyden and Kligman estimate that sensitization to it occurs no more frequently "than once in every 100,000 ordinary usages."[15] However, there is a considerably higher risk when neomycin is applied for long periods of time and particularly to such chronic lesions as stasis ("varicose") dermatitis or ulcers. The allergenicity of neomycin is potentiated by this antibiotic's considerable tissue persistence (skin substantivity), which also, of course, tends to prolong its local activity.

Disturbances of Skin Ecology

Superinfection. The application of potent antimicrobial agents to the skin surface is intended to produce radical changes in the local microflora. Ideally, all pathogenic organisms will be eliminated and the normal ecologic balance of skin commensals reestablished, both from surrounding untreated areas and from any residual members of the normal flora locally. Unfortunately, the temporary ecologic vacuum created by topical therapy provides optimal conditions for colonization and subsequent infection of the skin.[13] The risks increase in proportion to the duration of treatment and are particularly high in hospitals, due to the prevalance of considerable degrees of environmental contamination by potential pathogens.[28]

The principal types of microorganisms seen in superinfections are strains of *S. aureus* possessing reduced susceptibility to antibiotics, gram-negative bacilli, including *Pseudomonas aeruginosa, Klebsiella* and *Proteus* species, and *Candida albicans.*

Drug-resistant Microorganisms. Topical antimicrobial therapy can produce an increase in the prevalence of organisms resistant to a variety of drugs. Two separate mechanisms operate. One involves preexisting, microbial strains resistant to the topical agent in use and present either as a small minority population in a skin lesion or in the environment. Under the selective pressure of the antibiotic these strains readily overgrow more susceptible organisms. This is the main basis of superinfection. Since multiple drug resistance is a frequent feature in such minority populations, its incidence can also increase due to the use of unrelated topical antibiotics.

The evolution of neomycin resistance due to topical treatment of *S. aureus* infection is well documented.[15] This has occurred almost entirely in hospital and among bacteriophage-group III strains, which are fortunately considerably less virulent than the notorious group I (80/81) "hospital staphylococcus." Happily, neomycin resistance does not extend to the newer aminoglycosides, such as gentamicin.

The second mechanism involves genetic changes in bacteria, and apparently occurs to a significant extent outside hospital. It has been observed most alarmingly in relation to topical gentamicin, and although *S. aureus* has received most attention. gram-negative bacilli—notably *P. aeruginosa*—have evidently acquired gentamicin resistance by this mechanism in leg ulcers. burns. and otitis externa.[30] Just as epidemics of neomycin-resistant *S. aureus* infection followed topical application of this antibiotic—and not its wide use in preoperative bowel preparation—so most reports of outbreaks of gentamicin-resistant staphylococcal infection have been closely associated with the topical administration of this antibiotic rather than with its systemic use.[23]   This resistance is mediated by extrachromosomal plasmids and although often relatively labile. it can prove stable at a remarkably high level. Thus a minimum inhibitory concentration exceeding 3000μg/ml has been reported.[21] Apart from *S. aureus*, strains of skin staphylococci may act as reservoirs of resistance plasmids.

Bacteriophage-mediated transduction can readily occur on the skin surface, as shown in detail with neomycin.[22] The even more efficient form of resistance transfer, referred to as conjugation, may also take place in skin lesions. particularly among gram-negative bacilli.

Clearly, there is a serious risk of losing the use of such valuable antibiotics as gentamicin and newer aminoglycosides. It seems valid to conclude that the continued success of aminoglycoside treatment in severe systemic infection is endangered as long as gentamicin remains available for topical use (other than in gram-negative infections of the eye, which are usually posttraumatic). Similar constraints should apply to the antistaphylococcal antibiotic, fusidic acid, widely used in Europe systemically and topically in a variety of formulations, including tulle gras.

Drug Interactions and Microbial Stimulation

While antimicrobial agents used together can produce synergic, antagonistic, or merely additive interactions against microorganisms, the effects of combining different kinds of topical drugs cannot be classified so conveniently. For example the action of an antibiotic may initially be assisted by antiinflammatory and other desirable effects of a corticosteroid (as noted in the next section), but may then be hindered by the latter's depression of local immune defenses.[24] Conversely, the antibiotic can initially aid the effects of the corticosteroid, but it may subsequently promote superinfection and so lead to treatment failure.

A more subtle adverse interaction has recently been reported by Raab,[24] who observed that low concentrations of some corticosteroids stimulated metabolism in *C. albicans, S. aureus,* and *P. aeruginosa,* and by this means, opposed the antimicrobial effect of various agents, including hamycin and fusidic acid. However, the antimetabolic effect of higher concentrations depressed microbial respiration, except in the case of the pseudomonas, which was thought to have used the corticosteroid as a metabolic substrate.

Although the clinical significance of Raab's findings remains uncertain, it is noteworthy that corticosteroid concentrations in the depths of a skin lesion are within the stimulatory range, and indeed, Marples and his colleagues found that topical cortocosteroid in the absence of antibiotics was associated with moderately higher staphylococcal counts than the vehicle alone in human experimental infections.[19]

Practical Aspects of Treatment

Available Preparations

From the vast range of topical preparations used to treat skin infections (Table 34.1), only a modest selection is required to deal with most clinical problems.

Nonspecific Medicaments. When infected lesions are severely inflamed, edematous, exudative, or crusted, nonspecific measures are initially as important as antimicrobial therapy.[13,26]

Efficient local cleansing and gentle removal of accumulated debris (débridement) is necessary to permit adequate access of topical preparations. Mild keratolytic agents, such as salicylate ointments, are useful in chronic hyperkeratotic lesions, while grossly edematous, exudative areas respond well initially to hypertonic compresses of magnesium sulphate or sodium chloride. A helpful drying effect is produced by astringents such as aluminium acetate compresses or the antiseptic dyes discussed below.

Such treatment usually forms a brief prelude to more definitive anti-inflammatory and antimicrobial measures, but we have obtained satisfactory results in moderately severe primary skin sepsis and infected dermatoses using, as sole therapy, compresses followed by cream containing simple mixtures of inorganic salts and allantoin derivatives.[4]

Corticosteroids are, of course, the most active of the nonspecific topical preparations. Their anti-inflammatory effect usually brings rapid—if temporary—symptomatic relief in infected dermatoses. But while this promotes the onset of healing, it is to a varying extent opposed both by the simultaneous suppression of the local defenses and the possible stimulation of microbial metabolism at certain steroid concentrations that create favorable conditions for infection to persist and even extend.[24] The combined topical use of corticosteroid and appropriate antimicrobial agents is strongly recommended in the initial treatment of dermatoses that are either infected or colonized by significant numbers of pathogens.[4,15]

Topical Antimicrobial Agents. As a general rule, no anti-infective agent should be used topically if it has value as a systemic drug. The particular relevance of this to gentamicin and fusidic acid has already been discussed. Tetracyclines—although now less widely used topically than in the past —are an exception to the rule, since any tetracycline-resistant organisms selected in a skin lesion are not relevant to the infections treated systemically with antibiotics of this group. Two other exceptions are the polymyxins (including colistin) and amphotericin B. Both of these act on microbial cell membranes, and the acquisition of stable resistance to the drugs is rare.

Neomycin is justifiably the most popular topical antibiotic. It is often combined with bacitracin or related peptide antibiotics, which are also too toxic for systemic use. A polymyxin and an antifungal agent may be added when appropriate. The desirability of restricting topical chloramphenicol, penicillin, and sulfonamide to ophthalmic use has been discussed.

Several antiseptics are useful alternatives to topical antibiotics.[30] These include chlorhexidine, iodophors, halogenated quinolines, antiseptic dyes, nitrofurazone, and the organic mercurial, hydrargaphen.[27] Quaternary ammonium compounds (cationic detergents), such as benzalkonium and cetrimide, are valuable cleansing agents but have poor antimicrobial activity in vivo. Even the better antiseptics such as the iodophor, povidone-iodine, are unfortunately less active than topical antibiotics both in treating skin infection[15] and preventing wound infection.[30] Nevertheless, because such preparations are inexpensive and do not produce cross-sensitization or cross-resistance to any antibiotics, they are recommended for use in mild primary skin infections, and—with appropriate topical corticosteroids—in dermatoses, also prophylactically in minor injuries. Formulations of chlorhexidine (not to be confused with hexachlorophene) are nonstaining, and therefore usually more acceptable to patients than iodophors, quinolines, and traditional dyes, such as crystal violet, brillant green, or Castellani's paint.

Therapeutic Strategy

Because of the extremely high local concentrations of antimicrobial agents achieved during topical therapy, satisfactory results are often obtained even when pathogenic organisms are resistant to drug levels attained in the tissues during systemic therapy.

Spectrum of Activity. The antiseptics in general possess a broad range of action, extending from the fungi—including yeasts—through the gram-positive cocci to the gram-negative bacilli. The antiseptic dyes, and to a lesser extent chlorhexidine, however, possess reduced activity against gram-negatives, notably *Ps. aeruginosa.* Iodophors possess the broadest in vitro spectrum among the antiseptics, although body proteins rapidly inactivate iodine and its derivatives.

Antibiotics are all more selective. Bacitracin and gramicidin show good activity only against gram-positive bacteria, and polymyxins only against gram-negative bacteria (except *Proteus* sp.). Neomycin usually possesses high activity against *S. aureus* and most gram-negative bacteria except pseudomonads. *Streptococcus pyogenes* is less susceptible to neomycin than *S. aureus*.

The newer imidazole antifungals, clotrimazole, miconazole, and econazole, possess moderate antistaphylococcal activity in addition to a broad-spectrum of efficacy against pathogenic fungi.

Topical Regimens. Despite frequent assertions that skin infections should only be treated with systemic antibiotics, the case for rational topical therapy has recently been convincingly argued[15]. Table 34.2 summarizes a practical approach to the management of primary and secondary infection. After initial cleansing and débridement, well-localized primary infections which are judged to be staphylococcal on clinical grounds—notably bullous impetigo—can be treated twice daily with a topical antibiotic such as neomycin or tetracycline alone. A specimen for bacteriologic culture and subsequent antibiotic susceptibility testing should, if possible, be taken before applying the antibiotic. If the infection appears to be extending, systemic antibiotic such as a penicillinase-resistant penicillin or a cephalosporin should be administered. In addition, it is desirable to apply topical antibiotic, and so suppress accessible *S. aureus*.

The formation of a cutaneous abscess will necessitate an incision to allow drainage; systemic antibiotic is not then usually necessary in well-localized lesions, except facial abscesses.[29] Axillary abscesses often yield *Proteus* species; a recent report has indicated that anaerobic infections with *Bacteroides* species and other relatively antibiotic-resistant bacteria were also often present in axillary lesions.[12]

Table 34.2. Principles of Treatment of Cutaneous Infections

1. *Nonspecific measures*	
In all cases:	Initial. débridement, astringents, hypertonic soaks, as required.
II. *Specific measures in*	
Localized staphylococcal pyoderma	Topical antibacterial[a]
Extending staphylococcal pyoderma	Topical + systemic antibiotic
Streptococcal infection	Systemic antibiotic[b]
Infected dermatosis	
(a) first week	Topical corticosteroid + antibiotic[c]
(b) subsequently	Topical corticosteroid alone ± antiseptic

[a]Usually *antibiotic* (neomycin ± bacitracin or gramicidin; tetracycline); *antiseptic* adequate in minimal sepsis.
[b]Ideally combined with topical antibiotic; topical antibiotic adequate in minimal cases.
[c]e.g. neomycin + bacitracin (or gentamicin) + polymyxin; antifungal may also be appropriate.

In streptococcal infections—diagnosed clinically and confirmed, when possible, bacteriologically—systemic antibiotics are required if the lesions include cellulitis or extending ecthyma. The application of topical antibiotic in addition rapidly reduces the population of *S. pyogenes* in the lesion. This encourages early healing as well as preventing dangerous dispersal.[33] However, skillful topical treatment alone with a combination of neomycin, bacitracin (or gramicidin), and polymyxin has proved effective in more localized infections. Moreover, the incidence of poststreptococcal glomerulonephritis is the same in patients treated either topically or systemically.[15]

In secondarily infected dermatoses, topical treatment with a combination of antibiotic and corticosteroid is logical.[14,35] Microbiologic monitoring is advisable to provide data on the infecting organisms; these include possible gram-negative species, *Candida albicans*, and others that may resist standard antibiotics. Usually the antibiotic component of the treatment can be withdrawn within 1 week. It is indeed unwise to continue with topical antibiotics for long periods because of the risks of superinfection and resistance acquisition discussed earlier. In addition, because of the special hazard of allergic sensitization, neomycin should be avoided on ulcers or eczema of the lower leg.

Finally, attention should again be drawn to the value of a period of topical antibiotic treatment in clinically 'noninfected' but heavily colonized dermatoses.

References

1. Branemark PI, Ekholm R (1967) Tissue injury caused by wound disinfectants. J Bone Joint Surg (Am) 49:48–62
2. Chittasobohn N, Smith JMB (1979) The production of experimental dermatophyte lesions in guinea pigs. J Invest Dermatol 73:198–201
3. Committee on Safety of Medicines (United Kingdom) (1977) Adverse Reactions Series, No. 14. Department of Health, London
4. Copeman PWM, Selwyn S (1975) New non-steroid non-antibiotic skin medicaments. Br Med J 4:264
5. Crosfill F, Crosfill ML (1979) Antibiotic aerosols. J Antimicrob Chemother 5:486–487
6. Dajani AS, Hill PL, Wanamaker LW (1971) Experimental infection of the skin in the hamster simulating impetigo. II Assessment of various therapeutic regimens. Pediatrics 48:83–88
7. de Gruchy GC (1975) Aplastic anaemias. In: Drug-induced blood disorders. Blackwell Scientific Publications, Oxford, pp 39–75
8. Forsgren O, Schmeling D (1977) Effect of antibiotics on chemotaxis of human leukocytes. Antimicrob Agents Chemother 11:580–584
9. Hexachlorophene—yes or no? (1977) Br Med J 1:337–338
10. Holder IA, Schwab M, Jackson L (1979) Eighteen months of routine topical antimicrobial susceptibility testing of isolates from burn patients; results and conclusions. J Antimicrob Chemother 5:455–463
11. 2 Kings 5:1–14
12. Leach RD, Eykyn SJ, Phillips I, Corrin B, Taylor EA (1979) Anaerobic axillary abscess. Br Med J 2:5–7
13. Leyden JJ (1974) Antibiotic usage in dermatological practice. Int J Dermatol 13:342–352

14. Leyden JJ, Kligman AM (1977) The case for steroid-antibiotic combinations. Br J Dermatol 96:179–187
15. Leyden JJ, Kligman AM (1978) Rationale for topical antibiotics. Cutis 22:515–528
16. Leyden JJ, Stewart R, Kligman AM (1979) Updated *in vivo* methods for evaluating topical antimicrobial agents on human skin. J Invest Dermatol 72:165–170
17. Manten A (1975) Antibiotic drugs (excluding those in chapter 23). In: Dukes MNG (ed.) Meyler's side effects of drugs, vol 8. pp 603–647 Excerpta Medica, Amsterdam
18. Marples RR, Kligman AM (1972) Bacterial infection of superficial wounds: a human model for *Staphylococcus aureus*. In: Maibach H, Rovee DT (eds) Epidermal wound healing. Year Book, Medical Publishers, Chicago
19. Marples RR, Rebora A, Kligman AM (1973) Topical steroid-antibiotic combinations: assay of use in experimentally induced human infections. Arch Dermatol 108:237–240
20. McRipley RJ (1976) Topical treatment of microbial infections of skin and mucous membranes. In: Gadebusch HH (ed) Chemotherapy of Infectious Disease. CRC Press, Cleveland, pp 55–69
21. Naidoo J, Noble WC (1978) Acquisition of antibiotic resistance by *Staphylococcus aureus* in skin patients. J Clin Pathol 31:1187–1192
22. Naidoo J, Noble WC (1978) Transfer of gentamicin resistance between strains of *Staphylococcus aureus* on skin. J Gen Microbiol 107:391–393
23. Noble WC, Naidoo J (1978) Evolution of antibiotic resistance in *Staphylococcus aureus*: the role of the skin. Br J Dermatol 98:481–489
24. Raab W (1976) Effects of local corticosteroids in skin infections. Dermatologica 152 (Suppl 1): 67–79
25. Rebora AE, Marples RR, Kligman AM (1973) Experimental infection with *Candida albicans*. Arch Dermatol 108:69–73
26. Rook A, Wilkinson DS, Ebling FJG (1979) Textbook of dermatology. 3rd ed. Blackwell Scientific Publications, Oxford, pp 551–603, 811–867, 2233–2247, 2293–2328
27. Sandifer SH (1970) Clinical trial of topical nitrofurazone, with or without hydrocortisone, in 252 children with skin infections. JSC Med Assoc 66:363–365
28. Selwyn S (1965) The mechanism and prevention of cross infection in dermatological wards. J Hyg (Camb) 63:59–71
29. Selwyn S (1977) Cutaneous abscesses. Br Med J 2:1499
30. Selwyn S (1978) Antibiotic resistance and topical treatment. Br Med J 2:649–650
31. Selwyn S (1979) Pioneer work on the 'penicillin phenomenon,' 1870–1876. J Amtimicrob Chemother 5:249–255
32. Selwyn S (1979) Early experimental models of disinfection and sterilization. J Antimicrob Chemother 5:229–30
33. Selwyn S, Chalmers D (1965) Dispersal of bacteria from skin lesions: a hospital hazard. Br J Dermatol 77:349–356
34. Selwyn S, Ellis H (1972) Skin bacteria and skin disinfection reconsidered. Br Med J 1:136–140
35. Wachs GN, Maibach HI (1976) Co-operative double-blind trial of antibiotic/corticoid combination in impetiginized atopic dermatitis. Br J Dermatol 95:323–328

Chapter 35
Treatment of Serious Cutaneous Staphylococcal and Streptococcal Infections

STEPHEN N. COHEN

At least three-quarters of cutaneous infections are due to staphylococcal and streptococcal species. More than half of these infections are due to *Staphylococcus aureus*, which, whether acquired in the community or in hospital, is usually resistant to penicillin. Therapy for severe staphylococcal infection begins with any parenterally administered penicillinase-resistant penicillin or cephalosporin (see Table 35.1) unless the susceptibility of the specific isolate to penicillin is known. If possible, any predisposing factors should be removed. After a good clinical response is evident, oral therapy may be substituted. Penicillin G (or V for oral therapy) should always be given if the organism is not a penicillinase producer since it is 10–100 times more active than the penicillinase-resistant drugs. Ampicillin and car-benicillin, and other new anti–gram–negative penicillins, are susceptible to staphylococcal penicillinase and are not suitable for antistaphylococcal therapy of penicillin G-resistant infections.[6]

Tetracycline resistance is relatively common in *S. aureus*; erythromycin resistance emerges rapidly during therapy. We recommend clindamycin for the penicillin-allergic patient in whom bactericidal therapy does not appear critical, or vancomycin for such life-threatening infections as endocarditis. Clindamycin can cause neutropenia, apparently of an allergic and reversible type, and with variable frequency leads to an ulcerative colitislike picture, which is probably due to proliferation of the enterotoxigenic, clindamycin-resistant anaerobe *Clostridium difficile*.[3] Discontinuing the drug is usually sufficient to reverse the syndrome, but oral vancomycin and steroid therapy may be required to control a severe colitis.

Gentamicin resistance is becoming common in hospital staphylococci; Porthouse et al.[11] have described epidemics due to gentamicin-inactivating staphylococci. Gentamicin killing of staphylococci is inferior to that of penicillinlike drugs.[8,15] An increased speed of bacterial killing by the combination of penicillins and aminoglycosides[14] provides the potential for increased toxicity and has not been shown to attain superior clinical results in staphylococcal disease.

We do not recommend cephalosporin therapy for patients with allergy to

Table 35.1. Treatment Schedules for Severe Staphylococcal and Streptococcal Infection (see text for selection of antibiotic)[a]

Nafcillin	1–2 gms IV q4h
Dicloxacillin	250–500 mg po q4h
Aqueous Benzyl Penicillin G	1–2 m units IV q4h
Phenoxymethyl Penicillin (Penicillin V)	0.5–1.0 gm po q4h (the higher dose may be necessary with staphylococcal endocarditis or osteomyelitis).
Vancomycin	0.5–1.0 gm IV q12h
Clindamycin	600 mg IV q6h
	later 300 mg po q6h
Erythromycin	500 mg IV q6h
	later 500 mg po q6h

[a]Treatment of severe infection is generally continued for at least 10 days. If staphylococcal endocarditis or osteomyelitis is suspected, therapy should extend for 6 weeks (3 weeks suffices with vancomycin). Oral treatment should only be instituted after a good clinical response has been obtained, and, if osteomyelitis or endocarditis is suspected, only if administration of the drug can be guaranteed, usually by observation in hospital.
Combination therapy with aminoglycosides (usually gentamicin 80 mg IM q8h) is only recommended for endocarditis due to enterococcal species of streptococci.

penicillins. Many patients with a history of penicillin allergy do tolerate cephalosporins, but often the history of allergy is incorrect. Cephalosporins are often though not always cross-allergenic, and fatal anaphylaxis and other manifestations of cross allergenicity to cephalosporins have been reported. Oral cephalosporins are usually much less active against staphylococci than parenteral cephalosporins; the serum antibiotic levels attained with oral therapy may may be marginal against an established staphylococcal infection.

Methicillin resistance is widespread among *Staphylococcus aureus* in Europe, but uncommon in the United States. Vancomycin is the therapy of choice for methicillin-resistant staphylococci, including *Staphylococcus epidermidis*. Vancomycin toxicity is primarily manifested by phlebitis or by a reversible depression of of renal function. For technical reasons, probably reflecting the different proportion of organisms intrinsically resistant to the two drugs, a methicillin-resistant organism may appear sensitive to cephalothin by disc testing, yet appear cephalothin-resistant by both dilution techniques and respond poorly to a clinical trial of cephalosporin therapy. Neither methicillin-like nor cephalosporin drugs should be used for the treatment of methicillin-resistant staphylococcal infections.

Penicillin-tolerant staphylococci have recently been described. Organisms displaying "tolerance" are inhibited normally by penicillin or methicillin, but are poorly killed by these drugs due to an inhibitor of the intrinsic intrabacterial autolytic system.[13] Tolerance is a frequently demonstrable test tube phenomenon among clinical isolates of various staphylococcal phage types, but its clinical significance is unknown.

Staphylococcus epidermidis has become frequent in hospital-acquired infections,[2] but it only rarely causes severe cutaneous infections.

Staphylococcus epidermidis is an interesting bellwether for the presence of antibiotic resistance in *S. aureus*, frequently displaying such resistance before it becomes prominent in the commonly pathogenic species, e.g., gentamicin resistance is becoming widespread in *S. epidermidis* and is now present in 14% of strains compared with 4% of *S. aureus*.

Group A streptococcal infections remain a major source of cutaneous morbidity. Group B streptococcal cutaneous infections are particularly common in diabetics and alcoholics.[9] Groups C and G β-hemolytic streptococcal infections are less common, while many other varieties of streptococci, including anaerobes, are often recovered in mixed cultures from such sites as decubitus ulcers.

Penicillin G remains the drug of choice for treatment of streptococcal infections. Cephalosporins are equally effective, but at greater expense. In the United States, erythromycin and clindamycin are suitable alternatives in the penicillin-allergic patient and vancomycin can also be used, but tetracycline resistance is widespread, and in some areas of the world resistance to multiple antibiotics is prevalent.[10] Severe streptococcal infections are no exception to the general rule that severe illness should initially be treated with parenteral therapy. It is common for patients with severe streptococcal infection to remain acutely ill with high, swinging fevers and major clinical toxicity for several days after the institution of completely effective antimicrobial therapy.[12]

Strains of one special variety of streptococcus, *Streptococcus pneumoniae*, which are highly resistant to penicillin G and methicillin and to most other commonly used antibiotics have recently been isolated in South Africa.[7] Resistance in these strains is not due to penicillinase production but to a permeability barrier to the site of inhibition. The implication for the treatment of all types of streptococcal infections throughout the world is worrisome indeed.

References

1. Acar JF, Courvalin P, Chabbert A (1971) Methicillin-resistant staphylococcemia: bacteriological failure to treatment with cephalosporins. Antimicrob Agents Chemother 1970:280–285
2. Andriole VT, Lyons RW (1970) Coagulase-negative staphylococcus Ann NY Acad Sci 533–544
3. Bartlett JG, Onderdonk AB, Cisneros RL, Kasper DL (1977) Clindamycin-associated colitis due to a toxin-producing species of *Clostridium* in hamsters. J Infect Dis 136:701–705
4. Federal Register, Rules and Regulations (1972) Antibiotic Susceptibility Discs, 37:20525–20529.
5. Federal Register Rules and Regulations (1973) Antibiotic susceptibility discs (correction) 38:2757
6. Garrod LP, Lambert HP, O'Grady F (1973) Antibiotic and chemotherapy, 4th edn. Churchill Livingstone, Edinburgh
7. Jacobs MR, Loornhof HJ, Robins-Browne RM, Stevenson CM, Vermaak ZA, Freiman I, Miller GB, Witcomb MA, Isaacson M, Ward JI, Austrian R (1978) Emergence of multiply resistant pneumococci N Engl J Med 299:735–740

8. Klastersky J Hensgens C, Daneau D (1975) Therapy of Staphylococcal Infections: (A Comparative Study of Cephaloridine and Gentamicin) Am J Med Sci 269:201–207

9. Lerner PI, Gopalakrishna KV, Wolinsky E, McHenry MC, Tan JS, Rosenthal M (1977) Group B streptococcus (*S. agalactiae*) bacteremia in adults: analysis of 32 Cases and review of the literature Medicine 56:457–473

10. Nakae M, Murai T, Kaneko Y, Mitsuhashi S (1977) Drug resistance in *Streptococcus pyogenes*. Antimicrob Agents Chemother 12:427–428

11. Porthouse A, Brown DFJ, Smith RG, Rogers T (1976) Gentamicin resistance in *Staphylococcus aureus*. Lencet 1:20–21

12. Quintiliani R, Engh GA (1971) Overwhelming sepsis associated with Group A beta hemolytic streptococci J Bone Joint Surg (Am) 53:1391–1399

13. Sabath LD, Wheeler N, Laverdiere M, Blazevic, Wilkinson BJ (1977) A new type of penicillin resistance of *Staphylococcus aureus* Lancet 1:443–447

14. Steigbigel RT, Greenman RL, Remington JS (1975) Antibiotic combinations in the treatment of experimental *Staphylococcus aureus* infection J Infect Dis 131:245–251

15. Wilson SG, Sanders CC (1976) Selection and characterization of strains of *Staphylococcus aureus* displaying unusual resistance to aminoglycosides. Antimicrob Agents Chemother 10:519–525

Chapter 36
The Role of Ketoconazole in the Management of Mucocutaneous Candidiasis Syndrome

JOHN R. GRAYBILL AND DAVID J. DRUTZ

Candida species at the surface of the body, especially *Candida albicans*, are responsible for some of the most common nuisance infections to which humans are susceptible. Among these are thrush and vaginal candidiasis. Thrush is of potential life-threatening significance in the immunosuppressed patient where it may signal the presence of candidiasis throughout the gastrointestinal tract. Fatal *Candida* dissemination frequently arises from clinically inapparent intestinal colonization under conditions of immunosuppression.[6] Vaginal candidiasis has become a problem of major clinical significance since the advent of birth control steroidal medications.[14]

A surface infection of much less common occurrence, but of major immunologic and endocrinologic interest, is chronic mucocutaneous candidiasis. Also known as chronic mucocutaneous microbiosis,[4] this syndrome is characterized by fungal infection of the oropharynx, face and skin surface, anus, vagina, and nails.[9] Lesions may assume a granulomatous form, resulting in considerable disfigurement. Patients with this syndrome often manifest abnormalities of T-lymphocyte function, and may have a variety of endocrinologic abnormalities including hypoparathyroidism, hypothyroidism, or hypoadrenalism.[9] Treatment modalities employed in patients with this disease have ranged from topical antifungal agents to parenteral amphotericin B to transfer factor.[5] In no case has there been permanent improvement of the infection. However, ketoconazole, an orally absorbable imidazole compound, has shown considerable promise in the management of this disease state.

The Imidazoles

The antimycotic effects of imidazole derivatives having the general formula shown in Figure 36.1 have been known since 1967.[2] Essential parameters for the antimycotic effect include an unsubstituted imidazole ring that is bound to the residual molecule by a C-N bond.[2] The substituents 1, 2, and 3 can be widely varied and influence more the pharmacokinetics than their antifungal activity.

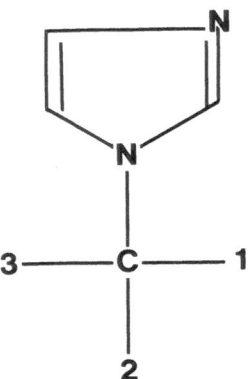

Fig. 36.1. General chemical structure of the imidazoles.

Imidazoles are remarkably versatile; a variety have been employed in the management of microbial and parasitic infections. Hints of their versatility first came in the 1960s when thiabendazole was introduced. This thiazolyl benzimidazole derivative has a wide antihelminthic spectrum and good activity against dermatophytes, *Cladosporium*, *Phialophora*, and *Madurella* species (7). Tetramisole and its enantiomer levamisole are phenylimidazothiazoles with considerable antinematode activity.[8]   However, they also possess potent immunomodulating activity including enhancement of graft rejection, restoration of defective cutaneous delayed type hypersensitivity, and enhancement of monocyte function.[5]   Mebendazole, another benzimidazole derivative, is a broad-spectrum oral anti-helminthic with some inhibitory capacity for *Aspergillus* species and some dermatophytes.[8]   Metronidazole, best known as an antitrichomonal agent, has potent activity in amebiasis and giardiasis, and in infections due to anaerobic gram-negative bacilli.[7]   Its antifungal activity is negligible.

Clotrimazole and miconazole were the first imidazole derivatives to have broad antifungal spectra and to possess high in vitro activity against most fungi of clinical interest.[8]   Despite early promise as an orally absorbed broad-spectrum antifungal agent, clotrimazole is now known to be erratically absorbed.   Moreover, there is a steady decline in achievable clotrimazole serum levels as a result of hepatic microsomal enzyme induction with progressively rapid drug inactivation.[3] Because of these problems, clotrimazole is principally useful as a topical preparation. Clotrimazole troches have shown clinical efficacy in the management of the oral lesions of mucocutaneous candidiasis syndrome.[10] Esophageal, vaginal, nail, and skin lesions were not eliminated by this therapy.

Miconazole is poorly absorbed when administered by the oral route, but is an extremely efficacious topical antimycotic agent with a spectrum of activity similar to that of clotrimazole.[14]   Miconazole can be administered intravenously and intrathecally and has shown considerable promise in the management of coccidioidomycosis and paracoccidioidomycosis,[13] *Alles-cheria* (*Petriellidium*) *boydii* infection,[11] and several other mycoses, including candidiasis. The side effects of parenteral miconazole therapy are prodigious and include phlebitis; the syndrome of inappropriate antidiuretic hormone secretion with hyponatremia; anemia, thrombocytosis or thrombocytopenia; and augmentation of coumarin drug effects.[13] In addition, patients develop a profound hyperlipidemia (cholesterol and triglycerides) apparently attributable to cremophor EL (polyethyoxylated castor oil), the lipid vehicle of intravenous miconazole formulations.[1] The drug produces an intense and persistent pruritus (formication) in about 25% of patients, in the presence or absence of a rash.[13] Miconazole has shown transitory efficacy in chronic mucocutaneous candidiasis.

The latest of the imidazoles to undergo clinical trials is ketoconazole. Like miconazole, this drug has excellent in vitro activity against *Candida albicans*, *Histoplasma capsulatum*, *Coccidioides immitis*, *Cryptococcus neoformans*, and many dermatophytes. However, unlike miconazole, ketoconazole has the distinct advantage of water solubility. There is ample evidence in laboratory animals and in man that the drug is readily absorbed after oral administration. As measured by Dr. Hillel Levine, Naval Biosciences Laboratory in Oakland, Ca., the drug is detectable at a concentration of about 1 μg/ml, 1 hr after a 200 mg oral dose. Peak levels of approximately 2 μg/ml are maintained for up to 4 h, and the concentration declines to about 1 μg/ml by 8 h after the dose. Higher concentrations are seen after a 400 mg oral dose. By 24 h following either dose, there is no detectable activity in the blood. The drug is not excreted significantly by either urinary or biliary routes, but is extensively catabolized by a number of metabolic pathways. In 24 h urine collections, the ketoconazole urine concentration is approximately 0.4 μg/ml.

The relationship of achievable serum concentrations to the minimum inhibitory concentration (MIC) for *Candida* species is difficult to assess because the MIC for imidazoles is extremely dependent upon the method used. Our *Candida* strains have been tested only by an agar diffusion method; this produces a zone of inhibition in an agar plate seeded with *Candida*. The MICs for five strains of *Candida albicans* have ranged from 0.6–3.0 μg/ml. In most cases, the zone of inhibition was found to contain *Candida* colonies, some apparently resistant to concentrations of 12 μg/ml of ketoconazole or more. This apparent in vitro resistance of a small number of *Candida* cells does not appear to correlate with clinical response to treatment, and its significance is unknown.

Clinical Experience with Ketoconazole

The first clinical trials of ketoconazole in candidiasis occurred in Europe and involved patients with gastrointestinal and vaginal *Candida* infections. Unpublished studies have suggested that *Candida* infections of these sites were rapidly cleared, usually within 10 days, on a course of only 200 mg once daily. A few patients did not improve, possibly because of failure of absorption of the ketoconazole. Ketoconazole base dissolves only in an acid environment, and at least some of these patients had achlorhydria (R. Legendre, Janssen R&D, personal communication).

In the United States, the experience with ketoconazole treatment of candidiasis has been more restricted. We have treated six patients with candidiasis, and Rosenblatt[12] and his colleagues have treated three others. The therapeutic results in our six patients are summarized in Table 36.1. The patients ranged in age from 14–36 years. Five had the syndrome of chronic mucocutaneous candidiasis, and the sixth had persistent thrush associated with immunosuppressive medication following renal transplantation. A

Table 36.1. Response of Six Patients with Candidiasis to Ketoconazole

Pt #	Age & sex	Dx	Age at onset (yrs)	Site and severity
1	26 M	CMC[b]	< 1	Hand, arm (severe); thrush (mild)
2	36 F	CMC	6	Thrush (mild); vaginal (resolved)[c]
3	21 M	CMC	< 1	Thrush; esophagitis, plaques over head, thorax (severe)
4	14 F	CMC	< 1	Thrush (mild)
			14	Disseminated histoplasmosis (relapse of disease originally treated at age 7)
			14	Cryptococcosis (bone marrow)
5	23 M	CMC	17	Thrush, face, neck, chest, groin, foot (severe)
6	29 M	Renal transplantation	28	Thrush (mild to moderate)

[a] M = topical miconazone; N = topical nystation; C = oral clotrimazole; C (IV) = intravenous clotrimazole; G = oral griseofulvin; F = oral flucytosine; A = intravenous amphotericin B; T = intramuscular transfer factor.
[b] CMC = chronic mucocutaneous candidiasis.
[c] hysterectomy for candida endometritis (vaginitis cleared).

single family was represented by a 36-year-old woman (Patient 2), her brother (Patient 3), and her daughter (Patient 4).

Candidiasis was severe in three patients and mild in the others. In all patients it had resisted topical treatment. Several patients had responded to intravenous amphotericin B, but relapsed after that drug was stopped. This tendency to relapse after termination of treatment is attributable to defective host cell—mediated immunity and repeated exposure to *Candida*, which can exploit this weakness of immune defense. The usual persistence of mucocutaneous candidiasis in the absence of treatment gives additional credence to the results of uncontrolled therapeutic trials such as ours.

None of our patients received other antifungal therapy during their treatment with ketoconazole. All were begun on 200 mg per day. A similar pat-

Prior therapy[a]	Ketoconazole therapy		
	Dose (mg/d)	Duration	Response and most recent culture status
M,N,C,G, F,A,T	200–400	8 mo	Thrush cleared < 1 wk hand, arm 90% cleared by 8 mo (Hand and mouth still culture + for Candida).
N	200	9 mo	Thrush cleared < 1 wk (culture −).
M,A,T, C (IV)	200–400	7 mo	Thrush cleared < 1 wk (culture +). Cutaneous lesions 80% cleared (culture − for Candida, but + for dermatophyte).
G,A	200–400	20 days	Thrush cleared < 1 wk (culture −). Nc improvement in histoplasmosis (amphotericin B required). ? improvement in cryptococcosis (marrow culture became negative after ketoconazole, and before amphotericin B).
N,A	200–400	5 mo (course #1)	Thrush improved in 21 days and cleared in 7 wks (culture −). Facial lesions gone in < 2 mo (see text).
		1 mo (course #2)	Thrush improved. Facial lesions improving (culture +).
N	200	21 days (course #1)	Clinical improvement only (culture +).
		10 days (course #2)	Clinical improvement only (culture +).

tern of response was seen in all. Within a week definite improvement was noted in the thrush. By 3–7 weeks, thrush was absent. In Patients 2, 4, and 5, oral cultures became free of *Candida* in the absence of mucosal lesions.

The large cutaneous plaques present in the more seriously affected patients responded much more slowly to treatment. As a result, all three patients with extensive disease (Patients 1, 3, and 5) eventually had an increase in ketoconazole dosage to 400 mg per day. There was no indication, however, that increasing the dose altered the rate of improvement. Of these three patients, Patient 5 had complete clearance of his skin lesions. Patients 1 and 3 had more than 80%–90% clearance. The skin of Patients 1 and 5 never became free of *Candida* by culture criteria. In Patient 3, skin cultures became negative for *Candida*, but a dermatophyte was later isolated from the remaining lesions.

The response of mucocutaneous candidiasis to ketoconazole is well illustrated in the case of Patient 5. This 23-year-old man first developed crusting plaques on his chin at age 17. A few months later he was hospitalized with diabetic ketoacidosis. The lesions were diagnosed as candidiasis and treated with topical nystatin with some response. Ultimately the candidiasis progressed to involve his entire face, neck, and most of his chest. Areas of healing were marked by loss of skin pigmentation. The process spread to involve his scrotum and inguinal area with an erythematous scaling rash. A similar process involved his feet, although the nailbeds were spared. His immune response was evaluated in 1977, and he was found to have defective skin-test responses to a battery of recall antigens, including *Candida* extract, and defective lymphocyte blastogenic response to phytohemagglutinin and *Candida* extract. Migration inhibition factor responses to *Candida* antigen were negative. A course of 2 gm of intravenous amphotericin B, given over 2 months, resulted in marked improvement. He relapsed 2 months later. By August, 1978, his disease was as extensive as it had been before the amphotericin B (Fig. 36.2a).

He was begun on 200 mg per day of ketoconazole on November 9, 1978. By November 30, 1978, his thrush had markedly improved, and the skin lesions had regressed moderately (Fig. 36.2b). His dose was reduced to 200 mg every other day, and the improvement of lesions appeared to be slowed. On December 15, 1978, culture of a facial lesion was still positive for *Candida*, although the groin was now negative. The dose was increased to 200, then 400 mg per day. In late December, his lesions on the face, feet, and groin were gone (Fig. 36.2c). From February 1–March 18, 1979, the dose was reduced to 200 mg per day. Cultures of the mouth and groin remained negative for *Candida*. His dose was then lowered to a maintenance of 200 mg three times per week. From April to August, 1979, the patient was lost to follow-up. In June, his mucocutaneous disease began to return. In August (Fig. 36.2d), his lesions were not as severe as at his initial presentation. Nevertheless, the face was extensively involved, and he had severe thrush. Cultures were positive for *C. albicans*, MIC 0.5 μg/ml. He was again treated with ketoconazole, 200 mg per day, and showed a good clinical response.

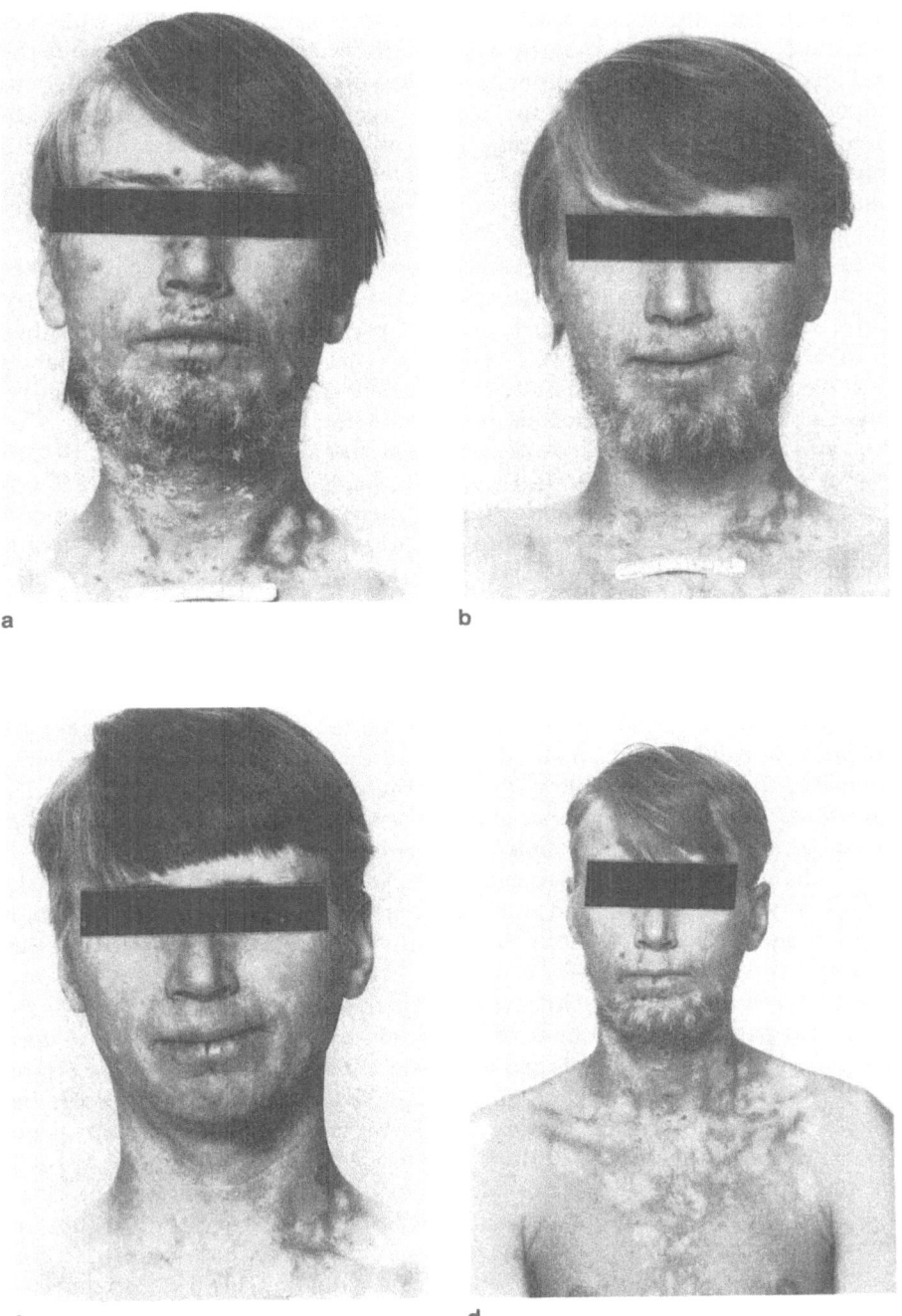

Fig. 36.2. a–d Patient 5: **a** Facial appearance just prior to initiation of ketoconazole therapy. **b** Facial appearance 3 weeks after initiation of ketoconazole. **c** Facial appearance 6 weeks after initiation of ketoconazole. **d** Appearance following discontinuance of ketoconazole for 3 months.

This patient illustrates the excellent response achievable with ketoconazole, though some months were required to do so. He illustrates the problems of discontinuing ketoconazole in a patient with chronic mucocutaneous candidiasis. Just as the disease recurred after stopping amphotericin B, it recurred after stopping ketoconazole. This is not unique and happened to Patient 6, who had received a renal transplant and immunosuppression. He also improved during a second course of ketoconazole.

What long-term alternatives are available for patients receiving ketoconazole? First, we cannot simply stop the drug. The disease clearly recurs, which is not surprising in light of previous experience with other therapeutic agents. Second, we can decrease the dose to a low maintenance level. This was done successfully with Patient 2, a woman who had active thrush and inactive fingernail lesions at the time of her initial treatment. Ketoconazole has been reduced, after an initial 200 mg per day course, to 200 mg three times weekly. This strategy, while successful in some patients, will not work in all. In a recent report,[12] one patient whose dose was so reduced had relapse of *Candida* esophagitis. Fortunately, this responded to an increase in dosage.

While ketoconazole appears very effective in this small series, its role in other mycotic infections is less well defined. This is dramatically illustrated in Patient 4, a 14-year-old girl who had concurrent thrush; disseminated histoplasmosis involving multiple lymph nodes, the oropharynx, and the bone marrow; and cryptococcosis involving the bone marrow. Three weeks of ketoconazole treatment dramatically cleared the thrush and may have been responsible for a negative follow-up bone marrow culture for *Cryptococcus neoformans*. However, the histoplasmosis was not affected at all, and she required amphotericin B for control of this infection.

Major benefits of ketoconazole therapy, in addition to its effectiveness, are ease of administration and minimal toxicity in patients treated up to 1 year. The most severe reactions thus far attributable to ketoconazole in the 31 patients we have treated for a variety of mycoses have been rash in one patient, nausea and vomiting in several at the higher 400 mg dose, and dizziness in two patients. In neither of the latter are we sure the reaction was drug-related. Patient 6 developed an acute cataract, but we are uncertain whether this was caused by ketoconazole or the large doses of steroids he was receiving. Monitoring of CBC, SMAC, urinalysis, audiograms, and ECGs showed no evidence for drug toxicity.

In summary, the imidazoles provide a possible alternative to amphotericin B in a variety of mycotic infections. Topical miconazole is very effective in many cutaneous mycoses, but does not benefit all patients with chronic mucocutaneous candidiasis. Intravenous use of miconazole is extremely expensive and may be complicated by significant toxicity. Ketoconazole appears to be the first orally absorbable, minimally toxic imidazole drug that can be chronically administered to patients with chronic mucocutaneous candidiasis. At the present time, we do not know whether indefinite use of any systemic antifungal drug is feasible or whether microbial resistance will develop; with ketoconazole, prolonged therapy is at least within the realm of consideration.

Acknowledgements

This work was supported by Janssen R&D, Inc., and by the General Medical Research Service of the Veterans Administration.

References

1. Bagnarello AG, Lewis LA, McHenry MC, Weinstein AJ, Naito HK, MC-Cullough AJ, Lederman RJ, Gavan TL (1977) Unusual serum lipoprotein abnormality induced by the vehicle of miconazole. N Engl J Med 296:497–499
2. Bennett J (1978) Chemotherapy of fungal diseases. In: Siegenthaler W, Luthy R (eds) Current chemotherapy: proceedings of the 10th international congress of chemotherapy, vol 1. American Society for Microbiology, Washington, pp 53–55
3. Burgess MA. Bodey GP (1972) Clotrimazole (Bay b 5097): In vitro and clinical pharmacological studies. Antimicrob Agents Chemother 2:423–426
4. Chappler R, Maibach H, Conant M, Aly R (1978) Mucocutaneous candidiasis or mucocutaneous microbiosis. JAMA 239:428–429
5. Edwards JE Jr (moderator) (1978) Severe candidal infections. Clinical perspective, immune defense mechanisms, and current concepts of therapy. Ann Intern Med 89:91–106
6. Eras P, Goldstein MJ. Sherlock P (1972) Candida infection of the gastrointestinal tract. Medicine (Baltimore) 51:367–379
7. Galgiani JN, Busch DF, Brass C, Rumans LW, Mengels JI, Stevens DA (1978) Bacteroides fragilis endocarditis, bacteremia, and other infections treated with oral or intravenous metronidazole. Am J Med 65:284–289
8. Holt RJ (1975) New antifungal drugs. Drugs 9:401–405
9. Kirkpatrick CH (moderator) (1971) Chronic mucocutaneous candidiasis: Model-building in cellular immunity. Ann Intern Med 74:955–978
10. Kirkpatrick CH, Alling DW (1978) Treatment of chronic oral candidiasis with clotrimazole troches. N Engl J Med 299:1201–1203
11. Lutwick LI, Galgiani JN, Johnson RH, Stevens DA (1976) Visceral fungal infections due to Petriellidium boydii (Allescheria boydii). In vitro drug sensitivity studies. Am J Med 61:632–640
12. Rosenblatt HM, Byrne W, Ament ME, Graybill JR, Stiehm ER (1979) Successful treatment of severe chronic mucocutaneous candidiasis (CMC) with a new oral agent, ketoconazole. 11th International Congress of Chemotherapy and 19th Interscience Conference on Antimicrobial Agents and Chemotherapy, October 1–5, 1979, Boston, abstract #144
13. Stevens DA (1977) Miconazole in the treatment of systemic fungal infections. (editorial) Am Rev Respir Dis 116:801–806
14. Trends in vaginal candidiasis. Proceedings of an international symposium held by Janssen Pharmaceutical Limited at the Royal Society of Medicine. 15 November 1976. (1977) Proc R Soc Med 70(supp14):1–32

Index

Page numbers in *italic* refer to illustrations or tables.